Chemistry for the Graphic Arts

Chemistry for the Graphic Arts

Third Edition

by
Dr. Nelson R. Eldred

GATF*Press*
PITTSBURGH

Library of Congress Catalog Card Number: 00-107663
International Standard Book Number: 0-88362-249-1

Printed in the United States of America

GATF Catalog No. 14013
Third Edition
First Printing, February 2001

INTERNATIONAL Ⓐ PAPER

Printed on Williamsburg offset, 60 lb., smooth finish by International Paper

GATF*Press*
Graphic Arts Technical Foundation
200 Deer Run Road
Sewickley, PA 15143-2600
Phone: 412/741-6860
Fax: 412/741-2311
Email: info@gatf.org
Internet: www.gatf.org • www.gain.net

Orders to:
GATF Orders
P.O. Box 1020
Sewickley, PA 15143-1020
Phone (U.S. only): 800/662-3916
Phone (Canada only): 613/236-7208
Phone (all other countries): 412/741-5733
Fax: 412/741-0609 • Internet: www.gain.net

GATF*Press* books are widely used by companies, associations, and schools for training, marketing, and resale. Quantity discounts are available by contacting Peter Oresick at 800/910-GATF.

Contents

Foreword

Chemical processes are basic to several graphic arts operations and play an important role in many others. For this reason *Chemistry for the Graphic Arts* is considered to be a fundamental text in the Graphic Arts Technical Foundation's family of educational books.

Chemistry for the Graphic Arts is an informational textbook designed for use both by industry and in the schools. It has been organized in such a manner that people with no background in chemistry can read the first three chapters to develop a working knowledge of the basics of chemistry. Then, the book progresses into the chemistry involved in various areas of the graphic arts.

This is the third edition of Chemistry for the Graphic Arts. The first edition, published in 1979 and reprinted with minor revisions in 1983, was written by the late Dr. Paul J. Hartsuch, a longtime GATF researcher who was at the forefront of the activities that changed lithography from a largely trial-and-error craft to a scientific technology.

The second edition, published in 1992, was written by Dr. Nelson R. Eldred, who used the first edition to provide a framework for the development of the second edition.

Dr. Eldred was again asked to write this third edition. After developing and appending the comprehensive chapter outlines, the author enlisted the assistance of various industry experts to ensure the technical accuracy of the material presented in this book (see acknowledgments).

Thomas M. Destree
Editor in Chief

Acknowledgments

Numerous industry people have assisted the author and GATF with this edition of *Chemistry for the Graphic Arts*. We would like to thank the following individuals for their assistance:

Mark Atchley, X-Rite, Inc.
Melissa Ayotte, X-Rite, Inc.
Tony Bart, E.I. DuPont de Nemours & Co.
Bob Bassemir, Sun Chemical Corp.
Bruce Blom, Mead Papers
Robert Booker, Rogers Corporation
Mike Bruno, consultant
Lucinda Cole, Flexographic Technical Association
Dutch Drehle, Screenprinting and Graphic Imaging Association
Rudiger Feldman, Voith Sulzer Finishing GmbH
Dr. Richard Fisch, Imation
Jeff Frazine, X-Rite, Inc.
David Gerson, Printers' Service Co.
Dr. Ralph W. Gordon, consultant
Henry Hatch, International Prepress Association
Terry Hutton, M & R Printing Equipment, Inc.
Greg Ikuta, Pacific Transducer Corp.
Liz Jones, Cole-Parmer Instrument Co.
Josephine Kirks, Newspaper Association of America
George Leyda, ColorInfo Technology
Ken Lowery, E.I. DuPont de Nemours & Co.
Bruce Lyne, International Paper Co.
John MacPhee, Baldwin Technology Systems
Donald Marsden, Ulano Corporation
David McDowell, Eastman Kodak Co.
Michelle Mercier, Creo Products

Ronald L. Mihills, Research & Engineering Council of the
    Graphic Arts Industry, Inc.
Martin Kent Miller, Hell Gravure Systems
David Muchorski, Thwing-Albert Instrument Co.
Susan Newcomb, Canadian Silver Recovery Systems
Richard Podhajny, consultant
Dr. Joseph Rach, MacDermid, Inc.
Neil Richards, Richards Research
Joel Schulman, Jelmar Publishing Co., Inc.
Kent Seibel, Ohio Electronic Engraving
John W. Shreve, Midwest Rubber Plate Co.
William Stephens, Cole-Parmer Instrument Co.
Bill Sunter, Gravure Association of America
Timothy Warriner, Beloit Corporation
Larry Warter, Fuji Photo Film
Michael Wiest, Flexographic Technical Association
Joe Zilka, Cole-Parmer Instrument Co.

The following GATF employees have also assisted in
ensuring the technical accuracy of this book:

Dr. Richard M. Adams II
Brad Evans
William Farmer, Jr.
Dee Gentile
Rick Hartwig
Gary Jones
John Lind
Raymond J. Prince
Bruce Tietz
Dick Warner

The author and GATF truly regret if we have failed to
recognize anyone who has assisted with this project. The
affiliations listed for some of the industry individuals may
have changed since the time that they provided us with
assistance.

1   An Introduction to Chemistry

The chemist attempts to find the relation of the structure of matter (chemicals) to its behavior in order to understand how things work:

- How do printing inks dry?
- How does water affect paper?
- What makes a good paper coating?
- What makes a good fountain solution?
- What makes a good printing plate?
- How can we make better photographic film?

Introduction

Chemistry involves the basic composition and structure of materials. Anything that has mass is a chemical: air, water, foodstuffs, paper, ink, and photographic developers. Things that have no mass—light, sound, and heat—are not chemicals, although they may result from chemical reactions.

In a ***chemical reaction,*** materials called ***reactants*** undergo changes (or react chemically) to form other materials called ***products.***

When gasoline burns, the reactants are gasoline and air, and the products are carbon dioxide and water (and also, unfortunately, some ***byproducts*** called oxides of nitrogen that result from the reaction of nitrogen and oxygen in the air). When the cook mixes baking soda with vinegar, the reactants are sodium bicarbonate (soda) and acetic acid (vinegar), and the products are carbon dioxide and sodium acetate. The chemist has learned that ice, water, and steam are all the same chemical, so that when ice melts and water boils, no chemical reactions are involved.

The change from ice to water is called a ***physical change*** since no reactants disappear and no products are formed.

Chemistry is involved in many areas of the graphic arts. Drying of ink sometimes involves a chemical change and sometimes a physical change. Several chemical reactions occur during the development and fixing of photographic films and papers. Chemistry is involved in the action of light and developers on the coatings of offset and relief plates. These examples represent only a few of the kinds of chemical change that are considered in this book.

Chemical reactions follow certain rules. Some things are possible, and some are not; the more you know about these rules, the better you can predict whether two materials are apt to react with each other and the better you can understand the chemical hazards of the workplace. For example, it is impossible to change aluminum into iron, or iron into gold. Adding drier will speed the drying of some types of inks, but it will not affect the drying of other kinds of inks. An understanding of chemistry makes sense of these things and many more.

Conservation of Mass

When a chemical reaction occurs, the total amount of matter remains exactly the same; matter cannot be created or destroyed. Of course, the reactants do disappear, but products are formed in their place. The weight of the products is exactly the same as the weight of the reactants that disappeared. This fact is known as the *law of the conservation of mass.*

This is not immediately obvious. When coal burns, a chemical reaction is certainly taking place, but all that is left is a small amount of ash. The rest seems to have disappeared. It appears that matter has been destroyed. However, such is not the case. When coal burns, gaseous products including water vapor and carbon dioxide are formed. These have weight even though they are invisible. The total weight of the ash plus the gaseous products is exactly the same as the weight of the coal and air that reacted during burning.

This fundamental law of chemistry also applies to ink drying, photographic processes, development of offset plates, and ultraviolet-curing coatings.

Conservation of Energy

A similar law applies to energy. It takes exactly as much energy to decompose a gram of water into hydrogen and oxygen as is liberated when hydrogen and oxygen combine to form a gram of water. Also, the amount of heat required to melt a gram of ice is exactly the same as the amount given up when one gram of water freezes.

At the beginning of the twentieth century, Einstein showed that matter and energy can be interconverted, and he summarized it in his famous equation $e = mc^2$, but even here, the amounts of energy and matter are exactly balanced.

Materials

Gases, Liquids, and Solids

All materials have chemical structure. Materials exist in three states, called *gas, liquid,* and *solid.* The term "gas" in chemistry means much more than natural gas or heating gas; it refers to any material that is a vapor.

You are familiar with the states of many materials at room temperature. Thus aluminum, iron, and table salt are solids. Water, alcohol, and naphtha are liquids; air is a gas. These materials are respectively solids, liquids, and gases at room temperature and the pressure of the atmosphere; but a change in temperature or pressure, or both, may cause a material to change to another state. Water is a good example. If the temperature is lowered enough, liquid water changes to a solid (ice). If the temperature is raised enough, water changes to a gas (steam or water vapor). If iron is heated high enough, it changes to a liquid. Ammonia is a gas at 77°F (25°C) and one atmosphere pressure, but it becomes a liquid at 77°F (25°C) and 10 atmospheres pressure. Here a change in pressure produces a change from gas to liquid, with no change in temperature. Even a gas like oxygen can be changed to a liquid, but it must be cooled to –297°F (–183°C) at one atmosphere to accomplish the change. (It can be accomplished with less cooling by increasing the pressure.)

Pure Substances and Mixtures

A material is either a pure substance or a mixture. A *pure substance* is defined as a material that has a constant, fixed composition by weight. When a pure substance is analyzed, it always has the same percentage of elements that make up the substance. When it is decomposed electrolytically, pure water always yields exactly two volumes of hydrogen for every volume of oxygen, and this fact is expressed in the chemical abbreviation H_2O. Other examples of pure substances are crystallized, white sugar, table salt, iron, copper, magnesium, and aluminum.

Besides having a fixed composition by weight, a pure substance has a fixed melting point and a fixed boiling point (at one atmosphere pressure). It is necessary to fix the pressure, since the boiling point of a substance increases as the pressure increases.

Mixing two or more pure substances creates a *mixture.* A mixture does not have a definite composition, as the substances may be mixed in different proportions, within limits. For example, seltzer water is a mixture of carbon dioxide and water, and household ammonia is a mixture of ammonia gas and water. If these materials are allowed to stand in the open, the gases evaporate, and the composition of the mixture gradually changes. It is impossible to tell by looking at a material whether it is a pure substance or a mixture.

Most of the materials used in the graphic arts are mixtures. Inks are a mixture of several substances. Gasoline, naphtha, and other petroleum materials are mixtures. A solid dissolved in a liquid is a mixture, called a *solution.* For example, a photographic fixing bath is a solution of sodium thiosulfate ("hypo") and a hardening agent in water.

Air is a mixture of gaseous substances, mostly oxygen and nitrogen, but including carbon dioxide, water vapor, argon, and very small amounts of other substances. By cooling air to a very low temperature, it can be changed to a liquid, and then distilled by fractional distillation, to separate it into oxygen, nitrogen, and argon. Fractional distillation is done commercially on a large scale. Tanks of oxygen are used in hospitals and certain welding operations. Nitrogen is also available in tanks and is used to promote drying of electron beam (EB) inks and coatings.

Elements and Compounds

So far, materials have been divided into pure substances and mixtures. In turn, pure substances are divided into two classes, called elements and compounds. Pure substances get their characteristics from basic particles (called atoms) of which they are composed. (Atoms are described in more detail later in this chapter.) *Elements* are substances made up of only one type of atom. Many familiar substances including iron, copper, oxygen, hydrogen, iodine, nitrogen, and sulfur are elements.

A chemical *compound* consists of two or more elements chemically combined in a fixed proportion by weight. As an example, the compound magnesium nitrate consists of 16.4% magnesium, 18.9% nitrogen, and 64.7% oxygen by weight. In this case, the three elements—magnesium, nitrogen, and oxygen—have combined to form the compound called magnesium nitrate.

Many chemical compounds occur in nature. Water is a chemical compound of the elements hydrogen and oxygen.

Common table salt has the chemical name sodium chloride and is a compound of the elements sodium (a metal) and chlorine (a highly toxic gas). Limestone is principally calcium carbonate, a chemical combination of calcium (a metal), carbon, and oxygen, all of which are elements.

Some chemical compounds do not occur in nature, but chemists have devised ways to chemically combine naturally occurring materials in new ways. If these materials do not readily react, the reaction can often be forced by the use of heat, electric current, or light.

Only a few of the elements occur as such in nature. Among the common ones are oxygen, nitrogen, and argon that occur in air, and sulfur and gold that occur in the earth. Diamond is pure carbon. Most of the common metals, such as iron, lead, zinc, aluminum, and copper, occur in nature only as compounds. By one method or another, these compounds are broken down to produce the metals in their elementary state (as free elements). Many big companies produce iron and steel, aluminum, magnesium, and other metals from ores (naturally occurring compounds) of these metals.

Figure 1-1 shows the relationship between materials, pure substances, elements, and compounds.

Figure 1-1.
The relationship between materials, pure substances, elements, and compounds.

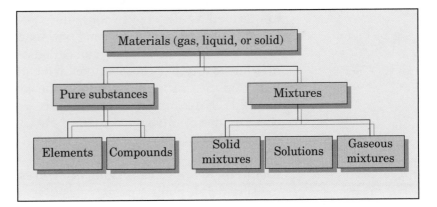

Metals and Nonmetals

Elements may be divided into **metals** and **nonmetals.** Table 1-I lists the common metals and nonmetals, many of which are discussed later. The abbreviations after the name of the element, such as Au for gold and Ag for silver, are the **symbols** of the element. Symbols are used when writing equations for chemical reactions.

Metals commonly react with nonmetals to form compounds. Sometimes a metal is already combined in nature with a nonmetal. For example, iron (a metal) combines with

Table 1-I.
Common metals and nonmetals, with symbols.

Metals		Nonmetals
Aluminum, Al	Mercury, Hg	Argon, Ar
Barium, Ba	Molybdenum, Mo	Bromine, Br
Cadmium, Cd	Nickel, Ni	Carbon, C
Calcium, Ca	Platinum, Pt	Chlorine, Cl
Chromium, Cr	Potassium, K	Fluorine, F
Cobalt, Co	Silver, Ag	Hydrogen, H
Copper, Cu	Sodium, Na	Iodine, I
Gold, Au	Tin, Sn	Nitrogen, N
Iron, Fe	Titanium, Ti	Oxygen, O
Lead, Pb	Tungsten, W	Phosphorus, P
Magnesium, Mg	Zinc, Zn	Silicon, Si
Manganese, Mn		Sulfur, S

oxygen (a nonmetal) to form magnetite, an iron ore. This is only a general rule, because a metal may not react with all of the nonmetals. Platinum and gold are very unreactive. Metals do not react with other metals under most conditions, but under certain conditions, so-called "intermetallic compounds" are formed as alloys of two metals.

Some of the nonmetals react with each other. Carbon forms a compound with oxygen called carbon dioxide and also forms another compound called carbon monoxide. Carbon combines with sulfur to form the compound carbon disulfide. Nitrogen forms compounds with oxygen. Some of these oxides of nitrogen contribute to the smog that forms over our big cities.

Physical Properties

Physical properties include color, crystalline form, density, melting point, boiling point, refractive index, surface tension (of liquids), tensile strength, and thermal expansivity. The term "physical properties" is used because these qualities are not concerned with the chemical reaction of a material with another material.

Typical physical properties of metals include a metallic luster, such as the characteristic reflective glow of clean, polished copper, silver, or gold. Many are malleable (capable of being hammered into thin sheets), and they are usually ductile (capable of being drawn into a wire). In addition, many are fair-to-excellent conductors of electricity and heat.

Law of Combining Proportions

Over the centuries, chemists have found that elements combine to form compounds in exact proportions. One volume of oxygen combines with exactly two volumes of hydrogen, and the more carefully the experiment is carried out, the more precisely the relationship is confirmed. Actually, there are two different compounds of hydrogen and oxygen, water and hydrogen peroxide. In hydrogen peroxide, one volume of oxygen combines with one volume of hydrogen, but the ratio is precisely one to one.

This is one of the most important facts supporting the atomic structure of matter. The fact that one volume of oxygen always reacts with exactly one or exactly two volumes of hydrogen can only be explained by the fact that two atoms of hydrogen react with one or two atoms of oxygen, never with one and nine-tenths of an atom or one and seven-sixteenths of an atom. The ***law of combining proportions*** results from this observation.

Chemical Measurements

This leads to the observation that careful measurements are necessary in order to establish the laws of science. Chemists have found that it is much simpler to carry out calculations using the metric system of meters, grams, and liters, instead of the clumsy imperial system that uses feet and yards, ounces and pounds, and (liquid) ounces and cups and gills and pints and quarts. In the metric (or SI) system, conversions from one measurement to a larger or smaller one are always done in factors of ten, and the conversion can be done without a calculator. The metric system is presented in appendix B. It is used throughout the world, except in the United States, and it is becoming increasingly familiar even here.

Chemical Symbols and Formulas

Every science and technology has its own terminology and abbreviations as scientists attempt to convey large amounts of information in a few words or symbols. Chemists frequently use letters and numbers to communicate the results of centuries of observation and experimentation. Some of these letters and numbers are familiar to the general public: H_2O (water), H_2SO_4 (sulfuric acid), and NaCl (table salt). But the observations and experiments are usually unfamiliar, and the symbols become confusing instead of helpful. The formula H_2O tells the chemist that one volume of water vapor can be decomposed (by electricity) to yield one volume of hydrogen gas and one half volume of oxygen gas, or one volume of oxygen and two volumes of hydrogen will react to give two volumes of water vapor. The entire reactions can be written:

$$H_2O \rightarrow H_2 + \tfrac{1}{2}O_2$$
$$\text{or}$$
$$O_2 + 2H_2 = 2H_2O$$

Use of chemical symbols should help the reader, not pose a problem. Chemical symbols become especially significant in the chemistry of carbon compounds. (See chapter 3, "Organic Chemistry.")

Because carbon atoms can form stable compounds by bonding to each other as well as to other atoms, the symbols indicate not only the number of atoms in a molecule, but the order in which they are connected. For example, C_2H_6O can be either:

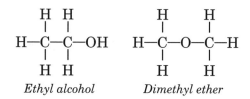

Ethyl alcohol   Dimethyl ether

These are two different compounds with different properties and different chemical behavior.

Atoms

All chemical elements consist of very tiny particles called **atoms.** A piece of aluminum weighing one ounce consists of 6.35×10^{23}, which is a short way of saying that it contains 635,000,000,000,000,000,000,000 atoms! Thus the weight of a single atom is very, very small. Yet each aluminum atom reacts chemically the same as every other aluminum atom.

The reason sulfur or oxygen is different chemically from aluminum is that sulfur and oxygen atoms are different from aluminum atoms. Indeed, the atoms of all the elements differ from one another in their weights. They also differ in the structure of the atoms, and this difference is very important because the structure of the atoms determines how those atoms react chemically.

Structure of Atoms

The atoms of all elements are made up of still smaller particles called protons, neutrons, and electrons. The atoms of all elements consist of some combination of protons, neutrons, and electrons. The number of protons in each atom makes the atoms of any element different from the atoms of another element.

The atoms of all elements consist of two main parts—the center, or **nucleus,** consisting of **protons** and **neutrons,** and the **electrons** outside of the nucleus. These three particles differ in their mass and electrical charge, as follows:

- Proton—"unit" mass and "unit" positive electrical charge.
- Neutron—"unit" mass and no electrical charge
- Electron—"unit" negative electrical charge

The protons carry an electrical charge, and the neutrons, which are electrically neutral, hold all of the protons together in the nucleus. The number of electrons exactly matches the number of protons to make the atom electrically neutral. (If there are more or fewer electrons, then the atom is known as an **ion.)**

The mass of an electron is only 1/1837 times as great as the mass of the proton and the neutron, and for most practical purposes it can be considered to be zero.

It is possible to give a number to every element, starting with one (hydrogen) and going up to 109 or higher (scientists are still making new elements in the laboratory). This number is called the **atomic number** and is equal to the number of protons in the nucleus of the atoms of that element. For example, the metallic element sodium has the atomic number 11. This means that all sodium atoms have eleven protons in the nucleus. And, since atoms have no net electrical charge, all sodium atoms have eleven electrons around the nucleus (extranuclear electrons).

The total weight of any atom is equal to the sum of the weights of the protons and neutrons in the nucleus. (Remember, the weight of the electrons is negligible.) The **mass number** of a particular kind of atom is the sum of the protons and the neutrons in the nucleus. The mass number of sodium is 23. Since there are eleven protons in the nucleus of sodium atoms, twenty-three minus eleven gives twelve neutrons in the nucleus.

The complete symbol for sodium is:

$$\text{Mass number} \rightarrow {}^{23}_{11}\text{Na} \leftarrow \text{Atomic number}$$

and the structure of a sodium atom is:

Nucleus	**Extranuclear electrons**
11 p	11 e
12 n	

Only one more thing is needed to complete the structure of sodium, and other, atoms. The extranuclear electrons have different energies and exist in different energy "levels" or "shells." Two electrons are the maximum number that can be accommodated in Energy Level 1. Level 2 holds a maximum of eight electrons. Level 3 holds a maximum of eight electrons with the lighter atoms, but with the elements heavier than calcium, it holds up to eighteen (see the periodic table in appendix A). A sodium atom has eleven extranuclear electrons. Two of these are in Level 1, eight (the maximum) in Level 2, leaving one electron in Level 3 (figure 1-2).

Figure 1-2.
Structure of a sodium atom.

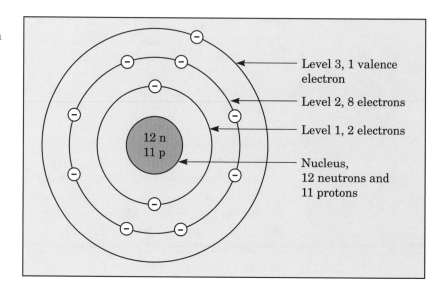

The number of electrons in the outermost energy level is especially important. These are the most loosely bonded electrons, and they can be lost (or shared) when atoms of one element combine with the atoms of another element. These electrons that can participate in forming chemical bonds are called ***valence electrons.*** Sodium atoms, for example, have only one valence electron.

With a little practice, and by using the rules given above, you can show the structure of the atoms of any element—at least up to element No. 20, calcium, if you know its atomic number and its mass number. Thus the structure of $^{32}_{16}$S (sulfur) is:

Nucleus	**Extranuclear electrons**		
	Level 1	Level 2	Level 3
16 p			
16 n	2 e	8 e	6 e

The atomic number of sulfur is 16; there are sixteen protons in the nucleus and sixteen extranuclear electrons. Of these sixteen electrons, two exist in Level 1 and eight in Level 2, leaving six in Level 3. Since the mass number is 32, the nucleus has sixteen neutrons in the nucleus (figure 1-3).

Figure 1-3.
Structure of a sulfur atom.

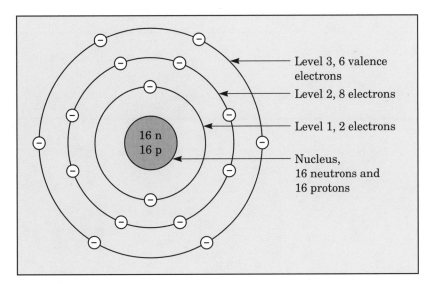

Atoms of a particular element must have as many protons in the nucleus and as many extranuclear electrons as the atomic number of that element. However, sometimes atoms of a particular element do not all have the same weight (mass). Some may have more neutrons in the nucleus than others. Atoms that have the same atomic number but have different numbers of neutrons in the nucleus are known as *isotopes.*

Hydrogen, the lightest of all the elements, has an atomic number of 1. Most hydrogen atoms also have a mass number of 1. Thus the structure of $^{1}_{1}$H is:

Nucleus	**Extranuclear electrons**
1 p	1 e

Hydrogen atoms have *one* valence electron.

A very small percentage of hydrogen atoms have a mass number of 2. These are called ***deuterium*** atoms (previously "heavy hydrogen"). The structure of deuterium, $_1^2H$, is:

Nucleus	**Extranuclear electrons**
1 p	1 e
1 n	

Deuterium is still hydrogen since there is one proton in the nucleus, but the atoms of deuterium are twice as heavy as those of $_1^1H$. Another isotope of hydrogen, ***tritium,*** has two neutrons in the nucleus (figure 1-4).

Figure 1-4.
Isotopes of hydrogen.

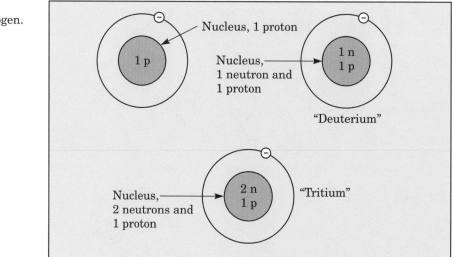

Many elements have several isotopes, but some have only one. Table 1-II gives the structure of the atoms of elements with atomic numbers 1 through 20. The electrons in the outermost energy level are the valence electrons. Thus carbon has four valence electrons, nitrogen has five, oxygen has six, and chlorine has seven. This difference in the number of valence electrons makes one element react differently from another.

The Periodic Table of the Elements

By the early 1800s, chemists had noted the similarity between lithium, sodium, and potassium, and between fluorine, chlorine, bromine, and iodine, and they had tried to show that all of the elements could be grouped according to their reactions and physical properties. In the mid-19th century,

Table 1-II.
Structure of the
principal atoms of
elements 1–20.

Element	Atomic No.	Mass No.	Nucleus Protons	Nucleus Neutrons	Extranuclear Electrons Level No. 1	2	3	4
Hydrogen	1	1	1	0	1			
Hydrogen	1	2	1	1	1			
Helium	2	4	2	2	2			
Lithium	3	6	3	3	2	1		
Lithium	3	7	3	4	2	1		
Beryllium	4	9	4	5	2	2		
Boron	5	11	5	6	2	3		
Carbon	6	12	6	6	2	4		
Nitrogen	7	14	7	7	2	5		
Oxygen	8	16	8	8	2	6		
Fluorine	9	19	9	10	2	7		
Neon	10	20	10	10	2	8		
Sodium	11	23	11	12	2	8	1	
Magnesium	12	24	12	12	2	8	2	
Magnesium	12	25	12	13	2	8	2	
Magnesium	12	26	12	14	2	8	2	
Aluminum	13	27	13	14	2	8	3	
Silicon	14	28	14	14	2	8	4	
Phosphorus	15	31	15	16	2	8	5	
Sulfur	16	32	16	16	2	8	6	
Chlorine	17	35	17	18	2	8	7	
Chlorine	17	37	17	20	2	8	7	
Argon	18	40	18	22	2	8	8	
Potassium	19	39	19	20	2	8	8	1
Calcium	20	40	20	20	2	8	8	2

a Russian chemist by the name of Mendeleev successfully arranged all of the elements into a scheme that is known as the ***periodic table.*** At the beginning of the 20th century, as the structure of atoms became known, chemists learned that elements in Group I all had one valence electron in the outer shell, elements in Group VII had seven valence electrons in the outer shell, and the noble gases (Group VIII) all had a completed outer shell.

The periodic table, reproduced in appendix A, presents the elements in the order of their atomic structure, and accordingly, they are grouped together according to their chemical behavior. This periodic table helps chemists remember the elements and understand their chemistry.

Four important families of elements are the alkali metals, alkaline earth metals, halogens, and noble gases.

Alkali metals. The common metals in this family are lithium, sodium, and potassium. The atoms of all these elements have *one* valence electron. They readily give up one electron so that the ions have a +1 charge: Na^+ or K^+.

Alkaline earth metals. The common metals in this family are beryllium, magnesium, calcium, strontium, and barium. The atoms of these elements have *two* valence electrons and produce ions with a +2 charge: Ca^{2+} or Mg^{2+}.

The atoms of most metals have one or two valence electrons although aluminum has three. So elements classified as metals usually have one, two, or at most three valence electrons. The heavy metals lead and tin are exceptions. Their atoms have four valence electrons.

Halogens. The halogens are nonmetals and include fluorine, chlorine, bromine, and iodine. The atoms of all these elements have *seven* valence electrons. These elements readily accept an electron to fill the outer shell, producing an ion with a −1 charge: Cl^- or Br^-.

Noble gases. The noble gases include helium, neon, argon, krypton, and xenon. In the atoms of all of the noble gases, the outermost energy level is filled with all of the electrons it can hold. Helium has two electrons in Level 1, all that Level 1 can hold. As general rule, the others have eight electrons, and that is all that the outermost energy level can hold.

Since the outermost energy levels of the atoms of the noble gases are filled, one can say that these atoms do not have any valence electrons. Therefore, the noble gases are extremely inert. Until 1962, no compounds of these gases had ever been produced.

Atomic Weights

Since the actual weight of atoms is such a very small figure, chemists have adopted the practice of using the relative weights of the atoms of elements. These are called the *atomic weights* of the elements even though they are expressed in relative, or unitless, numbers.

Atomic weights can be used to calculate weight relations in chemical reactions. By their use, you can calculate what weight of one substance will react with a given weight of another substance, and what the weights will be of the substances that are produced.

In originally making a table of atomic weights, it was necessary to pick one element as the standard. For many years oxygen was used as the standard and was assigned an atomic weight of 16.000. The number 16 was chosen so that the lightest element, hydrogen, would have an atomic weight close to one. By using chemical combining weights, it was possible to determine the relative weights of other elements, compared with 16 for oxygen.

Chemists used the table of atomic weights for many years before it was discovered that many elements consist of two or more isotopes. This discovery helped to explain why the atomic weights of the elements are not even multiples of the atomic weight of hydrogen. Actually the exact atomic weights are based on the combining weights of elements in chemical reactions that involve billions and billions of atoms. For example, the atomic weight of chlorine (35.5) represents the average weight of chlorine atoms. (An examination of table 1-II suggests that chlorine normally contains about three atoms of $^{35}_{17}Cl$ to one of $^{37}_{17}Cl$, and this is confirmed by experiment.) It is such average weights that determine the proportions by weight of substances that are involved in a chemical reaction. The approximate atomic weights of many of the common elements are given in table 1-III.

In recent years, the expansion of the metric system known as "SI" (for Systeme International) has provided a new standard for atomic weights. It is based on using 12.000 for the atomic weight of the principal isotope of carbon, with a mass number of 12. This new definition changes the former atomic weights by only 0.0045%. Although this is a small amount, it is important in the most precise scientific calculations.

Molecules

Molecules of the Elements

If two or more atoms, either alike or different, are held tightly together, the combination is called a ***molecule.*** Most elements consist of single atoms, but in a few cases, two atoms of an element join to form a molecule of the element. Hydrogen, for instance, exists as molecules having two atoms each. The molecular formula is H_2.

Why does hydrogen form such molecules? Any atom is more stable when it has a completed level of outer electrons, which make it more like the noble gases. Hydrogen has one valence electron, and the first energy level is complete with two electrons. When two hydrogen atoms join to form a molecule of hydrogen, they share two electrons, and this makes them more like the nearest noble gas, helium, which has a

Table 1-III.
Selected characteristics of common elements.

Element	Symbol	Atomic Number	Atomic Weight	No. of Valence Electrons
Aluminum	Al	13	27.0	3
Argon	Ar	18	39.9	0
Barium	Ba	56	137.3	2
Boron	B	5	10.8	3
Bromine	Br	35	79.9	7
Cadmium	Cd	48	112.4	2
Calcium	Ca	20	40.1	2
Carbon	C	6	12.01	4
Chlorine	Cl	17	35.5	7
Chromium	Cr	24	52.0	1 (T)
Cobalt	Co	27	58.9	2 (T)
Copper	Cu	29	63.5	1 (T)
Fluorine	F	9	19.0	7
Gold	Au	79	197.0	1 (T)
Helium	He	2	4.0	0
Hydrogen	H	1	1.008	1
Iodine	I	53	126.9	7
Iron	Fe	26	55.8	2 (T)
Krypton	Kr	36	83.8	0
Lead	Pb	82	207.2	4
Magnesium	Mg	12	24.3	2
Manganese	Mn	25	54.9	2 (T)
Mercury	Hg	80	200.6	2
Molybdenum	Mo	42	95.9	1 (T)
Neon	Ne	10	20.2	0
Nickel	Ni	28	58.7	2 (T)
Nitrogen	N	7	14.01	5
Oxygen	O	8	16.00	6
Phosphorus	P	15	31.0	5
Platinum	Pt	78	195.1	1 (T)
Potassium	K	19	39.1	1
Silicon	Si	14	28.1	4
Silver	Ag	47	107.9	1 (T)
Sodium	Na	11	23.0	1
Sulfur	S	16	32.1	6
Tin	Sn	50	118.7	4
Titanium	Ti	22	47.9	2 (T)
Tungsten	W	74	183.8	2 (T)
Xenon	Xe	54	131.3	0
Zinc	Zn	30	65.4	2

(T) = "transition" element

completed energy level of two electrons. A pair of electrons shared between two atoms, whether the two atoms are of the same element or two different elements, is called a ***covalent bond.*** Valence electrons are often pictured as small black dots. Thus a hydrogen atom is written as H·, and a molecule of hydrogen is written H:H. The sharing of two electrons between two hydrogen atoms is the chemical force that holds the molecule together (figure 1-5).

Figure 1-5.
Two hydrogen atoms forming a molecule of hydrogen.

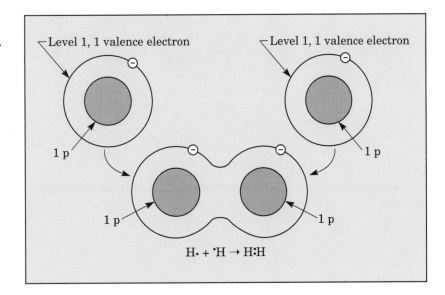

$$H\cdot + \cdot H \rightarrow H\!:\!H$$

Fluorine, chlorine, bromine, and iodine each have seven valence electrons. The noble gas nearest to each of them has completed energy levels of eight electrons. two atoms of these elements can share two electrons between them so that each atom has a complete energy level of eight electrons. Chlorine, for example, can be written:

$$:\!\overset{\cdot\cdot}{\underset{\cdot\cdot}{Cl}}\cdot$$

But when two of these combine to give a Cl_2 molecule, the structure is:

$$:\!\overset{\cdot\cdot}{\underset{\cdot\cdot}{Cl}}\!:\!\overset{\cdot\cdot}{\underset{\cdot\cdot}{Cl}}\!:$$

In a similar way, atoms of the other halogens combine to form diatomic molecules—F_2, Br_2, and I_2. Other elements that also form diatomic molecules include oxygen, O_2, and nitrogen, N_2.

Molecules of Compounds

Atoms of different elements combine chemically to form molecules of a compound. The atoms in thousands of different kinds of molecules are held together by covalent bonds. A simple example is the compound water. The ***molecular formula*** of water is H_2O, a shorthand expression meaning that a molecule of water consists of two atoms of hydrogen combined with one atom of oxygen.

Oxygen atoms have six valence electrons and require eight to complete the second energy level. (The nearest noble gas, neon, has a completed second energy level of eight electrons.) Because one hydrogen atom has only one valence electron, it takes two hydrogen atoms to combine with one atom of oxygen to form one molecule of H_2O. A pair of electrons is shared between the oxygen atom and each hydrogen atom. We can picture this as follows:

$$2H\cdot \text{ plus } \cdot \ddot{\underset{\cdot}{O}}: \text{ gives } H:\ddot{\underset{\cdot\cdot}{O}}:$$
$$\qquad\qquad\qquad\qquad\qquad H$$

Another compound of hydrogen and oxygen is the bleaching agent hydrogen peroxide—H_2O_2. Its structure is:

$$\begin{array}{c} H \\ :\ddot{\underset{\cdot\cdot}{O}}:\ddot{\underset{\cdot\cdot}{O}}: \\ H \end{array}$$

In molecules of hydrogen peroxide, each oxygen atom shares two electrons with one hydrogen atom. In addition, the two oxygen atoms share two electrons with each other. Such an arrangement is less stable than the H_2O arrangements. Hydrogen peroxide decomposes readily to make water and a free oxygen atom, which makes hydrogen peroxide an effective bleaching agent.

Gram Molecular Weight

The ***gram molecular weight*** is the molecular weight of a pure substance, compared to the weight of a carbon 12 atom. Thus, the gram molecular weight of hydrogen is 2.016 (H_2), the gram molecular weight of nitrogen gas is 28.02 (N_2), and the gram molecular weight of water is 18.016.

Before isotopes were discovered, chemists were puzzled by the fact that combining weights were usually close to, but never quite exactly whole numbers. Since we now know that oxygen in the air contains some oxygen with an atomic weight of 18, and it combines with hydrogen, most of which has an atomic weight of 1, but some of which weighs 2, it is apparent why the combining weights are not exactly 1:8 (H_2O).

The gram molecular weight of any gas has a volume of 22.41 L under standard conditions of temperature and pressure, a volume that contains 6.023×10^{23} molecules. (This is called ***Avogadro's number*** for the Italian chemist who first determined the number of molecules in a gram molecular weight of any pure substance.) It is the number of atoms in 16.00 g of oxygen 16, or 12.00 g of carbon 12. It is the number of molecules in 32 g of oxygen 16 (O_2), or the number of molecules in 18.016 g of water (H_2O) or in 44.01 g of CO_2.

The atomic weight of each element is given in appendix A.

The Difference Between Mixtures and Compounds

A vast difference exists between a mixture and a compound. A mixture of hydrogen and oxygen gases consists, of course, of hydrogen and oxygen. The compound water also consists of hydrogen and oxygen. Water consists of molecules of H_2O with the hydrogen and oxygen atoms chemically combined. A mixture of hydrogen and oxygen gases consists of H_2 molecules and O_2 molecules. Furthermore, since they are a mixture, the two gases can be present in any proportion and not necessarily in the proportion required to produce molecules of H_2O.

Suppose a mixture consists of 4 g of H_2 and 32 g of O_2. This is the correct proportion by weight required to form H_2O, because hydrogen molecules are much lighter than oxygen molecules, but they cannot react with each other until they are "activated." Activation requires an electrical spark or a lighted match, which provides what is called the ***energy of activation.*** The gases react so rapidly that there is an explosion. It can be expressed as follows: 4 g (0.14 oz) H_2 react with 32 g (1.12 oz) O_2 to produce 36 g (1.26 oz) H_2O plus 569,400 joules (136.0 kilocalories or 420,000 foot-pounds) of energy. This reaction is another example of the law of the conservation of mass. The weight of H_2 and O_2 that disappear (4 g plus 32 g) is exactly the same as the weight of water (36 g) that is formed. The reaction also liberates a large quantity of heat. All chemical reactions proceed either with emission or absorption of energy of one kind or another. Sometimes the amount is small. Other times, as with hydrogen and oxygen, it is sufficient to cause an explosion.

If a material is dissolved in water to enable the solution to carry an electric current, and two carbon electrodes are inserted into the water, it is possible to pass an electric current through the water and decompose the water molecules. Hydrogen gas is formed at one electrode and oxygen gas at the other electrode. Energy is being consumed to reverse the chemical reaction just discussed. It is found that the electrical energy required to form 4 g of H_2 and 32 g of O_2 is 569,400 joules. This is an example of the **law of the conservation of energy,** which states that energy cannot be created or destroyed. In this case, the amount of energy released when water is formed from hydrogen and oxygen gases is exactly equal to the amount of energy absorbed to convert the water back to hydrogen and oxygen gases.

The energy involved in changing a substance from one state to another is an additional example of the law of the conservation of energy. For instance, heat must be absorbed to change 18 g (0.63 oz) of ice into liquid water at 0°C. (32°F). This is called the **molar heat of fusion.** More heat must be absorbed to raise the temperature of the liquid water to 100°C (212°F), the boiling point. A considerable amount of heat is then absorbed at 100°C to convert the water into steam. This is called the **molar heat of vaporization.** Now, if this process is reversed, exactly the same amount of heat is emitted that was originally absorbed in each step.

Ionization and Conductivity

The atoms in the molecules of many compounds are held together by covalent bonds: bonds in which each atom shares electrons. Others are held together with electrostatic bonds. If such a compound is soluble in water, it may break apart to form electrically charged particles that are called **ions.** One kind of ion carries a positive charge and the other a negative charge. Thus the common **nitrate** ion is NO_3^-: the combination of three oxygen atoms and one nitrogen atom has an electrical charge of −1. In solution, negative ions are always balanced by positive ions so that the net electrical charge is zero.

Atoms of the elements are uncharged, with as many extranuclear electrons as there are protons in the nucleus. Some kinds of atoms, however, tend to lose electrons while other kinds tend to gain them. Consider table salt, sodium chloride, NaCl, as an example. Sodium ions (that have one valence electron in the outer shell) can achieve a stable configuration (eight electrons in energy Level 1) by losing one electron. Chlorine atoms have seven valence electrons and can achieve a stable shell by gaining one electron. As a consequence, each sodium atom donates its valence electron to a chlorine atom. Since the electron donated carries a negative charge, the chlorine atom is changed to a Cl^- ion, called a *chloride* ion. The sodium atom, which lost an electron, becomes a Na^+ ion (figure 1-6).

Figure 1-6.
Sodium atom donating an electron to a chlorine atom.

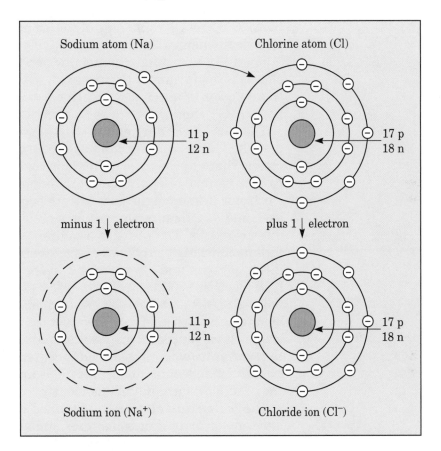

The stability of the ions with complete outer shells accounts for the fact that sodium chloride is a very stable and neutral compound while sodium metal and chlorine gas are very reactive.

The tendency for this electron transfer is so great in the case of sodium chloride, NaCl, that even solid sodium chloride consists of a network of sodium and chloride ions; even in the solid state there are no molecules of NaCl. So when NaCl dissolves in water, the Na^+ and Cl^- ions merely separate and start moving around among the water molecules. A solid compound with this characteristic is called an *ionic solid.*

The compound calcium chloride has the formula $CaCl_2$. Calcium atoms have two valence electrons. When calcium chloride dissolves in water, each calcium atom donates an electron to each of two chlorine atoms. The calcium atom minus two electrons becomes the Ca^{2+} ion. A solution of calcium chloride in water consists of billions of Ca^{2+} ions and twice as many Cl^- ions. This of course makes the solution electrically neutral.

Similarly, an aluminum atom, which has three valence electrons, can transfer the electrons to three chlorine atoms, so a solution of aluminum chloride in water consists of billions of Al^{3+}, and three times as many Cl^-.

The same procedure occurs with more complicated compounds. Magnesium nitrate has the formula $Mg(NO_3)_2$. This means that one magnesium atom is combined with two nitrate groups, NO_3^-. Magnesium atoms have two valence electrons. When magnesium nitrate is dissolved in water, each magnesium atom contributes an electron to each of two nitrate groups. The result is a solution with billions of magnesium ions, Mg^{2+}, and twice as many nitrate ions, NO_3^-. Since nitrogen and oxygen atoms in the nitrate group are held together by shared electrons (covalent bonds), they do not ionize in water. The magnesium and nitrate groups are held together with ionic or electrostatic bonds, and they do ionize. More precisely, magnesium nitrate is an ionic solid that *solvates* when it is placed in water. In an aqueous solution, the magnesium and nitrate ions are separated, and they can move around independently.

The *transition elements* (identified with "T" in table 1-III) have one or two valence electrons and are metals. Examples are iron, copper, silver, and gold. What makes the transition elements special is that they can lose not only their valence electrons, but they can also lose an electron from the energy level next to the outermost energy level.

Copper, for example, can form *cuprous* chloride, $CuCl$, where the one valence electron of copper atoms is involved, or it can form *cupric* chloride, $CuCl_2$. In water solution, each $CuCl_2$ molecule gives a cupric ion, Cu^{2+}, and two chloride ions, Cl^-. In this case, each copper atom has lost its one valence electron and also one from the energy level next to the outermost one. In the same way, iron atoms can lose their two valence electrons to form ferrous ions, Fe^{2+}, or they can lose three electrons to form ferric ions, Fe^{3+}. So one chloride compound is ferrous chloride, $FeCl_2$, and another is ferric chloride, $FeCl_3$.

Tables 1-IV and 1-V list the common ions with their positive or negative charge. These ions cannot exist alone. In any water solution, or in any ionic solid, there must be exactly as many positive charges as negative. Certain ions, such as oxygen ions, O^{2-}, are not found in nature. Oxygen forms compounds with other elements that are held together by covalent bonds. It is merely convenient to consider oxygen as having a –2 charge when writing formulas for compounds of oxygen with another element.

Table 1-IV.
Some positively charged ions.

Name	Symbol	Name	Symbol
Aluminum	Al^{3+}	Lithium	Li^+
Ammonium	NH_4^+	Magnesium	Mg^{2+}
Barium	Ba^{2+}	Manganese (II),	Mn^{2+}
Calcium	Ca^{2+}	or manganous	
Chromium (II),	Cr^{2+}	Mercury (I),	Hg_2^{2+}
or chromous		or mercurous	
Chromium (III),	Cr^{3+}	Mercury (II),	Hg^{2+}
or chromic		or mercuric	
Cobalt	Co^{2+}	Nickel (II),	Ni^{2+}
Copper (I),	Cu^+	or nickelous	
or cuprous		Nickel (III),	Ni^{3+}
Copper (II),	Cu^{2+}	or nickelic	
or cupric		Potassium	K^+
Hydrogen,	H^+, H_3O^+	Silver	Ag^+
or hydronium		Sodium	Na^+
Iron (II),	Fe^{2+}	Tin (II),	Sn^{2+}
or ferrous		or stannous	
Iron (III),	Fe^{3+}	Tin (IV),	Sn^{4+}
or ferric		or stannic	
Lead	Pb^{2+}	Zinc	Zn^{2+}

Formulas, Valence, and Combining Power

It has already been explained why the formula of sodium chloride is NaCl, calcium chloride is $CaCl_2$, and aluminum chloride is $AlCl_3$. Formulas for many other compounds can be written by using the proper combination of the positive and negative ions given in tables 1-IV and 1-V, based on the fact that the molecules of any compound are *uncharged*.

Chemists use the term **valence** to indicate the combining power of an atom. Sodium readily loses its outermost electron and gains one positive charge, while chlorine readily accepts one electron and gains one negative charge. Thus,

Table 1-V.
Some negatively
charged ions.

Name	Symbol	Name	Symbol
Acetate	CH_3COO^-	Hydroxide	OH^-
Aluminate	AlO_2^-	Hypochlorite	ClO^-
Bromide	Br^-	Iodide	I^-
Carbonate	CO_3^{2-}	Metaborate	BO_3^{3-}
Chlorate	ClO_3^-	Metasilicate	SiO_3^{2-}
Chloride	Cl^-	Monohydrogen	HPO_4^{2-}
Chlorite	ClO_2^-	phosphate	
Chromate	CrO_4^{2-}	Nitrate	NO_3^-
Citrate	$C_6H_5O_7^{3-}$	Nitrite	NO_2^-
Cyanide	CN^-	Oxalate	$C_2O_4^{2-}$
Dichromate	$Cr_2O_7^{2-}$	Oxide	O^{2-}
Dihydrogen phosphate	$H_2PO_4^-$	Perchlorate	ClO_4^-
Ferricyanide	$Fe(CN)_6^{3-}$	Permanganate	MnO_4^-
Ferrocyanide	$Fe(CN)_6^{4-}$	Phosphate	PO_4^{3-}
Fluoride	F^-	Sulfate	SO_4^{2-}
Hydrogen carbonate,	HCO_3^-	Sulfide	S^{2-}
or bicarbonate		Sulfite	SO_3^{2-}
Hydrogen sulfate,	HSO_4^-	Tartrate	$C_4H_4O_6^{2-}$
or bisulfate		Tetraborate	$B_4O_7^{2-}$

sodium has a valence of +1 and chlorine has a valence of –1. Calcium has a valence of +2, and it reacts with two chlorine atoms. Sometimes, chemists refer to the electrons in the outer shell of the atom as valence electrons, because these electrons are lost or gained in order to make a complete energy level or shell.

The total positive charge of the positive ions must be the same as the total negative charge of the negative ions. Here are some examples:

- **Sodium sulfate.** Formula is Na_2SO_4. (Two sodium ions each with a +1 charge are needed to balance the –2 charge of the sulfate ion.)
- **Ammonium chloride.** Formula is NH_4Cl.
- **Potassium dichromate.** (Printers often call this "bichromate.") Formula is $K_2Cr_2O_7$. (Two potassium ions each with a +1 charge are needed to balance the –2 charge of the dichromate ion.)
- **Calcium phosphate.** Formula is $Ca_3(PO_4)_2$. (Three calcium ions each with a +2 charge are needed to balance two phosphate ions each with a –3 charge.)
- **Iron (III) oxide.** Formula is Fe_2O_3. (Two ferric ions each with a +3 charge are balanced with three oxygens, with a charge of –2 each.)

Tables 1-IV and 1-V can be used to assist in writing the formulas of other compounds.

Most of the chemical elements occur in nature as compounds. The principal ore of silver is silver sulfide, Ag_2S. Zinc occurs in nature as zinc sulfide, ZnS. Aluminum occurs as bauxite, $Al_2O_3 \cdot 2H_2O$. The dot in the center of the preceding formula marks the division between different molecules in combination. In bauxite, two molecules of water are combined with one molecule of aluminum oxide, Al_2O_3. Calcium occurs in nature as calcium carbonate, $CaCO_3$, and also as calcium phosphate, $Ca_3(PO_4)_2$.

When some compounds that are dissolved in water form solid crystals as the solution is evaporated or cooled, they carry with them a certain amount of water, called the ***water of crystallization.*** Sodium carbonate forms:

$$Na_2CO_3 \cdot 10H_2O$$

This is called ***hydrated*** sodium carbonate because of the associated water molecules. It is often possible to drive off this water by heating the crystals or sometimes merely exposing them to dry air. Then the crystalline form disappears and the compound becomes a powder, called an ***anhydrous*** compound. The hydrated and anhydrous forms behave the same in a chemical reaction. The only difference is that a greater weight of the hydrated form must be used because of the water present in it.

Sometimes when two compounds are dissolved together in water, and the solution is evaporated or cooled, solid crystals form that contain both compounds. The ***alums*** are examples of this. Here is the formula of potassium aluminum alum, a common alum:

$$K_2SO_4 \cdot Al_2(SO_4)_3 \cdot 24H_2O$$

The two compounds K_2SO_4 and $Al_2(SO_4)_3$ do not combine chemically. They simply crystallize out together. When the crystals are dissolved in water, they merely produce a mixture of the two compounds. (Incidentally, papermakers' alum is aluminum sulfate, a different substance.)

Inorganic and Organic Compounds

Most of the compounds that have formulas that can be written with the aid of tables 1-IV and 1-V are called ***inorganic compounds. Organic compounds*** are compounds of the element carbon. In fact, all compounds containing carbon are called organic except for carbon dioxide, CO_2; carbon monoxide, CO; and a few compounds such as sodium carbonate, Na_2CO_3 and sodium cyanide, $NaCN$. There are so many organic compounds and so many types, they are covered in a separate chapter.

Acids and Bases

The subject of acids and bases is treated at length in chapter 2. These materials react with each other in a reaction that is called neutralization. The reaction can be expressed in a chemical *equation.*

$$HCl + NaOH \rightarrow NaCl + H_2O$$

This means that a molecule of HCl (acid) reacts with a molecule of $NaOH$ (base) to form a molecule of $NaCl$ (a salt) and a molecule of water. This does not completely describe all that happens, but it is a convenient equation for calculating what weight of HCl will react with a given weight of $NaOH$.

If the atomic weights in HCl, approximately 1.0 and 35.5, are added, the total, 36.5, is called the *molecular weight* of HCl. In the same way, the atomic weights 23, 16, and 1 are added to give a total of 40, the molecular weight of $NaOH$. The molecular weight of $NaCl$ is 58.5 and H_2O is 18. These numbers show the relative reacting weights of HCl and $NaOH$, and the relative weights of the products of the reaction, $NaCl$ and H_2O. The quantity of substance having a weight in grams numerically equal to the molecular weight of the substance is called a *mole.* Now we can write:

$$HCl + NaOH \rightarrow NaCl + H_2O$$
$$\textit{36.5 g} \quad \textit{40.0 g} \quad \textit{58.5 g} \quad \textit{18 g}$$
$$\textit{1 mole} \quad \textit{1 mole} \quad \textit{1 mole} \quad \textit{1 mole}$$

HCl and $NaOH$ do not have to be mixed in this proportion, but if they are not, the one in excess will be left over. If these weights are used, then 36.5 plus 40.0 g, or 76.5 g. of the reactants disappear, and 58.5 plus 18.0 g, or 76.5 g, of new substances (products) are formed. This is an example of the law of the conservation of mass, and any equation that is written must show such an equality or it is not "balanced."

Consider the reaction between nitric acid, HNO_3, and sodium carbonate, Na_2CO_3:

$$HNO_3 + Na_2CO_3 \rightarrow NaNO_3 + CO_2 + H_2O$$

The equation as written is not balanced, because it shows two sodium atoms disappearing and only one remaining in the product. This imbalance can be remedied if two molecules of $NaNO_3$ are formed. However, there are now two NO_3 groups on the right-hand side of the equation. This new imbalance can be corrected if two molecules of HNO_3 react. So the balanced equation is:

$$2HNO_3 + Na_2CO_3 \rightarrow 2NaNO_3 + CO_2 + H_2O$$

126 g	106 g	170 g	44 g	18 g
2 moles	1 mole	2 moles	1 mole	1 mole

The weights can be verified by adding the atomic weights as they are given in table 1-III. To see if the equation is balanced, you can check the number of each kind of atom on each side. In this case, there are one carbon, two hydrogen, two nitrogen, two sodium, and nine oxygen atoms on each side. Also the total weight of reactants (232 g) is the same as the total weight of products (232 g). Every equation that is written must show a similar balance between the reactants that disappear and the products that are formed.

The reactions between HCl and NaOH and between HNO_3 and Na_2CO_3 are typical of acid-base reactions. The acid is said to ***neutralize*** the base, and the base is said to neutralize the acid.

Sometimes an acid-base reaction can be carried only part way. If only one mole of nitric acid reacts with one mole of sodium carbonate, the reaction is:

$$HNO_3 + Na_2CO_3 \rightarrow NaNO_3 + NaHCO_3$$

63 g	106 g	85 g	84 g
1 mole	1 mole	1 mole	1 mole

Although equations like this are fine in determining what weight of a particular acid will react with a certain weight of a particular base, it is not necessary to use molar weights. You may, for example, want to neutralize 500 g of NaOH with the acid HCl. To do this will require:

$$500 \text{ g NaOH} \cdot \frac{(36.5 \text{ g/mole HCl})}{(40.0 \text{ g/mole NaOH})} = 456.3 \text{ g HCl}$$

To understand the mechanism of typical acid-base reactions, another kind of equation is required. Substances such as HCl, HNO_3, and Na_2CO_3 are almost completely ionized in an aqueous solution. So what really happens when HCl reacts with NaOH is:

$$H^+ + OH^- \rightarrow H_2O$$

In other words, hydrogen ions from the HCl react with the OH$^-$ ions from the NaOH to form molecules of water. The Na$^+$ and Cl$^-$ ions do not enter into the reaction. They are present at the start and are still present after the reaction has taken place. In graphic arts, the subject of relative weights of reacting chemicals is the more important consideration, so we will use molecules in equations even though they do not react as such.

Hydrolysis

When dissolved in water, salts of a strong base and a weak acid or salts of a strong acid and a weak base form a solution that is basic or acidic. This results from a reaction, called *hydrolysis,* between water and ions of the salt. Examples of such compounds are sodium acetate, CH_3COONa, ammonium chloride, NH_4Cl, and sodium carbonate (washing soda), Na_2CO_3. The equations show what happens during a hydrolysis reaction:

- For sodium acetate: $CH_3COO^- + H_2O \rightarrow CH_3COOH + OH^-$
  (In this case the OH$^-$ ions make the solution alkaline.)
- For ammonium chloride: $NH_4^+ + H_2O \rightarrow NH_4OH + H^+$
  (The H$^+$ ions make the solution acid.)
- For sodium carbonate: $CO_3^{2-} + H_2O \rightarrow HCO_3^- + OH^-$
  (Solutions of sodium carbonate are alkaline, due to the formation of hydroxide ions, OH$^-$.)

Equilibrium Reactions

Many chemical reactions do not proceed to completion, but reach what is called an *equilibrium.* At equilibrium, some of the reacting substances are still present, along with the products that are formed. Such an equilibrium mixture could be left for a long time and would still contain the same percentage of reactants and product. Often it is possible to change these percentages by adding an excess of one of the reactants, by heat, or by pressure. Then a new equilibrium is reached, with different percentages of reactants and products.

The hydrolysis reactions above provide good examples of equilibrium. With some of them, less that 0.1% of the total amount of ions in the solution react with water. Then the solution reaches an equilibrium and nothing more happens, but even this small amount of reaction is enough to generate a small concentration of either H$^+$ or OH$^-$ ions, making the solution either slightly acid or slightly alkaline.

Chemical Kinetics

Some chemical reactions, like those that occur in aqueous solution, occur very rapidly. Others take place slowly. Some inks take several hours to dry, but the hardening of light-sensitive coatings on lithographic plates requires only 1–3 min. The study of the speed of chemical reactions is called *chemical kinetics.*

Solubility

If any solid compound such as calcium sulfate (gypsum or plaster of paris) is added to water or another solvent, some of it dissolves. If such a mixture is stirred for a long time at a particular temperature, the amount of the compound that dissolves will reach a fixed concentration and no more will dissolve. This forms a *saturated solution.* The compound is said to have a certain *solubility* in that solvent. Tables are available that list the solubility of many compounds in a variety of different solvents. The concentration may be expressed in different ways, but a common expression of concentration is the grams of the compound that can be dissolved in 100 g of solvent. The temperature must also be given, because compounds are usually more soluble as the temperature increases. A few compounds, like calcium sulfate, have a lower solubility as the temperature is raised.

The situation is somewhat different if a liquid is mixed with water. Some liquids, like ethyl alcohol (grain alcohol), are completely soluble in, or miscible with, water. With others, only a certain amount dissolves in the water; at the same time a certain amount of water may dissolve in the liquid.

Compounds vary widely in their solubility in water. Only 0.615 g of $Ca(OH)_2$ will dissolve in 100 g of water at 68°F (20°C), but the amount is 11 g for Na_3PO_4 and 36 g for NaCl at the same temperature. Some compounds dissolve so little in water that they are called "insoluble." Even the "insoluble" compounds have a very small solubility. The solubility of silver chloride, AgCl, is 0.00015 (1.5×10^{-4}) g per 100 g of water. (Silver chloride is used in the light-sensitive emulsion of photographic films and papers.)

Since compounds are formed by a combination of positive and negative ions, it is possible to construct a table that lists, in general, the solubility of many compounds by referring to various combinations of positive and negative ions. Table 1-VI is such a table. It makes a useful reference table.

To determine if ammonium chloride, NH_4Cl, is soluble in water, look for the NH_4^+ ion in the column of positive ions and find that almost all compounds of NH_4^+ with negative

Table 1-VI.
Solubility of common
compounds in water.

Negative Ions	Positive Ions	Compound of Positive and Negative Ions
Almost all	NH_4^+, Li^+, Na^+, K^+	Soluble
NO_3^-, CH_3COO^-	Almost all	Soluble
OH^-	Li^+, Na^+, K^+, NH_4^+, Sr^{2+}, Ba^{2+} All other positive ions	Soluble Slightly soluble
Cl^-, Br^-, I^-	Ag^+, Pb^{2+}, Hg_2^{2+}, Cu^+ All others	Slightly soluble Soluble
SO_4^{2-}	Ca^{2+}, Sr^{2+}, Ba^{2+}, Pb^{2+} All others	Slightly soluble Soluble
PO_4^{3-}, CO_3^{2-}, SO_3^{2-}	Li^+, Na^+, K^+, NH_4^+ All others	Soluble Slightly soluble

Compounds vary greatly in their solubility in water, but those that are
fairly soluble are listed as "soluble," and those with low to very low
solubility are listed as "slightly soluble."

ions are soluble. So ammonium chloride must be fairly soluble in water. To determine if ferric chloride, $FeCl_3$, is soluble
in water, look for Cl^- in the column of negative ions. The
positive-ion column shows that compounds of Cl^- with silver,
lead, mercurous mercury, and cuprous ions have low solubility, while all other chloride compounds are soluble. This
means that $FeCl_3$ is soluble. Other compounds can be
checked in a similar way for their solubility in water. Use
table 1-VI to check the correctness of the following: $Mg(OH)_2$,
low solubility; $BaSO_4$, low; KBr, soluble; and $CaCO_3$, low.

Precipitates

If a clear aqueous solution of one compound is mixed with an
aqueous solution of another compound, one of two things will
happen. Either the mixture of the two solutions will remain
perfectly clear, or a cloudiness will appear, and sooner or
later a solid will settle to the bottom of the container. This
sediment is called a ***precipitate.***

When two such solutions are mixed, a precipitate forms
and settles out if the positive ions of one compound can combine with the negative ions of the other compound to form a
new compound that has low solubility in water. Suppose one
solution contains silver nitrate, $AgNO_3$, and the other con-

tains potassium chloride, KCl. Both of these compounds are fairly soluble in water, However, when they are mixed, the silver ions of the silver nitrate combine with the chloride ions of the potassium chloride to form silver chloride, which is very insoluble in water. The ionic equation is:

$$Ag^+ + Cl^- \rightarrow AgCl \text{ (solid)} (\downarrow)$$

In the above equation, the downward-pointing arrow ($\downarrow$) indicates that the product of the reaction precipitates—that is, the product is insoluble and comes out of solution.

If one solution contains water-soluble lead acetate, $Pb(CH_3COO)_2$, and the other contains water-soluble magnesium sulfate, $MgSO_4$, a precipitate of water-insoluble lead sulfate, $PbSO_4$, forms when the two solutions are mixed. The ionic equation is:

$$Pb^{2+} + SO_4^{2-} \rightarrow PbSO_4 \text{ (solid)} (\downarrow)$$

On the other hand, if a rearrangement of the positive and negative ions of two compounds gives two other compounds that are also soluble in water, then nothing happens. The mixture of the two solutions merely contains two kinds of positive ions and two kinds of negative ions. In this case, the mixture remains clear.

Oxidation and Reduction

Oxidation and reduction are two interdependent chemical concepts: it is impossible to have one without the other. An *oxidation-reduction reaction* is one in which some atoms lose electrons to other atoms. The atoms that lose electrons are said to be *oxidized;* the atoms that gain electrons are said to be *reduced.* For example, when oxygen from the air combines with iron to form iron oxide, or rust, the iron is oxidized (it loses electrons to the oxygen), and the oxygen is reduced (it gains electrons from the iron). However, "oxidation" (the taking of electrons from some atoms by others) does not always involve oxygen.

To determine if any oxidation-reduction reaction has occurred, it is necessary to define "oxidation number." The *oxidation number* is the number of electrons that must be lost or gained by an atom in a combined state to convert it to its elemental form. The oxidation number of an uncombined element is therefore 0.

Water, H_2O, was established as the reference compound for determining oxidation number of elements in compounds.

The arithmetic sum of the oxidation numbers of any compound must equal zero, because a compound has no net charge. Therefore, oxygen was given the oxidation number of –2, and hydrogen was given the oxidation number of +1. The oxidation numbers of other atoms in combined form are calculated in relation to the oxidation number of oxygen and hydrogen. Now we can say that oxidation occurs when there is an increase in the oxidation number of some element, and reduction occurs when there is a decrease in the oxidation number of another element.

When magnesium burns in air, the compound magnesium oxide, MgO, is formed. In this reaction, the oxidation number of oxygen has gone from zero (free oxygen) to –2 in MgO, so oxygen has been reduced. At the same time, the oxidation number of magnesium has gone from zero (Mg metal) to +2 in MgO, so magnesium has been oxidized. A substance that oxidizes something else is called an ***oxidizing agent,*** and a substance that reduces something else is called a ***reducing agent.***

The atoms in multiatomic ions, such as sulfate, SO_4^{2-}, can be assigned an oxidation number. For multiatomic ions, the sum of the oxidation number of the four oxygen atoms is $4 \times (-2)$, or –8. Since the sulfate ion has a charge of –2, the sulfur atom in it must have an oxidation number of +6. Using the same kind of arithmetic, the oxidation number of sulfur in the sulfite ion, SO_3^{2-}, is +4. This means that a compound with sulfite ions must be oxidized to convert it to a sulfate compound. In the same way, a nitrite ion, NO_2^-, must be oxidized to convert it to a nitrate ion, NO_3^-.

Sodium hypochlorite, NaOCl, is a good oxidizing agent. This means that some atom in this compound must be easily reduced. This is the Cl atom. In NaOCl, the Cl has an oxidation number of +1. When this compound oxidizes something else, the Cl^+ changes to chloride ion, Cl^-, so there has been a reduction in the oxidation number of Cl from +1 to –1. That is, the Cl in NaOCl has been reduced.

Conductive Solutions

Compounds that form positive and negative ions when the compounds are dissolved in water form solutions that conduct an electric current. Such compounds are called ***electrolytes.*** There are both strong electrolytes and weak electrolytes. A strong electrolyte is a compound that ionizes almost completely when dissolved in water. Strong acids, for example, are strong electrolytes. If only a small percentage of the molecules of a compound ionize when the compound is dissolved in water, then it is a weak electrolyte (meaning that the solution does not conduct current as well). Weak acids, for example, are also weak electrolytes. Other compounds, such as sugar, do not ionize when dissolved in water. A solution of sugar does not conduct current any better than water, which is a very poor conductor. Substances like sugar are called **nonelectrolytes.**

In order to measure the ability of a solution to carry electricity, it is necessary to immerse two **electrodes** into the solution; one electrode is connected to a positive pole of a battery and the other is connected to the negative pole of a battery. If an electrolyte is dissolved in water, a current will flow. The current flows through the solution because the positive ions of the electrolyte move toward the negative electrode and the negative ions move toward the positive electrode. In the wires that go from the electrodes to the battery, the current is due to a flow of negatively charged electrons through the wires.

Figure 1-7 shows a complete electrical circuit that includes a battery and an electrolytic cell that consists of two electrodes immersed in a cupric chloride, $CuCl_2$, solution. The Cu^{2+} ions move toward the negative electrode (called the **cathode)** and the Cl^- ions move toward the positive electrode (called the **anode).** Accordingly, ions with a positive charge are called **cations** (pronounced cat-ions) and ions with a negative charge are called **anions.**

A chemical reaction occurs at each electrode:
- Positive electrode: $2Cl^- \rightarrow Cl_2 + 2e^-$
- Negative electrode: $Cu^{2+} + 2e^- \rightarrow Cu^\circ$ (metallic Cu)

The Cl^- ions lose electrons (e^-), and therefore oxidation occurs. The Cu^{2+} ions gain electrons, and they are therefore reduced. This is a case where oxidation and reduction occur simultaneously but at different places.

A battery can be considered to be an electron pump. It pulls in the electrons liberated when chloride ions change to chlorine gas, then pumps the electrons around to the negative electrode to supply the electrons necessary to reduce cupric ions to metallic copper.

Electrolytic cells can be used to produce some metals in the pure metallic state. In the electrolytic method for the recovery of silver from a photographic hypo fixing bath, a current is passed through the bath, depositing pure silver onto the negative electrode: silver ions in the bath are

Figure 1-7.
Electrolysis of cupric
chloride.

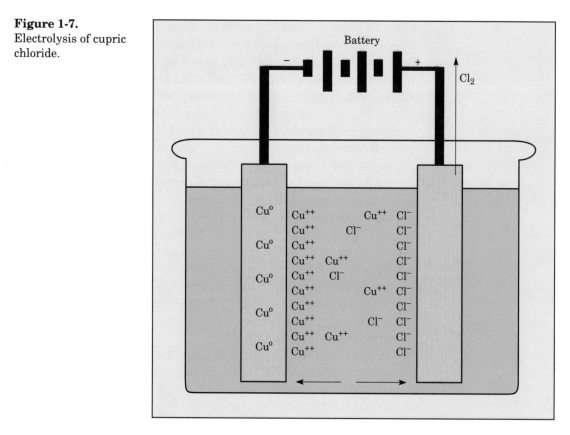

reduced to uncharged silver atoms, or metallic silver. A similar reaction occurs in electroplating a gravure cylinder with copper or chrome.

Relative Activity of Metals

Some metals are more active than others. For instance, some metals react readily with the oxygen in the air (in the presence of some water or water vapor) to form *oxides* of the metals. Others have little tendency to do this. Some metals react rapidly with strong acids, while others react slowly or not at all.

If metallic zinc is placed in a water solution of silver nitrate, the following ionic reaction occurs:

$$Zn° + 2Ag^+ \rightarrow Zn^{2+} + 2Ag° \text{ (metallic silver)}$$

The zinc metal goes into solution as zinc ions, and the silver ions in solution are precipitated as solid metallic silver.

A similar reaction is used to recover silver from a hypo fixing bath. In this reaction, zinc atoms lose electrons, and therefore zinc is oxidized. The silver ions gain electrons, so the silver ions are reduced. From this result, one concludes that zinc is a more active metal than silver. Zinc has a greater tendency to become oxidized than does silver. Further evidence: if a piece of silver is placed into a solution of zinc nitrate, nothing happens.

Based on experiments with other combinations of metals and experiments with electrolytic cells involving two different metals, the metals can be arranged in descending order of their tendency to react with water and acids. The metals at the top of this list (called the *electromotive series,* table 1-VII) are the most active.

Table 1-VII.
Electromotive series.

Potassium Barium Calcium Sodium	Very active metals; react very rapidly with acids, and even react with water
Magnesium Aluminum Zinc	Fairly active metals
Chromium Iron Cadmium Cobalt Nickel Tin Lead	Moderately active metals; usually less active than magnesium, aluminum, and zinc
Copper Mercury Silver	Rather inactive metals; react only with an "oxidizing" acid such as strong nitric acid
Gold Platinum	Very inactive metals; called "noble" metals

Magnesium, a fairly active metal, is used in the graphic arts to make magnesium relief plates. A solution of hydrochloric acid, HCl, is used to etch away the metal in the nonprinting areas. The molecular equation for this reaction is:

$$Mg° + 2HCl \rightarrow MgCl_2 + H_2 (\uparrow)$$

(The upward pointing arrow indicates that gaseous hydrogen is generated.)

The metallic magnesium is oxidized to Mg^{2+} ions, and the H^+ ions of HCl are reduced to hydrogen gas, H_2.

The metal copper is much lower in the list of table 1-VII, which means that copper is much less active than magnesium. Copper will not react with hydrochloric acid. It is necessary to use nitric acid, which is sometimes called an oxidizing acid. When nitric acid dissolves copper in the nonprinting areas of a copper photoengraving, oxides of nitrogen, such as NO and NO_2, are evolved as gases, instead of hydrogen. (The production of these toxic gases in the photoengraving of copper is one of the reasons that mechanical engraving has largely replaced chemical engraving.)

While the list of table 1-VII is useful, some metals at times appear to be less active than suggested by their position in this list. For instance, aluminum is the most widely used metal for lithographic printing plates. In table 1-VII, aluminum is listed as a "fairly active metal." Yet aluminum resists water and reacts very slowly with hydrochloric acid. In fact, concentrated nitric acid can be shipped in an aluminum tank car. Why don't these acids affect the aluminum? The answer is that a tightly adhering film of aluminum oxide covers the metallic aluminum used for lithographic plates or aluminum cans: the surface of the aluminum is already oxidized. This film keeps water or certain acids from reaching the active aluminum underneath the film.

Hydrofluoric acid, HF, is able to react with the film of aluminum oxide, to form aluminum fluoride, AlF_3, and then attack the aluminum metal underneath.

Corrosion

If two different metals are fastened together and exposed to a humid atmosphere, immersed in water, or immersed in a dilute solution of an electrolyte in water, one of the two metals usually corrodes (oxidizes) at the expense of the other metal. As might be expected, the metal that corrodes is the one that is higher in the list of table 1-VII.

The corrosion of metals in contact with each other is the principle of the protection of iron objects by galvanizing. Galvanizing is accomplished by dipping the iron object into a bath of molten zinc. When galvanized objects are subjected to the weather, the zinc, rather than the iron, gradually oxidizes.

Iron is often coated with a thin layer of tin to produce tinplate. Reference to Table 1-VII shows that tin is less active than iron, which means that iron is more apt to oxidize than is tin. In this case, the tin offers protection only if it covers all of the iron.

The principle of an active metal protecting a less-active metal is used to help keep steel storage tanks from corroding. One method is to bury large pieces of magnesium with a cable to the tank. Under moist conditions, the magnesium gradually oxidizes. The oxidation keeps the steel of the tank from corroding, or at least it greatly retards corrosion of the steel.

Certain active metals, particularly alloys of iron with other metals such as chromium and nickel, can be rendered passive by treatment with various chemicals or by electrolysis. The position of a metal or alloy in the active state is considerably higher on such a list as that of table 1-VII than the position of the same metal or alloy in the passive state. No explanation is offered here for these two states because differing theories exist about them.

Radioactivity and Atomic Energy

In the chemical reactions discussed earlier in this chapter, neutral atoms were sometimes converted to electrically charged ions, or the reverse. Other atoms formed compounds held together by covalent bonds. In neutralization reactions, the hydrogen ions of an acid combined with various ions to form molecules of water or other ions such as HCO_3^- or HPO_4^{2-}. In all of these reactions, each kind of atom remained the same. Hydrogen remained hydrogen, oxygen remained oxygen, and chlorine remained chlorine, although the atoms might be charged or uncharged.

Nuclear reactions involve the conversion of one element into another by changes in the nucleus of the atom. These reactions are accompanied by radioactivity and often evolve large amounts of energy.

Naturally radioactive radium, whose atomic number is 88, has several isotopes. The atoms of one of these, with mass number 226, emit alpha particles (helium ions) from the nucleus when it disintegrates. As a result, what is left are no longer radium atoms but atoms of *another* element, radon, with an atomic number of 86 and a mass number of 222.

The disintegration can be expressed with the following equation of radioactive decay:

$$\ _{226}^{88}\text{Ra} \rightarrow \ _{222}^{86}\text{Rn} + \ _4^2\text{He}$$

Radium atoms disintegrate one at a time. In any given length of time, only a certain number of atoms disintegrate. The time required for half the atoms in any quantity of the element to disintegrate is called the **half-life.** The half-life of radium-226 is 1,620 years. In this time, half of a sample of radium will be gone—changed into radon. In the next 1,620 years, half of what had been left will be gone. This process continues indefinitely. Many radioactive elements are known. Their half-lives vary from a fraction of a second up to thousands of years.

Several other naturally radioactive elements also emit alpha particles. These include bismuth-211, polonium-208, and uranium-238.

It is possible to bombard elements—with high-speed neutrons, for example—and make them radioactive. Over 2,000 such radioactive isotopes have been produced. Many of these emit **beta** particles (electrons) when they disintegrate. What happens is that one of the neutrons in the nucleus of one of these atoms is converted to a proton and an electron. The proton remains in the nucleus, but the electron (beta particle) is ejected. This leaves the nucleus with one more proton than before. So it is now a nucleus of an atom of a different element with an atomic number one greater.

A good example of such transformation is given by a radioactive isotope of phosphorus. A phosphorus atom has 15 protons; its atomic number is 15. Ordinary phosphorus has a mass number of 31, but it is possible to produce phosphorus with a mass number of 32, which has an extra neutron in its nucleus. Phosphorus-32 is radioactive with a half-life of 14.3 days. It emits beta particles (electrons), so that one of the neutrons becomes a proton, and the phosphorus atom turns into a sulfur atom! (Because electrons have negligible mass, the actual atomic weight remains almost the same.)

A very small quantity of H_3PO_4 containing radioactive phosphorus-32 atoms can be mixed with a larger quantity of ordinary H_3PO_4 to serve as a **tracer.** Whatever happens to the ordinary H_3PO_4 also happens to the phosphoric acid containing the atoms of phosphorus-32. For example, a lithographic plate is scrubbed with a dilute solution of phosphoric acid, and the acid is then rinsed from the plate with water several times. Has any phosphoric acid remained attached to the surface of the plate? This determination would be very difficult to make by other means, but by the use of a Geiger counter you can ascertain if beta particles are coming off the litho plate. If any are detected, they indicate that some phosphoric acid remains adsorbed to the surface of the plate, even after several rinsings with water.

In the radioactive changes mentioned above, some energy changes are involved, but they are very small. There are other changes that involve tremendous amounts of energy.

Most uranium atoms have a mass number of 238; a small percentage have a mass number of 235. When uranium-235 is separated from uranium-238 and is bombarded by neutrons, the neutrons first combine with the nucleus of the uranium-235 atoms. Then the atoms split apart to form atoms of two other elements, such as barium and krypton, whose atomic numbers total 92. In this process, three neutrons are evolved for every neutron that initially combined with a uranium-235 nucleus. In addition, a large amount of energy is liberated. The neutrons evolved react with the nuclei of more uranium-235 atoms. Since three neutrons are evolved for every one that reacts, this process builds up extremely rapidly and, if uncontrolled, results in an atomic explosion releasing a tremendous amount of energy almost instantaneously, as in a bomb. When one mole (235 g or about 8 oz) of uranium-235 reacts, about 4.5×10^{12} calories of energy are released. It is difficult to visualize how much energy that is; if controlled, it would be enough to operate an average automobile for about 100 years.

In a nuclear reactor (nuclear power plant), this reaction is controlled by the use of rods of cadmium metal located between rods that contain uranium-235. Cadmium is a good absorber of neutrons, and the amount of cadmium used is just enough so that some neutrons are left to keep the uranium-235 splitting reaction going at a steady rate. The result is a continuous evolution of energy that can be used to generate electrical power.

Presented earlier in this chapter, the law of the conservation of mass states that no mass (weight) can be lost or gained in chemical reactions. Also discussed was the law of the conservation of energy, which states that energy cannot be created nor destroyed. Both of these laws are essentially true for ordinary chemical reactions. However, Einstein postulated in 1905 that mass and energy are related. He theorized that if a certain amount of mass actually disappeared, a certain amount of energy would be created in its place. This was a radical assumption at the time, but it has turned out to be correct. The reason uranium-235 liberates a tremendous amount of energy is that the products of its fission (breaking apart) weigh less than the uranium that disappears, converted into an enormous quantity of energy. When only one gram of mass is lost, it is converted into 2.14×10^{13} calories of energy.

Summary

In ordinary chemical reactions, the weight of the products formed is exactly the same as the weight of the reactants that disappear. This is the law of the ***conservation of mass.*** Likewise, if a reaction or a change of state (gas to liquid, etc.) evolves a certain amount of energy, then the same amount must be employed to reverse the process. This is the law of the ***conservation of energy.***

Elements consist of atoms, which, in turn, consist of a nucleus containing protons and neutrons, with enough extranuclear electrons to balance the positive charge of the protons. The atomic number of an element is the number of protons in the nucleus. This is what makes the atoms of one element different from the atoms of another element. The number of electrons in the outermost shell or energy level (valence electrons) determines the chemical nature of an element. Most metallic elements have one to three valence electrons while some nonmetals have six or seven. The noble gases have the outermost energy level filled with as many electrons as that level can hold.

Isotopes of an element consist of atoms that have the same number of protons in the nucleus but a different number of uncharged neutrons and therefore a different atomic weight.

Atomic weights are the average relative weights of atoms of any element when compared with the atomic weight of 12.000 for the principal isotope of carbon.

In chemical reactions, atoms of one element may lose valence electrons to atoms of another element, thus forming positive and negative ions. In other cases, the atoms of a compound are held together by the sharing of two electrons, called a covalent bond.

When atoms of different elements combine chemically, molecules of a compound are formed. In some cases the compound consists of positive and negative ions instead of molecules. In many chemical reactions, one element is oxidized and another is reduced at the same time.

By using a balanced chemical equation and by calculating the weight of a mole of each reactant and product, it is possible to calculate what weight of one substance will react with a given weight of another substance. It is also possible to calculate the weights of the products of the reaction.

Many chemical reactions do not proceed to completion. Instead, they reach an equilibrium. Salts that hydrolyze in water provide examples of reactions that proceed only a little way before reaching an equilibrium. However, the amount is enough to make water solutions of these compounds either alkaline or acid, depending on the compound.

Pure substances consist of elements or compounds. A compound consists of two or more elements combined in a fixed proportion by weight.

Compounds vary widely in their solubility in water. When two solutions containing two different soluble compounds are mixed, a precipitate (solid) will form if another compound that has low solubility can be formed.

If a compound ionizes when dissolved in water, the solution will carry an electric current. The positive ions (cations) move toward the negative electrode; the negative ions (anions) move toward the positive electrode. Reactions occur at each electrode; an oxidation reaction occurs at the one electrode, and a reduction reaction occurs at the other.

Metals vary in their activity, including their tendency to react with the oxygen of the air, or to react with acids. Metals can be listed in the order of their decreasing activity. Such a list is helpful in determining which of two metals will corrode if they are in contact.

Ordinary chemical reactions involve changes in the outer electrons—those in the outer shells or energy levels. Nuclear reactions involve the conversion of one element into another by changes in the nucleus of the atom. These reactions are accompanied by radioactivity and often evolve large amounts of energy.

Most of the material in this chapter is concerned with inorganic chemistry. The chemistry of carbon is called organic chemistry. Organic chemistry is covered in chapter 3.

Further Reading

Seager, Spencer L. and Michael R. Slabaugh. *Chemistry for Today: General, Organic, and Biochemistry.* 3rd ed. 1997 (West Wadsworth).

2   Water, Acids, and Bases

Water, which is necessary for the survival of all plants and animals, is also essential in industrial operations. In graphic arts, water is the main ingredient in lithographic dampening solutions, photographic developers, and fixing baths. It is used in flexographic inks for package printing and light-sensitive coatings and developers for lithographic plates. Paper absorbs and gives off water vapor, and controlling water content of paper is essential to good paper performance.

Pure water is never found in nature. Water flowing from melting snow fields or glaciers is probably the purest natural water, but it contains large amounts of air. As the water flows from the ice field or spring, it picks up minerals from the rocks and land. It may also pick up raw sewage, agricultural wastes and runoff, and industrial wastes—treated or untreated. Water purification plants remove bacteria from natural water, but hardness and many industrial wastes remain in the water.

Most of the dissolved minerals that occur naturally are highly ionized. The principal ones are shown in table 2-I.

Table 2-I.
The principal ions found in tap water.

Positive Ions (Cations)	Negative Ions (Anions)
Calcium, Ca^{2+}	Carbonate, CO_3^{2-}
Magnesium, Mg^{2+}	Bicarbonate, HCO_3^{-}
Sodium, Na^{+}	Chloride, Cl^{-}
Iron, Fe^{2+}	Sulfate, SO_4^{2-}
Manganese, Mn^{2+}	Nitrate, NO_3^{-}

Conductivity

The purity of water can be judged by its electrical conductivity. When ionizable materials are dissolved in water, the water conducts electricity more readily that does pure water. The conductivity of the water can be measured with a conductivity meter, and the number can be used to indicate the purity of the water or a change in its composition.

The unit of conductivity is the micromho (pronounced "micro-mo") and abbreviated μmho.

Because ions have different conductivities, the conductivity reading does not indicate exactly the total dissolved solids (TDS) in tap water or any other aqueous solution, and accordingly no clear-cut numbers can be given for the proper level of conductivity of process water. Conductivity below 200 or 250 in the raw water is generally satisfactory for plant operations, while numbers like 1000 or 1250 indicate the presence of a high levels of dissolved solids.

Effect of Non-ionized Diluents

The effect of non-ionized diluents on conductivity must be considered when conductivity is used to measure the concentration of electrolytes in lithographic fountain solutions containing alcohol. Because the alcohol dilutes the solution, it changes the concentration of dissolved acids and other ions, but it also affects the electrical environment of the ions in the water. (See chapter 9, "Chemistry in the Pressroom" for a detailed discussion of lithographic dampening solutions.)

Raw Water and Hardness

"Raw water" refers to the water that comes into the printing plant, whether it comes from a water treatment plant, from wells, or from another source.

The impurities most commonly found in raw water are listed in table 2-II.

Table 2-II.
Common impurities in raw water.
(Source: J. A. MacPhee, TAGA Proceedings, 1988.)

Liquids	**Dissolved Gases**
Free acids	
Detergents	**Biological Growths**
Insecticides	Plant types
	Algae
Solids	Fungi
Dissolved	Bacteria
Suspended	Aerobic
Coarse particles	Anaerobic
Colloids	Slime

The impurities that are of most importance in the printing plant are the dissolved solids, especially the salts of calcium and magnesium that cause hardness. As a rule of thumb, total dissolved solids (TDS) in parts per million (ppm) times 1.5 is about equal to the conductivity of water (μmho/cm).

Water with a hardness greater than 220 ppm (conductivity greater than 330 μmhos/cm) can be considered to be very hard. If it has a hardness less than 50 ppm (75 μmhos/cm), it is considered soft.

Hardness is expressed as the total concentration of all calcium and magnesium salts calculated as calcium carbonate. That is, the total concentrations of calcium and magnesium ions are determined and reported as if they were calcium carbonate. For example, if a solution contains 40 grains of calcium ion, the concentration is reported as 100, because the ionic weight of calcium is 40, and the molecular weight of calcium carbonate is 100. (One grain/gal. equals 17.1 ppm or 17.1 mg/L.) Water with a hardness of 200 ppm is as hard as if it contained 200 parts of calcium carbonate in a million parts of water.

Hard Water

Calcium, magnesium, and iron ions make water hard, but most others do not, so that TDS is not, by itself, a good indication of the hardness of the water.

When soap is dissolved in hard water, a reaction such as the following takes place:

$$2RCOONa \text{ (aq)} + Ca^{2+} \text{ (aq)} \rightarrow (RCOO)_2Ca \text{ (}\downarrow\text{)} + 2Na^+ \text{ (aq)}$$

Soap   *Calcium soap*

The letter "R" in the above equation represents a hydrocarbon chain, in this case one such as $C_{17}H_{35}$, which is found in stearate soaps, or $C_{17}H_{33}$, which is found in oleate soaps. The "(aq)" in the equation indicates that the initial soap and Ca^{2+} ions and the final Na^+ ions are in aqueous solution. The "($\downarrow$)" indicates that the product has been deposited out of solution.

Soap is usually a sodium salt of a high-molecular-weight fatty acid. The soap reacts with the calcium and magnesium ions in the water to form calcium and magnesium soaps that are sticky materials and insoluble in water. They cause the bathtub ring that forms when soap is added to hard water. In fact, soap will not form lasting suds with hard water until enough soap has been consumed to react with all the calcium and magnesium ions in the water.

Water in different parts of the country varies greatly in hardness, from very low amounts up to as much as 1,000 ppm or more. Local water authorities can give the hardness figure for the water in their area. Sometimes, when a city has two different sources of water, the sources may differ significantly, and if the water company switches from one source to another, a printing plant may find the hardness of its water supply changes from month to month and even day to day.

Although the hardness of raw water is sufficient to change the pH of fountain solution by one to three units, the consistency of the raw water hardness is more important than the actual hardness. Manufacturers can prepare dampening solution concentrates to work with raw water of any hardness, but they cannot cope with water that varies in hardness. If the raw water has a conductivity higher than about 300, purification is recommended. Very soft waters (waters with very low conductivity) are not known to cause problems with lithographic dampening solutions.

The hardness of water also affects the way that some inks take up water. If the hardness of the water changes from time to time, the entire lithographic process may be thrown out of balance. This is a sound reason for careful control or purification of the water used in the pressroom.

Purity and Purification

The question of how pure the water must be to work well in the plant is not easily answered. It depends on the nature of the impurities and the use that is to be made of the water. Often, no treatment is required. In other cases, filtration may be necessary to remove suspended matter. In one lithographic plant, it was almost impossible to apply a good light-sensitive coating to the litho plates because there was so much suspended matter in the city water. Here, a good water filter was the solution. Filtration alone will not remove any of the compounds that are dissolved in water.

A reduction of dissolved compounds is often desirable, depending on what must be accomplished. For example, only a water-softening unit is needed to soften the water for washing clothes or to prevent buildup of scale in a boiler, but to remove all solids, distillation, deionization, or reverse osmosis must be employed.

The tap water of many printing plants is purified simply as a precaution against unknown contaminants and variations in the quality of incoming water.

Water Softening

When water is softened, the calcium, magnesium, and iron ions from the water are replaced with sodium ions. Even though the total amount of dissolved solids remains almost the same, the water has been softened, since the ions that make water hard have been replaced by others that do not.

Water is softened by passing it through a tank filled with tiny beads of a cation exchange resin. These beads are initially charged with replaceable sodium ions.

As the hard water passes through the bed of resin particles, the calcium and magnesium ions (which have a higher affinity for the ion exchange resins than the sodium ions) are retained by the resin and removed from the water. When the calcium, magnesium and iron ions become attached to the resin, a chemically equivalent number of sodium ions are released into the water. For simplification, a cation exchange resin can be expressed by the general formula Na_2R. A typical exchange reaction is:

$$Ca^{2+} (aq) + Na_2R \rightarrow 2Na^+ (aq) + CaR$$

$$\begin{matrix} \textit{Cation} & \textit{Exhausted} \\ \textit{exchange} & \textit{resin} \\ \textit{resin} & \end{matrix}$$

A similar reaction occurs with magnesium and iron ions. As a result, the ions responsible for making water hard are removed; the water is softened. Stain-forming iron ions are also removed, along with most suspended matter.

This process continues automatically until the supply of sodium ions in the exchange resin is almost depleted. When this happens, the cation exchange resin has lost its power to soften more water and is exhausted. To restore the resin to its original condition, a strong solution of common salt, sodium chloride ($NaCl$), is passed through the bed. The high concentration of sodium ions in the sodium chloride solution reverses the ion exchange reaction. A typical reaction is:

$$CaR + 2Na^+ \rightarrow Na_2R + Ca^{2+}$$

$$\begin{matrix} \textit{Exhausted} & \textit{Strong} & \textit{Regenerated} & \textit{(in the NaCl} \\ \textit{resin} & \textit{NaCl} & \textit{resin} & \textit{solution—goes} \\ & \textit{solution} & & \textit{to sewer)} \end{matrix}$$

The calcium, magnesium, and iron ions in the original water go to the sewer. The regeneration process can be described as one in which common salt is used to soften water. This cannot be accomplished by adding salt directly to the water to be softened, but it can be done indirectly by the process just described.

Millions of such water softeners are used in homes and industries. The energy cost for this process is very low, and the sodium chloride consumed is an inexpensive natural material.

Removal of Dissolved Solids

For some industrial uses, it is desirable to have purer water, with most of the dissolved solids removed. Three ways to accomplish this are distillation, demineralization or deionization, and reverse osmosis.

Distillation

Distilled water is produced by boiling ordinary tap water in a still. It is also available as the condensate from industrial boilers. The steam that rises from the boiling water is almost free of the mineral and organic matter that is present in tap water. This steam is fed through condenser coils where it is cooled and converted back into liquid water, called distilled water.

Figure 2-1.
Laboratory apparatus for distilling water.

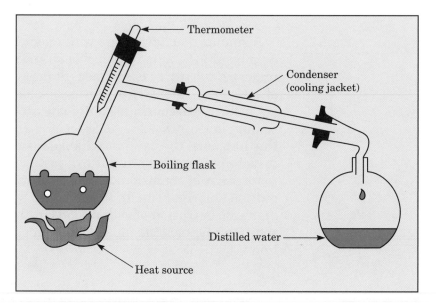

One might assume that distilled water would be free of all mineral matter because the mineral matter is not volatile. However, a small amount of mineral matter is carried over as small droplets in the steam, and of course, a tiny amount of the glass dissolves in the water. If the water initially contained about 200 ppm of dissolved solids, the distilled water will often contain 10 to 20 ppm, a level that is suitable for most applications. If still purer water is required, the distilled water can be distilled again and sometimes a third time. Energy costs are high for this process except when steam from an industrial boiler is available.

Deionization

The deionization, or demineralization, process requires two ion exchange resins. The cation exchange resin is initially charged with hydrogen ions; the anion exchange resin is initially charged with hydroxyl ions.

The water is passed first through a bed of cation exchange resin. Here almost all of the positive ions (cations) in the water become attached to the resin, and an equivalent amount of hydrogen ions leave the resin and enter the water. A typical reaction is:

$$(CR)–H + Na^+ \text{ (aq)} \rightarrow (CR)Na + H^+ \text{ (aq)}$$

Cation	*Exhausted*
exchange resin	*resin*

A similar ion exchange takes place with all of the other positive ions in the water.

As the water leaves the bed of cation exchange resin, it still contains all of the negative ions (anions) and hydrogen ions. In other words, it is a dilute solution of acids, such as HCl, HNO_3, H_2CO_3, and H_2SO_4. This water solution now passes through a bed of anion exchange resin. Here almost all of the negative ions become attached to the resin, and an equivalent amount of hydroxyl ions leaves the resin and enter the water. In the water, these hydroxyl ions combine with the hydrogen ions to form molecules of water. Here is a typical set of reactions:

$$(AR)–(OH)_2 + SO_4^{2-} \text{ (aq)} \rightarrow (AR)SO_4 + 2OH^- \text{ (aq)}$$

Anion	*Exhausted*
exchange resin	*resin*

$$2OH^- + 2H^+ \rightarrow 2H_2O$$

As the water issues from the second bed, most of the positive and negative ions have been removed. However, this process does not remove un-ionized, dissolved solids (or liquids).

Both ion exchange resins finally become exhausted and must be regenerated. The cation exchange resin is regenerated with a solution of a strong acid, such as hydrochloric acid or sulfuric acid. Regeneration removes the calcium, magnesium, and other positive ions and replaces them with hydrogen ions, H⁺, thus returning the resin to its initial condition.

It is important to mix the regeneration wastes from the cation and anion exchange resins, so that the acid from the one will be neutralized by the base from the other before the mixture goes to the sewer.

The purest water is produced by mixing the beads of cation and anion exchange resins in a single tank. Water issuing from such a tank is superior to triple-distilled water. The two resins must be separated before they can be regenerated, since one requires an acid for regeneration and the other requires a base. This separation can be accomplished easily because of a difference in the density of the two resins.

Reverse Osmosis   Most of the dissolved solids can be removed, at a cost less than that of producing distilled water, by a process called ***reverse osmosis.*** Reverse osmosis not only removes most of the positive and negative ions in the initial water, but also removes un-ionized dissolved solids (such as oils or liquid waste), suspended matter, and even bacteria.

To understand this process, it is necessary first to describe osmosis. When two aqueous solutions of different concentrations are separated by a semipermeable membrane, water passes through the membrane from the weaker solution to the stronger solution, as shown in figure 2-2.

Reverse osmosis is the opposite of osmosis. Applying pressure to the more concentrated solution forces water through the semipermeable membrane into the more dilute solution.

Figure 2-2.
Osmosis.

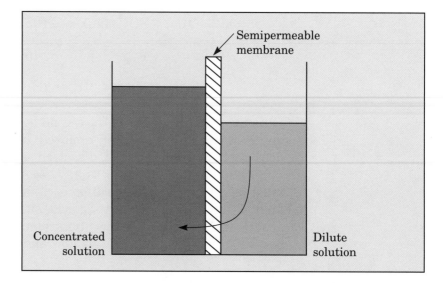

The membrane allows water to pass through, but only a small amount of materials dissolved in the water pass through as shown in the figure 2-3.

Figure 2-3.
Reverse osmosis.

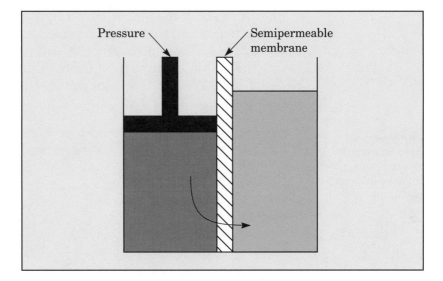

For water purification for printing or film processing, the "more concentrated solution" is the initial tap water. The liquid forced through the membrane is water that contains very little dissolved solids. Some positive or negative ions are held back better than others. Approximately 90% to 95% of dissolved solids are held back with an applied pressure of 50 lb./sq.in. (3.5 kg/cm^2 or 3.4 atmospheres). As the pressure is increased above 50 lb./sq.in., the purity of the water increases because more dissolved solids are rejected by the membrane.

The membrane is such a good filter that water containing suspended matter must first be passed through a regular filter, to prevent clogging the semipermeable membrane. With a hard-scale-forming water supply, it may be necessary to pretreat the water using a water softener or deionization unit.

The only energy required with reverse osmosis is that required to produce the necessary pressure. Unlike demineralization, no regenerating chemicals are needed. Because it is a continuous process, storage capacity is required for the treated water. The semipermeable membranes must be replaced about every two years.

Obtaining a reasonable amount of purified water requires a large semipermeable membrane area. One manufacturer uses special cellulose acetate tubing for the membrane, wrapping it into a spiral. Several tubes of this kind can be hooked together in parallel to give the desired rate of flow (figure 2-4). Another method makes use of long membranes in the form of tiny hollow fibers of a special nylon. These fibers are about the diameter of a human hair. The water to be treated enters a housing and surrounds the outside of these fibers. Under pressure, the water penetrates the fibers into the hollow centers, from which it flows to the product outlet. An outlet for the wastewater stream is required with both the spiral-wound module and the hollow-fiber systems.

Figure 2-4.
Reverse osmosis process using a spiral-wound semipermeable membrane.

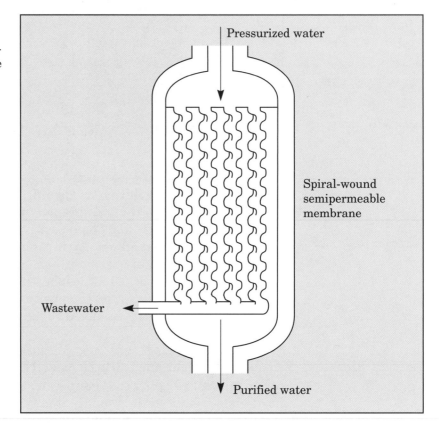

Since only a certain percent of the dissolved solids are removed by reverse osmosis at any given pressure, the quality of the purified water depends on the amount of dissolved solids in the initial water supply (and the applied pressure). If the water has low TDS initially, the effluent from a reverse osmosis unit is purer. For example, one town's water supply contains 1,440 ppm of dissolved solids, Reverse osmosis reduces this to 228. In another town, water with 1,150 ppm is reduced to 71.

Absolute and Relative Humidity

Press operators are often interested in the relative humidity of the air in the pressroom, as this determines whether paper will pick up moisture from the air and grow or lose moisture to the air and shrink, either of which causes several printing problems.

If the relative humidity is too low, as it often is in winter, the crew will have problems with static electricity and tight-edged sheets, which cause wrinkles and misregister. To avoid these problems, moisture must be added to the air. High relative humidity causes wavy-edged sheets, and it can retard the drying of ink. Therefore it is important to understand relative humidity and its relation to absolute humidity.

The amount of water vapor present in the air is measured in grains of water vapor per cubic foot (gr./cu.ft.) of air. (There are 437.5 grains in one avoirdupois ounce.) Using metric units, water vapor is measured as the number of grams of water vapor per cubic meter of air (g/m^3).

The maximum amount of water vapor that air can hold depends on the temperature of the air. It increases rapidly as the temperature increases. A table in appendix C gives the maximum amount for many temperatures. Table 2-III shows a few examples.

Table 2-III.
Maximum water vapor in air.

Temperature °F	°C	Grains/ Cubic foot	Grams/ Cubic meter
20	−6.7	1.236	2.938
32	0	2.119	4.849
50	10	4.108	9.401
60	15.6	5.798	13.27
68	20	7.561	17.30
70	21.1	8.064	18.45
86	30	13.28	30.39

Only saturated air contains the maximum amount of water vapor for a particular temperature. The amount of moisture in the air is called the ***absolute humidity.*** The ***relative humidity*** is the ratio of the absolute humidity to the maximum for that temperature, expressed as a percentage. For example, suppose that on a certain day, the moisture content of pressroom air at 20°C (68°F) is 7.785 g/m^3. The maximum amount of water vapor that air can hold at 20°C is 17.30 g/m^3. The relative humidity of this air is:

$$\frac{7.785}{17.30} \times 100\% = 45.0\%$$

In printing plants, the relative humidity is usually measured. One method uses a device with two thermometers called a **sling psychrometer.** One thermometer is the "dry bulb," which records the room temperature. The other, the "wet bulb," has, over the bulb, a cotton sleeve that is saturated with water. The device is swung around rapidly for a couple of minutes. Some of the water on the wet bulb evaporates, dropping the temperature of this thermometer. The less water vapor in the air (the lower the relative humidity), the more water evaporates from the wet bulb, and the more this temperature drops. The temperatures of the dry bulb and wet bulb are used to determine the relative humidity. (See appendix C.)

Figure 2-5.
Sketch of a sling psychrometer.

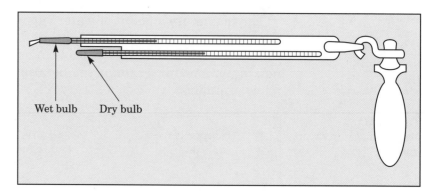

Wet bulb   Dry bulb

Effect of Temperature on Relative Humidity

The relative humidity of the air can be controlled with an air conditioner. This unit must do more than merely chill the air (which removes most of the water vapor). It is important that the unit add enough water vapor to bring the relative humidity up to the desired level.

In theory, the relative humidity can be increased by reducing the air temperature, provided that water is not removed by the chilling. However, as can be seen from the stream of water that flows from room cooling units, cooling removes water from the air, changing the absolute humidity as well as the relative humidity. In fact, these room cooling units (sometimes called "air conditioners") are responsible for very dry air that, in the summer time, causes tight-edged paper and static problems in pressrooms that are improperly humidified.

A related problem occurs in winter. Suppose the winter air is at 20°F (–6.6°C) and the weather bureau announces a relative humidity of 50%. The furnace in the printing plant raises the temperature of this air to 68°F. The following calculation shows that the air that had a relative humidity of 50% at 20°F has a relative humidity of only 8.5% at 68°F. Saturated air at 20°F holds 2.938 g/m³ of water, and 50% relative humidity means that the air on the cold winter morning holds 2.94 × 0.50 or 1.47 g/m³ of water. Saturated air at 68°F (20°C) holds 17.30 g/m³ of water. Therefore, the relative humidity of the 68°F air is 1.47/17.30 × 100% or 8.5%.

The best way to increase humidity is to inject a fine mist of water into the air. The apparatus that does this is called a **humidifier.** Compared to the cost of wasted paper, the cost of such equipment is very small. Automatically controlled air conditioning (temperature and humidity) is a good investment in the printing plant.

Acids and Bases

Materials that conduct electricity when they are dissolved in water are **electrolytes.** Acids, bases, and salts (the products of reaction of an acid and a base) are electrolytes. The ability of a solution to carry electricity is related to the concentration of electrolytes.

Acids

An acid is defined as a compound that generates hydrogen ions, H^+, when dissolved in water. Table 2-IV lists the names and formulas of some common acids.

Table 2-IV.
Common acids.

Acetic acid	CH_3COOH	Metaboric acid	H_3BO_3
Carbonic acid	H_2CO_3	Nitric acid	HNO_3
Citric acid	$H_3C_6H_5O_7$	Oxalic acid	$H_2C_2O_4$
Hydrochloric acid	HCl	Phosphoric acid	H_3PO_4
Hydrocyanic acid	HCN	Sulfuric acid	H_2SO_4
Hydrofluoric acid	HF	Sulfurous acid	H_2SO_3
Hydrogen sulfide	H_2S	Tartaric acid	$H_2C_4H_4O_6$

All of the formulas of these acids can be written by combining hydrogen ions with the appropriate negative ions (see table 1-IV). For example: $H^+ + CH_3COO^-$ (acetate) is acetic acid, and $H^+ + NO_3^-$ (nitrate) is nitric acid.

Properties of acids. Acids have certain properties.
- They increase the H^+ or H_3O^+ concentration in water.
- They react with metals such as zinc, magnesium, and iron.
- They conduct an electric current when dissolved in water.
- They taste sour in dilute solution. (**Warning!** Some acids—such as HCN—are highly toxic.)
- They turn blue litmus paper red. (Litmus paper contains a dye that turns red when the paper is immersed into an acid solution.)
- They react with bases to neutralize them.

Because all acids supply hydrogen ions in a water solution, one concludes that their acid properties are due to the hydrogen ions. Actually, hydrogen ions in aqueous solution are hydrated, that is, they are attached to a molecule of water. The resulting ion, H_3O^+, is called the *hydronium ion.* Consider the following chemical equation in which hydrochloric acid (HCl) is dissolved in water:

$$HCl + H_2O \rightarrow H_3O^+ + Cl^- (aq)$$

When HCl, which is a gas, is dissolved in water, the HCl molecules ionize to form hydronium ions and chloride ions. The hydrochloric acid that is sold commercially is a 40% solution of HCl in water.

For convenience, we refer to the hydrogen ion concentration of a solution, but it should be kept in mind that the hydrogen ions (H^+) are combined with water molecules to form hydronium ions (H_3O^+).

Strong and weak acids. Some acids are "strong." Unless they are dissolved in a large amount of water, they will attack skin or mouth and other tissues on contact. Others are "weak." Typical examples of weak acids are acetic acid, which is the acid present in vinegar, and phosphoric acid which is present in cola drinks.

A strong acid attacks some metals rapidly, while a weak acid reacts very slowly or not at all. The difference is due to the percentage of the acid molecules that ionize when the acid is dissolved in water. If most of the molecules ionize, the acid is strong. If only a small percentage of the molecules ionize, the acid is weak. Table 2-V shows the percent ionization of some common acids.

Table 2-V.
Percent ionization of
common acids.

Hydrochloric acid (HCl)	about 100%
Nitric acid (HNO$_3$)	about 100%
Sulfuric acid (H$_2$SO$_4$)	about 100%
Oxalic acid (H$_2$C$_2$O$_4$)	17.6%
Phosphoric acid (H$_3$PO$_4$)	8.3%
Tartaric acid (H$_2$C$_4$H$_4$O$_6$)	3.3%
Hydrofluoric acid (HF)	2.6%
Acetic acid (CH$_3$COOH)	0.43%
Carbonic acid (H$_2$CO$_3$)	0.06%
Metaboric acid (H$_3$BO$_3$)	0.003%

The above figures apply to dilute solutions of the acids. At low concentrations, 99.57% of acetic acid molecules, for example, remain in solution as molecules, and only 0.43% ionizes to give H$_3$O$^+$ and CH$_3$COO$^-$ ions. This percent ionization means that the hydronium (H$_3$O$^+$) ion concentration of the acetic acid solution is very much less than produced by an equivalent amount of hydrochloric acid.

Bases

One definition of a base is a substance that produces hydroxyl ions (OH$^-$) ions when dissolved in water. Another definition is that a base reacts with the hydrogen ions of an acid to produce a salt and water. Typical bases are:
- Sodium hydroxide, NaOH or lye
- Potassium hydroxide, KOH
- Calcium hydroxide, Ca(OH)$_2$ or quick lime
- Sodium carbonate, Na$_2$CO$_3$ or washing soda
- Sodium bicarbonate, NaHCO$_3$ or baking soda
- Sodium phosphate, Na$_3$PO4
- Ammonia, NH$_3$

Properties of bases. Bases have certain properties:
- They decrease the H$^+$ or H$_3$O$^+$ concentration of water.
- They taste bitter
- They feel slippery
- They conduct an electric current when dissolved in water
- They turn red litmus paper blue
- They react with acids to neutralize them

Strong and weak bases. Alkaline materials, too, can be strong or weak. Ammonium hydroxide (ammonia gas dissolved in water) is a weak base with a pH of around 11. A one molar* solution of the strong base sodium hydroxide in water is much more alkaline, with a pH of around 14. Calcium hydroxide is a stronger base than ammonia, but not as strong as sodium hydroxide.

Buffers

If a solution contains a weak acid together with its salt or a weak base together with its salt, the pH of the solution is not easily changed by addition of either acid or base. A solution of acetic acid and sodium acetate, for example, contains a high concentration of non-ionized acetic acid and a high concentration of acetate ion. If an acid is added to this solution, its H^+ ion is immediately captured by the acetate ion to form non-ionized acetic acid. If an alkali is added, its OH^- ion combines with the small amount of H^+ present, to form water, and further ionization of the acetic acid immediately restores the old value for the H^+ ion.

The same situation applies to disodium hydrogen phosphate (Na_2HPO_4) and sodium dihydrogen phosphate (NaH_2PO_4), salts of phosphoric acid, a weak acid. These materials exist in alkaline lithographic dampening solutions, so addition of an acid or base has little effect on the pH. Likewise, the concentration of these salts has little effect on the pH, and it is impossible, by measuring the pH, to determine the concentration of salts in the dampening solution. Concentration of alkaline dampening solutions is determined by measuring conductivity, discussed earlier in this chapter.

Ionization of Water

Even the purest water conducts electricity to a very small extent because a very small percentage of water molecules ionize to give hydrogen ions (H^+) and hydroxyl ions (OH^-). A mole of hydrogen ions is one gram (1 g). A mole of OH^- ions is 17 g. (A mole is one "gram molecular weight"; that is, it is the number of grams that matches the molecular weight as defined in chapter 1.) The ***concentration*** of these ions is measured in moles per liter. In pure water, the molar concentration of hydrogen ions is the same as the concentration of hydroxyl ions, but on a weight basis the hydroxyl ions weigh 17 times as much as the hydrogen ions. In pure water, the concentration of H^+ ions and the concentration of OH^- ions are each 10^{-7} moles per liter at 75°F (24°C). (The number 10^{-7} is a tenth of one-millionth.)

---

*A one molar solution contains one mole of the dissolved material in one liter of solution.

The number "–7" above and to the right of "10" is called an exponent. The expression 10^{-2} equals 0.01 while 10^{-7} is 0.0000001, etc. The exponent of 10 representing either the hydrogen ion or hydroxyl ion concentration in pure water (a neutral solution) is –7.

Logarithms

The exponent of 10 that gives a certain number is called the *logarithm** of that number. Therefore, pH can be defined precisely by the simple formula:

$$pH = -\log(H^+)$$

where "(H^+)" represents the concentration of hydrogen ions in moles per liter. This explains the notation of pH, the "power of hydrogen" ion concentration. The pH is the negative logarithm of the hydrogen ion concentration—negative because the H^+ concentration is almost always less than one, and the logarithm is accordingly negative. The definition of pH means that we do not have to write a minus sign every time we write the pH.

Measurement of Acidity and Alkalinity: pH

The acidity or the alkalinity of an aqueous solution—a solution in water—is indicated by the pH. Operators of lithographic presses often measure the pH of dampening solutions. Such measurements are particularly helpful when the composition of the raw water supply varies. Aqueous inks, used in flexography, will "sour" or precipitate if the pH gets too low. Effluents from a plant can be measured to see if they satisfy safety and health requirements related to acidity. These examples are only some of the applications of pH measurement in the graphic arts.

The degree of acidity or alkalinity of a solution depends on the comparative concentration of H^+ and OH^- ions. It has been found by experiment that the product of the concentrations of the H^+ and OH^- ions is always 10^{-14} at 24°C (75°F). This number is called the *ionization constant*. If the hydrogen ion concentration increases ten fold, then the hydroxyl ion concentration decreases ten fold. Thus, if one concentration is known, the other can be quickly calculated. Table 2-VI illustrates the relationship between pH and acidity or alkalinity.

---

*A detailed discussion of logarithms is beyond the scope of this book.

Table 2-VI.
Relationship between pH and the concentration of hydrogen ions and hydroxyl ions.

Hydrogen Ion Concentration (moles /liter)	Hydroxyl Ion Concentration (moles /liter)	pH
$1\ (10^0)$	10^{-14}	0
$1/10\ (0.1\ \text{or}\ 10^{-1})$	10^{-13}	1
$1/1{,}000\ (0.001\ \text{or}\ 10^{-3})$	10^{-11}	3
10^{-7}	10^{-7}	7
10^{-13}	10^{-1}	13
10^{-14}	1	14

If the pH of an aqueous solution is 7, the solution is neutral; it is neither acid nor alkaline. A solution with a pH of 5 is slightly acid, while one with a pH of 3 is considerably more acid. The lower the pH reading, (the greater the H^+ concentration), the more acid a solution is. As the pH rises above 7, the solution becomes alkaline. A solution with a pH of 8 is slightly alkaline, while one with a pH of 10 is considerably more alkaline. It is possible to make pH determinations in the printing plant and interpret them correctly without knowing more than is covered in the preceding two paragraphs, but effective use of the information is improved by deeper understanding.

Table 2-VII lists the approximate pH values of some common solutions. The pH values will vary somewhat, depending on how much of the material is dissolved in a specified quantity of water.

Table 2-VII.
Approximate pH of some common solutions.

Solution	pH
Hydrochloric acid, HCl (1 fl oz/gal, 7.8 mL/L)	1.2
Acetic acid, CH_3COOH (3 fl oz/gal)	3.0
Acetic acid (6 fl oz/gal)	2.8
Gum arabic solution (not acidified)	4.2–4.3
Ammonium hydroxide, NH_4OH (household ammonia) (1 fl oz/gal)	11.2
Trisodium phosphate, Na_3PO_4 (TSP)	12
Sodium hydroxide (1 molar: 40 g/L)	14

The pH of a water solution of HCl (a strong acid) is lower (more acid) than a solution of acetic acid (a weak acid.) If the acetic acid solution is made with 6 fl.oz./gal., instead of 3, the pH drops slightly—from 3.0 to 2.8—but it is still much higher than the pH of a solution of HCl made with 1 fl.oz./gal.

Measuring the pH of Solutions

Two methods are used to measure the pH of a solution. The *colorimetric* method depends on the color change of materials called *indicators,* which are added to the solution. The other method is electrical, and it depends on the change of the voltage of a small electrical cell.

Colored Indicators

One of the early indicators was a vegetable dye known as litmus, and the term "litmus paper" is still heard occasionally in referring (incorrectly) to pH papers. Indicators useful for determining pH are complicated organic compounds like bromcresol green, bromphenol blue, and thymol blue. Each of these indicators changes color over a range of about two pH units. Bromcresol green, for example, is yellow at a pH of 4.0, and it changes through various shades of green until it is blue at a pH of 5.6. Such an indicator is useful only in this narrow range of pH. At any pH below 4.0, it remains yellow, and at any pH above 5.6 it remains blue. Measuring the pH of a solution outside of this range requires another indicator. Thymol blue, which is sensitive to a pH range from 8.0 to 9.6, can be used for these values.

A few drops of an indicator can be added to a solution in a test tube. The color is compared with a series of control test tubes that contain the same indicator in solutions of different but known pH values. Such control test tubes can be purchased for each indicator. The color of the solution of unknown pH is matched with the color of one of the solutions in the series of test tubes. Its pH is the same as that of the matching control test tube.

Paper strips for the colorimetric measurement of pH are much more convenient (figure 2-6). These paper strips are impregnated with a solution of one of the indicators previously mentioned and then dried. The pH of the unknown solution is determined by immersing a strip of test paper in it and comparing the color developed in the test strip against a pH color chart that comes with the paper strips. The pH is given under each of the colors on the chart.

Short-range pH papers of this kind are useful for approximate pH measurements, such as the pH of a lithographic dampening solution. The accuracy is usually not closer than 0.3 to 0.5 pH to the actual pH value. If the color developed in the test strip is the same as the color at either end of the color chart, the solution must be retested with another short-range pH paper containing a different indicator. Several of

Figure 2-6.
Determination of
dampening solution
pH using pH paper
strips.

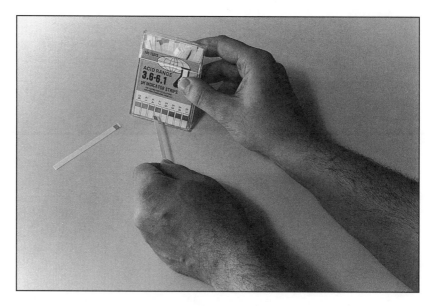

these papers are available; they cover a considerable part of
the pH scale when used properly.

pH meters

Electrically operated pH meters are offered by several manufacturers. Models are available that operate on 110- to 120-volt,
50- to 60-cycle alternating current. There are also portable,
battery-operated pH meters. Modern pH meters include
solid-state electronics and are much more stable than older
meters. Some show pH with a dial; others have a digital
readout (figure 2-7). Depending on the instrument, pH readings can be obtained with an accuracy of 0.01 to 0.05 pH.

Before a pH meter can be read correctly, it must be calibrated. Standard buffers can be obtained for any of several
pH values. They come either in liquid form, ready to use, or
in tablet form, to be dissolved in a specified quantity of distilled water.

A pH meter will read very accurately if it is calibrated
with a standard buffer solution that has a pH fairly close to

Figure 2-7.
A pH meter for use in
laboratory or plant.
*(Courtesy Cole-Parmer
Instrument Company)*

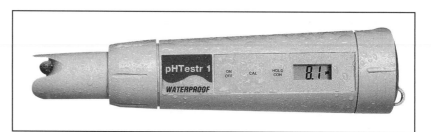

the pH readings to be made. (These buffer solutions can be purchased from a chemical supply company.) For example, if the pH readings are in the range of 3–5, one should use a standard buffer solution that has a pH of 4.00, or one close to that value.

Once the instrument has been calibrated, the pH of any solution is determined by immersing the electrode in a sample of the solution. The meter shows the pH value of this solution almost immediately. After each use, the electrode is washed off to prevent one solution from contaminating the next one to be tested.

A pH meter actually measures changes in voltage of the pH cell. It is a simple matter to produce a scale that reads in pH units rather than voltage. The term "volts per unit change in pH" is called the slope of the response curve. It is 0.059 volt/pH unit at 25°C. Some pH meters allow the operator to calibrate the slope of the response curve. Two standard buffer solutions are used for this calibration. One may have a pH of perhaps 3.00, and the other may have a pH of perhaps 8.00. By a proper adjustment of the meter, it is possible to get the meter to read both pH values correctly.

How an Electrical Cell Measures pH

Every electrical cell has two electrodes. The voltage of any cell is the sum of the voltages produced at each electrode. For example, a cell can be made using a piece of zinc immersed in a solution containing zinc ions for one electrode. The other electrode can be a piece of copper immersed in a solution containing copper ions. A voltage is produced at both the zinc electrode and the copper electrode. The voltage of the cell is the sum of these voltages.

A cell for the measurement of pH also has two electrodes. One is the glass electrode; the other is the reference electrode. Figure 2-8 shows a typical cell. In practice, the glass electrode and the reference electrode are assembled into one unit that can be easily immersed into any solution whose pH is to be measured.

The glass electrode is made of a special glass, as described later. Inside the glass is electrode A, which is immersed in solution B. Similarly, the reference electrode consists of electrode D, immersed in solution E. There is a small opening at the bottom of the reference electrode that allows a small electric current to pass from the reference electrode into solution C, whose pH is being measured.

Figure 2-8.
Electrodes for
measuring pH.

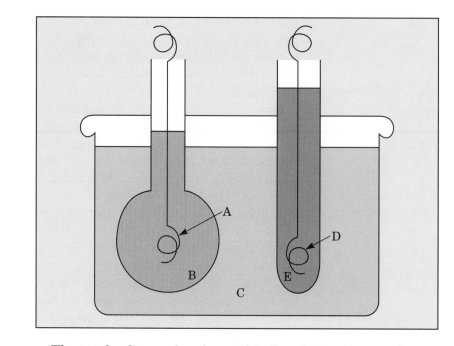

The total voltage of such a cell is the algebraic sum of several partial voltages as follows:

1. Voltage between electrode A and solution B
2. Voltage between solution B and the inside wall of the glass electrode
3. Voltage between solution C and the outside wall of the glass electrode
4. Voltage between electrode D and solution E.

Since solutions B and E always remain the same, the partial voltages 1, 2, and 4 never change, but due to the special glass used for the glass electrode, the partial voltage 3 does change, and it changes 0.059 volts for every 1.00 change in the pH of solution C.

The total voltage of a pH cell is immaterial. Only its property of changing voltage with a change in the pH of solution C is important. Manufacturers of pH meters incorporate the proper electrical circuits to produce instruments that read pH values directly.

For an electric current to pass from electrode A to electrode D, it must first pass through the wall of the glass electrode. This introduces a very high resistance, approximately 100 million ohms, into the cell. Such a cell is not intended to produce an appreciable current, Instead, measurement of the *voltage* of the cell is desired. With modern electronic circuits, it is possible to measure the voltage of pH cells in spite of their very high internal resistance.

The voltage developed between solution C and the outer wall of the glass electrode depends only on the hydrogen ion concentration of that solution. It does not change if other ions are present—unless, for some reason, those ions affect the H^+ concentration. Strongly alkaline solutions usually contain a high concentration of sodium or potassium ions which create an "alkali error" in the measurement of pH. If pH measurements are to be made in this region, the manufacturer of the pH meter should be consulted regarding a correction for the alkali error.

Reference and Glass Electrodes

The metal used for electrode A and the solution B, in which electrode A is immersed, vary from manufacturer to manufacturer. Even the electrodes offered by one manufacturer can vary. The same is true of the reference electrode, electrode D, and solution E.

Different combinations of electrodes give different total voltages when immersed in a solution (C) of a particular pH. As mentioned before, this total voltage is not important since the meter reading can be calibrated to read the correct pH value.

- **Calomel electrode.** The electrode metal is mercury. It is immersed in a saturated solution of potassium chloride, KCl. The solution is also saturated with slightly soluble mercurous chloride, Hg_2Cl_2, which is commonly called "calomel." This combination is often used for the reference electrode in a pH cell.

- **Silver-silver chloride electrode.** The electrode metal is silver that is coated with a thin layer of silver chloride, AgCl. The solution is usually saturated with KCl and with the very slightly soluble compound AgCl. This combination is often used inside the glass electrode.

- **Thalamid electrode.** The electrode metal is mainly mercury that contains 40% thallium (Tl) by weight. The solution is saturated with KCl and with slightly soluble thallium chloride, TlCl.

Conductivity Meters

A conductivity meter, such as that shown in figure 2-9, is used to measure the conductivity of water and aqueous solutions.

These instruments are used to check on the concentration of salts and other conducting materials in water, and in the pressroom they are especially useful for monitoring alkaline and neutral dampening solutions.

Figure 2-9.
Conductivity meter.
(*Courtesy Cole-Parmer Instrument Company*)

Summary

Raw water contains many impurities. They change the pH of lithographic dampening solutions. They change the way ink takes up water, and they create a film or scum over equipment. Ca^{2+} and Mg^{2+} are the ions that make water hard and cause the most trouble in the printing plant.

Conductivity measurements are useful for indicating the concentration of ionized materials. Unlike pH, the conductivity reading is directly proportional to the concentration of the dissolved material. Conductivity is also related to the hardness of raw water.

A common water softener removes hardness but does not remove dissolved solids from the water. Several methods are used to reduce dissolved solids, including distillation, deionization or demineralization, and reverse osmosis. These methods produce water containing very small amounts of dissolved solids.

Air always contains some water vapor. The amount that air can hold increases rapidly with an increase in temperature. The relative humidity is the ratio between the amount of water vapor actually present in the air and the maximum amount of water vapor that the air can hold at a specified temperature. Pressroom air is best controlled with an air conditioner that automatically controls both the temperature and the moisture content.

Acids are materials that produce hydronium ions (H_3O^+) when they are dissolved in water. Alkalies produce OH^- ions in water, and they neutralize acids.

The hydrogen or hydronium ion concentration of an aqueous solution can be expressed as a certain pH value. If the hydrogen ion concentration is 10^{-7} moles per liter, the pH is 7, and the solution is neutral. As the pH decreases, the H^+ (or H_3O^+) concentration increases 10 times for each pH unit (100 times for 2, 1000 times for 3, etc.). As the pH increases above 7, the solution becomes more alkaline in the same ratio.

The pH of a solution can be obtained conveniently and inexpensively by the use of special colored indicator papers that change color over a pH range of about 2 units. Short-range pH papers contain such indicators. At best, colored indicators are accurate to about 0.5 pH unit.

Modern pH meters give pH values rapidly and very accurately. They make use of the voltage generated between a glass electrode and a reference electrode. For each pH unit, this voltage changes 0.059 volt. The scales of the meters are so calibrated that changes in voltages are read directly as pH values.

3 Chemistry of the Compounds of Carbon

Organic chemistry is a major branch of chemistry, comprising the compounds of the element carbon. *Inorganic chemistry,* the chemistry of the metals, water, acids, bases, and the many compounds that are formed from various positive and negative ions, was the subject of chapters 1 and 2.

Carbon is a great "joiner." Carbon atoms link with each other to form straight-chain, branched, or cyclic molecules. They also combine with atoms of other elements, notably hydrogen, oxygen, nitrogen, sulfur, and chlorine. Paper, ink pigments and vehicles, solvents, and coatings are composed chiefly of compounds containing carbon that are included in "organic" chemistry.

Forms of Elementary Carbon

Carbon exists alone in several forms. Diamonds, graphite, lampblack, wood charcoal, and coke consist mostly of elementary carbon. A diamond is pure carbon in a crystalline form. The other materials are largely carbon but contain a small percentage of impurities.

Coke. When soft coal is heated in the absence of air, it changes chemically. This process is called *destructive distillation.* Some of the complicated molecules break down into simpler molecules with fewer atoms. Some of these products are gases—called *coal gas*—that can be burned to produce heat. Others change to a thick liquid called *coal tar.* Most of the coal remains as *coke,* a solid consisting mainly of elementary carbon and the mineral matter that was present in the soft coal. Coke can be burned as a fuel; it is also used in the reduction of iron ore (iron oxide) to metallic iron.

Charcoal. When wood is heated in the absence of air, it decomposes as soft coal does, but the products are different.

The recovered liquid is a mixture of wood tar, wood alcohol (methyl alcohol), acetone (a common solvent), and acetic acid. The solid that remains is *charcoal,* which is principally elementary carbon.

Activated charcoal. Activated charcoal (also called activated carbon) is obtained by heating granulated charcoal to remove adsorbed gaseous materials. Activated charcoal is used to adsorb other gaseous materials from a stream of air. In the graphic arts, air containing solvents from gravure inks is passed through tanks containing activated charcoal; the solvents are adsorbed onto the charcoal and removed from the air stream. Later, the bed of charcoal is regenerated by passing steam through it. Hydrocarbon solvents from heatset lithographic inks can also be adsorbed onto activated charcoal. To regenerate the charcoal, the hydrocarbon solvents are removed by vacuum distillation.

Graphite. Graphite, like diamond, is a crystalline form of carbon. It is occasionally found in nature, but it is usually made artificially by heating coke in an electric furnace to about 5,000°F (2,760°C).

Graphite is a fair conductor of electricity. It is used for electrodes in certain electrochemical processes. It is also used in some lubricants. Graphite is mixed with clay to form the lead in lead pencils.

Reaction of Carbon with Oxygen

All forms of elementary carbon react with oxygen. Under the proper conditions, they burn in an atmosphere of oxygen gas or of air, which contains oxygen, to form carbon dioxide gas. The reaction is:

$$C + O_2 \rightarrow CO_2$$

It is much harder to get a diamond to burn than it is to get coke or charcoal to burn. When it does burn, it forms carbon dioxide just as coke and charcoal do.

If carbon is burned with an insufficient supply of oxygen, carbon monoxide is formed. The reaction for producing carbon monoxide gas is:

$$C + \tfrac{1}{2}O_2 \rightarrow CO$$

Carbon monoxide is very poisonous. A person who breathes air containing only one part of CO in 750 parts of air will die in a short time. It is the CO in the exhaust fumes of an automobile that kills people when an automobile engine is operated in a closed, unventilated area.

Neither carbon dioxide nor carbon monoxide is usually included among the organic compounds, but carbon dioxide plays an important role in life processes in what is called the "carbon dioxide cycle" in nature. People and animals eat food, which is composed mostly of organic chemicals and water, and they inhale air containing oxygen. Inside the body, a reaction between the food and the oxygen occurs, supplying the energy that keeps the body active. The principal products of the reaction are water and carbon dioxide gas, which are exhaled into the air.

The air contains about 0.04% CO_2. Even this small amount of carbon dioxide in the air is very important. Plants take it through their leaves and convert it, together with water and other materials taken from the soil, into organic compounds such as starches, sugars, proteins, fats, and cellulose. This completes the "carbon dioxide cycle"—CO_2 in the air goes to organic compounds in plants, which are eaten by people and other animals. These produce CO_2 that goes back into the air again.

Carbon dioxide plays another important role in the atmosphere. By trapping the sun's rays, it causes the temperature

Figure 3-1.
The carbon dioxide
cycle.

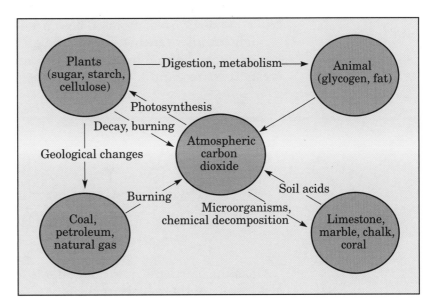

of the air to rise. Scientists are concerned that if the temperature rises a few degrees, serious changes in climate will occur. For example, if the Greenland ice cap were to melt, oceans would rise some 50 feet, covering low islands and cities near the ocean.

Raw Materials for Organic Compounds

All living things are composed of organic compounds. They are a source for a wide variety of raw materials. Coal, petroleum, and natural gas (which are derived from living material) are other sources. For example, trees supply cellulose fibers that are used to make paper for printing and other uses. The flax plant supplies flaxseed, which is pressed to make linseed oil. Toluene, xylene, and many other organic chemicals are produced from petroleum.

There are more than one million known compounds of carbon. They are organized into classes. Those classes and compounds having graphic arts uses are given principal attention here.

Hydro-carbons

Organic compounds that contain only carbon and hydrogen atoms are called **hydrocarbons.** There are two main classes of hydrocarbons: aliphatic and aromatic.

Aliphatic hydrocarbons. Aliphatic hydrocarbons include those correctly called alkanes but commonly called paraffins with no unsaturated carbon atoms, alkenes (olefins) with double-bonded carbon atoms, and alkynes (acetylenes) with triple-bonded carbon atoms. These are further divided into straight-chain, branched-chain, and alicyclic hydrocarbons. For example, normal butane (usually written "n-butane") is a straight-chain paraffin and isobutane is a branched paraffin.

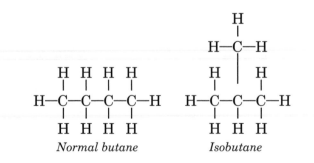

Normal butane   *Isobutane*

Gasoline, kerosene, fuel oil, and oils for heatset inks consist largely of these hydrocarbons.

Alicyclic hydrocarbons include cycloparaffins with no unsaturated carbon atoms and the complex terpenes that include double bonds. Turpentine is the best known of the terpenes. Alicyclic hydrocarbons have at least one carbon ring in their structure.

Aromatic hydrocarbons. These are unsaturated hydrocarbons that contain at least one six-carbon "aromatic" ring. In general, the solvent power of aromatic hydrocarbons is stronger than that of the aliphatics. Common compounds in this class are benzene, toluene, and xylene (see page 89).

Saturated Hydrocarbons

A *saturated* hydrocarbon is a hydrocarbon (either aliphatic or alicyclic) having only single bonds between carbon atoms. Every carbon-to-carbon bond in a saturated hydrocarbon comprises a single pair of valence electrons. (The carbon atoms are saturated with hydrogen.)

The simplest saturated hydrocarbon is *methane* (once called "marsh gas"). Its formula is CH_4. Natural gas consists largely of methane.

The element carbon has an atomic number of 6. This means that the atoms of carbon have two electrons in the first energy level and four valence electrons in the second energy level.

Carbon atoms do not gain or lose the valence electrons to form stable ions. Instead, they share a pair of electrons with another carbon atom or another element to form a covalent bond. Because of the number of valence electrons, carbon can form four covalent bonds. In methane, for example, each hydrogen atom furnishes one valence electron to form a covalent bond, and four hydrogen atoms combine with one carbon atom to form a molecule of CH_4. This molecule can be pictured as follows:

$$\begin{array}{c} H \\ \cdot\cdot \\ H\!:\!\overset{\cdot\cdot}{\underset{\cdot\cdot}{C}}\!:\!H \\ \cdot\cdot \\ H \end{array}$$

There are thus four covalent bonds in a CH_4 molecule. It is simpler to write the structural formula of methane as:

One of the reasons that there are so many compounds of carbon is that one carbon atom can share a pair of electrons with another carbon atom, the second carbon atom with a third, etc. This characteristic can lead to a long chain of carbon atoms.

The first example of such sharing is ***ethane,*** C_2H_6. Its structural formula is:

Ethane, like methane, is a gas and occurs in natural gas.

The next saturated hydrocarbon is ***propane,*** C_3H_8. Bottled gas is largely compressed propane. Its structural formula is:

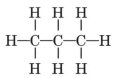

The next is ***butane,*** C_4H_{10}. Both propane and butane are gases at room temperature and pressure.

The next twelve members of this series are liquids at room temperature; those with more than 16 carbon atoms are solids. The general formula for any saturated aliphatic hydrocarbon is:

$$C_nH_{2n+2}$$

where "n" is the number of carbon atoms. For example, ***hexane*** is a liquid with six carbon atoms in each molecule, so its formula must be C_6H_{14}. It is used as a solvent in some flexographic and screen printing inks.

The gaseous saturated hydrocarbons occur in natural gas, but crude oil is the source for most of the remainder of the series. That is, petroleum is a mixture of many saturated hydrocarbons. It even contains some propane and butane, because gases are soluble in liquids that are chemically similar. Also some solid hydrocarbons are present in liquid petroleum; solids as well as gases dissolve in solvents that are chemically similar.

**Unsaturated
Hydrocarbons**

In the saturated hydrocarbons, each carbon atom holds as many hydrogen atoms as possible. In the unsaturated hydrocarbons, fewer than the maximum number of hydrogen atoms are attached to the carbon atoms; there are multiple bonds between carbon atoms. There are two pairs (or even three pairs) of electrons shared between the bonded carbon atoms.

Olefins or alkenes. Olefins are compounds with a carbon-to-carbon double bond. For example, the saturated hydrocarbon ethane, C_2H_6, has the structure:

A two-carbon unsaturated hydrocarbon, called *ethylene* (ethene), has the formula C_2H_4, with the structure:

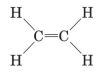

In this compound, the two carbon atoms are held together with a *double bond.* Two pairs of electrons are shared between the two carbon atoms.

Double bonds are more reactive than single bonds, and unsaturated hydrocarbons are more reactive than saturated ones. Under proper conditions, ethylene will react chemically

Figure 3-2.
Ethylene (ethene).

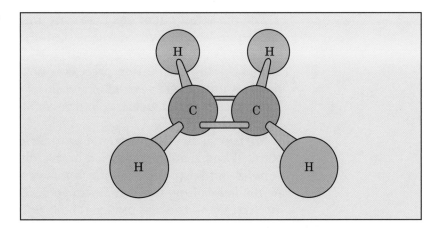

with hydrogen, bromine, hydrochloric acid, and water. Ethylene molecules also link with each other, forming a long chain. This action is called ***polymerization,*** and the result is the plastic called polyethylene. A lot of printing, particularly by flexography, is done on films of polyethylene.

Figure 3-3.
The polymerization
of ethylene to
polyethylene.

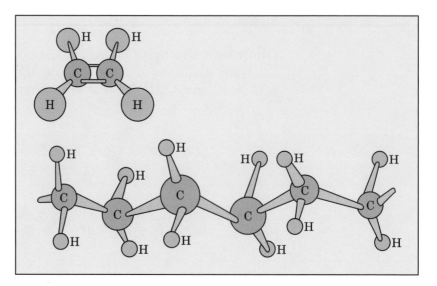

Alkynes or acetylenes. In oxyacetylene welding, one of the materials used is acetylene gas. It is a highly unsaturated hydrocarbon, with the formula C_2H_2. Its structure may be written H:C:::C:H or CH≡CH. As the structural formula indicates, there is a ***triple bond*** between the carbon atoms in this compound. This means that three pairs of electrons are shared between the two carbon atoms.

Hydrocarbons in Petroleum

At an oil refinery, petroleum is boiled (distilled) and separated into different fractions as diagrammed in figure 3-4.

The part that vaporizes first contains more of the lower-molecular-weight hydrocarbons. Later fractions contain more and more of the heavier hydrocarbons. The separation obtained by this distillation process is not complete; each fraction is a mixture of several of the saturated hydrocarbons. The fraction called gasoline contains mostly the hydrocarbons from C_5H_{12} to C_9H_{20}.

Mixtures such as this do not have a single, fixed boiling point. When gasoline is heated in the absence of air (to prevent fire and explosion), it starts to boil at about 140°F (60°C). As the low-boiling components evaporate, the boiling point increases, and the last material boils at about 375°F (190°C).

Figure 3-4.
Single-stage crude
distillation process.

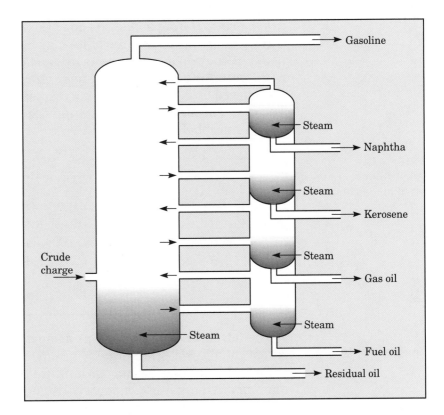

Solvents and fuels produced by the fractional distillation of petroleum include petroleum ether, gasoline, naphtha, mineral spirits, kerosene, fuel oil, and heatset ink oils. Some of the physical properties of these and other liquids are presented later in this chapter.

When crude oil is separated by fractional distillation, the amount of each fraction does not necessarily correspond to the market demand. There is less of the gasoline fraction than is needed, and too much of the kerosene and fuel oil fractions. The oil refineries compensate for this by "cracking" the heavier fractions, such as kerosene and fuel oil, using catalysts under high temperature and pressure (catalysts are materials that promote specific chemical reactions). The heavier (high-boiling or high-molecular-weight) hydrocarbon molecules are broken into two or more parts to give smaller molecules, many of which are in the boiling range of gasoline. Higher boiling fractions are suitable for use as heatset oils used in web offset or sheetfed inks.

Alkyl Groups

If one hydrogen atom were removed from a methane molecule, the group $-CH_3$ would remain. Such a group is called an *alkyl group.* Groups like this do not exist alone, because one carbon atom has only three of its four bonding forces satisfied, but they do occur in many organic compounds bonded with something else. Their names, which usually end in "yl," are derived from the names of the corresponding hydrocarbons. The ones encountered most often in the graphic arts are shown in table 3-I.

Table 3-I.
Alkyl groups encountered in the graphic arts.

Hydrocarbon		Organic Group	
CH_4	Methane	$-CH_3$	Methyl
C_2H_6	Ethane	$-C_2H_5$	Ethyl
C_3H_8	Propane	$-C_3H_7$	Propyl
C_4H_{10}	Butane	$-C_4H_9$	Butyl
		$-C_nH_{2n+1}$	(general formula)
		$-R$	(general symbol)

The letter "R" is often used as a shorthand notation to represent an alkyl or other organic group.

The names of the organic groups are often used in the naming of an organic compound. Thus the compound

$$\overset{\displaystyle O}{\overset{\displaystyle \|}{CH_3-C-C_2H_5}}$$

is called methyl ethyl ketone, or MEK.

Alkyl Halides: R–X

Organic compounds in which a halide—fluorine, chlorine, bromine, or iodine—has replaced one or more of the hydrogen atoms in a saturated hydrocarbon are called alkyl halides and often abbreviated R–X. If one hydrogen atom of methane is replaced with a chlorine atom (Cl), the product is methyl chloride whose structure is shown in figure 3-5.

If three hydrogen atoms of methane are replaced with Cl, the product is trichloromethane (commonly called chloroform), $CHCl_3$, with the structural formula:

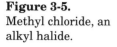

Figure 3-5.
Methyl chloride, an alkyl halide.

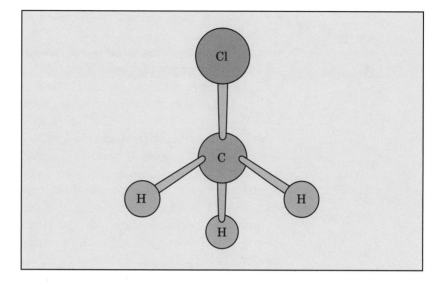

If all of the hydrogen atoms in methane are replaced with chlorine atoms, the product is carbon tetrachloride, CCl_4. Methane burns in air, but carbon tetrachloride does not. The vapor of carbon tetrachloride is highly toxic. CCl_4 is a common solvent, and its use is controlled by federal regulations.

An alkyl halide that is not nearly as toxic as carbon tetrachloride is bromotrifluoromethane, CF_3Br. It is used in hand-held fire extinguishers and for larger systems. If air contains as much as 5% of this compound, a fire will be extinguished rapidly. It is claimed that people can breathe in such an atmosphere without toxic effects for up to five minutes.

The formula for ethane is C_2H_6. If the three hydrogen atoms on the first carbon atom are replaced with chlorine atoms, the compound is called 1,1,1-trichloroethane and has the structural formula:

$$\begin{array}{c} \quad\;\; Cl \;\; H \\ \quad\;\;\; | \quad\; | \\ Cl-C-C-H \\ \quad\;\;\; | \quad\; | \\ \quad\;\; Cl \;\; H \end{array}$$

It is nonflammable, much less toxic than carbon tetrachloride, and a good substitute for it in liquid cleaners.

Alcohols: R–OH

If an organic group is combined with an –OH group, the compound is called **alcohol.** The formulas and names of the four lowest-molecular-weight alcohols are:

- CH_3OH, methyl alcohol, also called methanol or wood alcohol.
- C_2H_5OH, ethyl alcohol, also called ethanol or grain alcohol. (This is beverage alcohol.)
- C_3H_7OH, propyl alcohol, also called propanol.
- C_3H_7OH, isopropyl alcohol, also called 2-propanol.

Methyl alcohol, once obtained by the destructive distillation of hardwood, is now produced synthetically, from either coal or natural gas. Methyl alcohol has been used in some printing inks for food wraps and in some screen printing inks, but its use has been virtually eliminated in pressroom applications because of its toxicity. It causes blindness or even death if enough is taken internally by breathing or by mouth. Its structure is depicted in figure 3-6.

Figure 3-6.
Methyl alcohol, or methanol.

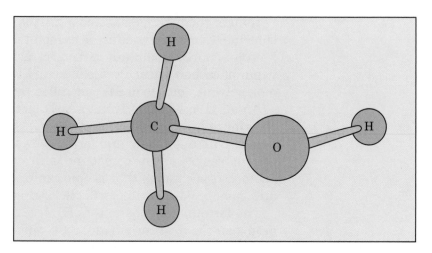

Various alcoholic beverages made by fermentation of grapes, grain, or potatoes contain ethyl alcohol. Ethyl alcohol produced in this way is too expensive for industrial uses. Most industrial ethyl alcohol is made synthetically from petroleum gases. Ethyl alcohol is used in both flexographic and gravure inks. Its structural formula is:

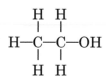

The molecular formula of propyl alcohol is C_3H_7OH. Since there are three carbon atoms in each molecule, it is possible to have the –OH group attached to a carbon atom at the end or to the one in the middle. Both compounds are produced commercially. The names and structures of the two compounds are:

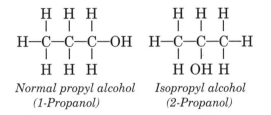

Normal propyl alcohol   *Isopropyl alcohol*
(1-Propanol)   *(2-Propanol)*

Each compound has exactly the same kind and number of atoms (in this case, three carbons, eight hydrogens, and one oxygen). Such compounds are called **isomers.** They are similar, but have somewhat different physical properties, such as boiling point and evaporation rate. Normal propyl alcohol is an example of a **primary** alcohol. This is an alcohol that has the –OH group attached to a carbon atom having two or three hydrogen atoms also attached. Isopropyl alcohol is an example of a **secondary** alcohol. In secondary alcohols, the –OH group is attached to a carbon atom that has one and only one hydrogen atom also attached. If the –OH group is attached to a carbon atom with no hydrogen atoms, it is a tertiary alcohol.

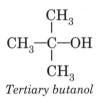

Tertiary butanol

Normal propyl alcohol and isopropyl alcohol are both used in the graphic arts for flexographic and screen printing inks. Isopropyl alcohol is also used in lithographic fountain solutions. Isopropyl alcohol is significantly more toxic than the familiar ethyl alcohol, and it creates air-pollution problems.

Other Alcohols

Alcohols can contain more than one –OH group as in ethylene glycol, with the structural formula:

$$
\begin{array}{cc}
\text{H} & \text{H} \\
| & | \\
\text{H—C—C—H} \\
| & | \\
\text{OH} & \text{OH}
\end{array}
$$

This compound is produced synthetically from petroleum gases. It is used as a permanent type of antifreeze. It is also employed in some screen printing inks and in the manufacture of polyester film base. Propylene glycol has low toxicity, and it is used in inks for food packaging. Its formula is $C_3H_6(OH)_2$, and its structure is:

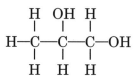

Glycerine (or more properly "glycerol") has the formula $C_3H_5(OH)_3$, with the structural formula:

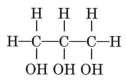

Glycerol is often used in the dampening solution for lithographic duplicator machines.

Ethers: R–O–R'

Ethers include diethyl ether, the one commonly called "ether." Its formula is $C_2H_5–O–C_2H_5$. The ethers are characterized by an oxygen atom that is bonded to two carbon atoms. Ether is a good solvent for oils, but it must be used carefully because it is volatile and very flammable. Diethyl ether (figure 3-7) is made commercially from ethyl alcohol.

Figure 3-7.
Diethyl ether.

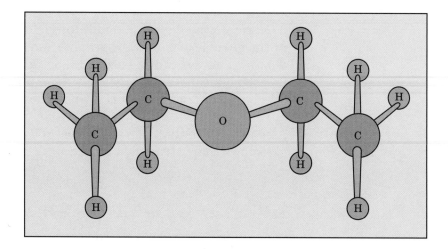

The glycol ethers have both alcoholic and ether groups, and this combination gives them unusual solvent properties. A common one is ethylene glycol monoethyl ether. Long names like this are often encountered in organic chemistry; they accurately tell how the molecules are constituted. In this case, the "R" of the general formula represents ethylene glycol, and the "R'" represents an ethyl group. The structural formula is:

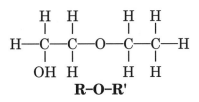

Ethylene glycol monoethyl ether is usually referred to by a familiar trade name: "Cellosolve," the registered trademark of Union Carbide Corp.

The glycol ethers are used as solvents in lacquers and in some gravure, flexographic, and screen printing inks. Glycol ethers are also used as alcohol substitutes in lithographic dampening solutions.

Organic Acids: R–COOH

If an organic "R" group is combined with a **_carboxyl_** group, –COOH, a simple organic acid results. A carboxyl group has the structural formula:

$$\begin{array}{c} \text{O} \\ \| \\ \text{—COH} \end{array}$$

One oxygen atom is double-bonded to a carbon atom and a hydroxyl group (–OH) is bonded to the same carbon atom. Acetic acid (CH_3COOH) consists of a methyl group attached to a carboxyl group. Acetic acid is the weak acid present in vinegar. Its structure is shown in figure 3-8.

When an organic acid is dissolved in water, the hydrogen ion that dissociates from the acid is the hydrogen atom of the –COOH group. Organic acids are weak acids: when they are dissolved in water, only a small percent of the molecules ionize to give a H^+ ion and an organic negative ion.

Some organic acids consist of molecules with two or three carboxyl groups; some have hydroxyl groups, –OH, instead of hydrogen atoms attached to certain carbon atoms. Formulas

Figure 3-8.
Acetic acid.

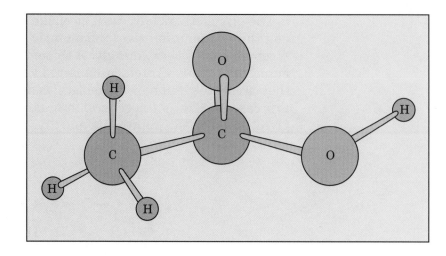

and structures of three common organic acids are presented below:

- Oxalic acid, $H_2C_2O_4$, with structure:

$$
\begin{array}{c}
\text{COOH} \\
| \\
\text{COOH}
\end{array}
$$

- Tartaric acid, $H_2C_4H_4O_6$, with structure:

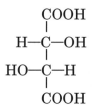

- Citric acid, $H_3C_6H_5O_7$, with structure:

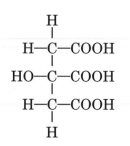

Only the hydrogen atoms that form part of a carboxyl group can ionize to furnish H^+ ions in solution. Thus, tartaric acid has two ionizable hydrogen atoms, and citric acid has three.

**Esters:
R–CO–OR'**

Esters are formed by the reaction of an organic acid with an alcohol. The organic group of the alcohol replaces the hydrogen atom of the –COOH. Most of the esters used in the graphic arts are formed from acetic acid and an alcohol, and they are called acetates. A typical reaction for the formation of an ester is:

$$CH_3COOH + CH_3OH \rightarrow CH_3COOCH_3 + H_2O$$

| *Acetic* | *Methyl* | *Methyl acetate* | *Water* |
| *acid* | *alcohol* | *(an ester)* | |

Methyl acetate is a solvent used sometimes in screen printing. Other esters used as solvents in the graphic arts are:
- Ethyl acetate, $CH_3COOC_2H_5$ (gravure, flexographic, and screen printing inks)
- Normal-propyl acetate, $CH_3COOC_3H_7$ (gravure and flexographic printing inks)
- Isopropyl acetate, $CH_3COOC_3H_7$ (gravure and screen printing inks)
- Normal-butyl acetate, $CH_3COOC_4H_9$ (screen printing inks)

The structure of methyl acetate is shown in figure 3-9.

Figure 3-9.
Methyl acetate.

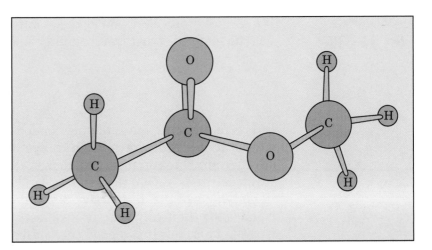

**Aldehydes:
R–CHO**

If a primary alcohol is oxidized carefully, an ***aldehyde*** can be produced. An aldehyde has the general structural formula:

$$\overset{\displaystyle O}{\underset{\displaystyle R-C-H}{\|}}$$

where an oxygen atom is double-bonded to a carbon atom. The simplest aldehyde is formaldehyde, HCHO, in which the "R" is a hydrogen atom. The structure of formaldehyde is shown in figure 3-10. Formaldehyde is a preservative and also a hardening or tanning agent.

Figure 3-10.
Formaldehyde.

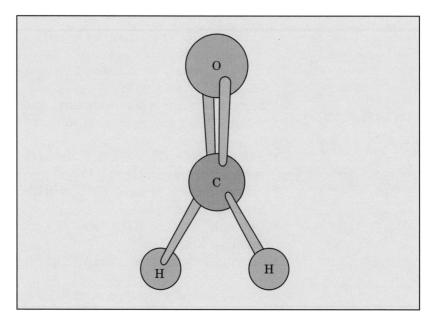

Ketones: R–CO–R'

If a secondary alcohol is oxidized, a ***ketone*** is produced. It has the general structural formula:

$$\overset{\overset{\displaystyle O}{\displaystyle \|}}{R—C—R'}$$

where an oxygen atom is double-bonded to a carbon atom and where "R" and "R'" are organic groups that may be the same or different. One common ketone is acetone, $CH_3–CO–CH_3$. Its structure is shown in figure 3-11. This solvent is one of the products of the destructive distillation of wood. Acetone is made synthetically from propylene or by oxidizing isopropyl alcohol. It is a good solvent for certain resins and is used in some screen printing inks.

Another ketone is methyl ethyl ketone, often referred to as MEK. Its formula is $CH_3–CO–C_2H_5$. It is produced by the oxidation of secondary butyl alcohol. It is a good solvent and is used in some gravure and screen printing inks. Their high solvency limits their use in flexographic inks that are applied from rubber or plastic plates.

Figure 3-11.
Acetone, a common
ketone.

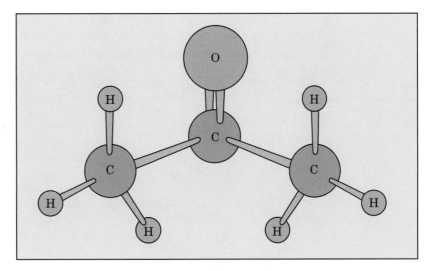

Aromatic Compounds

Benzene

Benzene is a volatile solvent with the formula C_6H_6. Benzene is an example of an aromatic hydrocarbon, and it is one of the materials obtained when soft coal is heated in the absence of air. Aromatic compounds have at least one unsaturated ring in them, and benzene is the simplest aromatic compound with a ring of six carbon atoms, each of which is attached to a hydrogen atom.

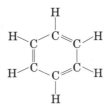

In formulas involving benzene, its "six-sided" carbon skeleton is sometimes represented by a hexagon with three double bonds. However, the bonds in benzene do not resemble the double bonds in the alkenes, and modern practice is to show it and other aromatic compounds as a hexagon with a circle.

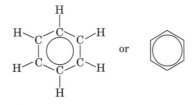

Benzene Derivatives

Benzene is the "parent" of thousands of compounds called ***benzene derivatives.*** These include solvents, dyes, pigments, medicines, photographic developers, and insecticides.

Many familiar compounds are benzene derivatives, in which one or more of the hydrogen atoms in benzene are replaced with another atom or groups of atoms.

Table 3-II.
Benzene derivatives.

Group or atom	Product
$-OH$	Phenol
$-NH_2$	Aniline
$-COOH$	Benzoic acid
$-CH=CH_2$	Styrene
$-Cl$	Chlorobenzene
$-CH_3$	Toluene
$-NO_2$	Nitrobenzene
$-CH_2OH$	Benzyl alcohol
$-SO_3H$	Benzene sulfonic acid

Phenol. Phenol, once called carbolic acid, is a benzene derivative in which one of the hydrogen atoms has been replaced with a hydroxyl group, $-OH$. The formula of phenol is C_6H_5OH and its structure is:

Phenol is not a carboxylic acid, but it is very weakly acidic, hence its name carbolic acid. It is toxic to bacteria and fungi, and has been used as a preservative and disinfectant. It is highly corrosive to the skin, and modern disinfectants are much more effective and less corrosive.

Aniline. Aniline has the formula $C_6H_5NH_2$ and the structure:

Because the amino group accepts hydrogen ions, aniline is a base. Aniline is the starting chemical for the production of aniline dyes. It is also used in making some ink pigments.

Benzoic acid. The formula for this carboxylic acid is C_6H_5COOH, and its structure is:

COOH

One hydrogen atom of benzene has been replaced with a –COOH group. When this acid reacts with the base NaOH, the main product is a compound called sodium benzoate that is widely used as a food preservative.

Toluene. Toluene, sometimes called "toluol," is the common name of methyl benzene. Its formula is $C_6H_5CH_3$, and its structure is:

CH_3

An important organic solvent, it is used in gravure and screen printing inks.

Xylenes. The xylenes have the formula $C_6H_4(CH_3)_2$. Two of the hydrogen atoms of benzene have been replaced with methyl groups, –CH_3. The xylenes are used in some gravure and screen printing inks.

Three compounds have the xylene formula, differing in regard to the position of the methyl groups on the benzene ring. These three compounds are examples of *isomers*—compounds with the same molecular formula but with different molecular structures.

To distinguish one isomer from another, it is convenient to number the carbon atoms around the benzene ring:

If adjacent hydrogen atoms (in "1,2" position) are replaced, the product is called an **ortho** compound. If the substitution is on carbon atoms separated from each other by one carbon atom (in the "1,3" position), the product is a **meta** compound, and if the substituted carbon atoms are across the ring (in the "1,4" position), the product is called a **para** compound. Commercial xylene, sometimes called xylol, is a mixture of these isomers.

The structures of the three xylenes are as follows:

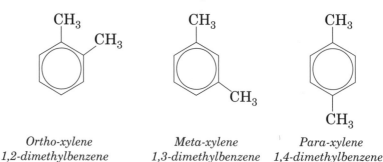

Ortho-xylene   *Meta-xylene*   *Para-xylene*
1,2-dimethylbenzene   *1,3-dimethylbenzene*   *1,4-dimethylbenzene*

Aromatic compounds may have two, or three, or more rings such as naphthalene ($C_{10}H_8$) or anthracene ($C_{14}H_{10}$):

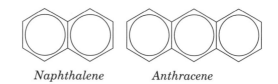

Naphthalene   *Anthracene*

(Chemists often write these structures without showing the location of the hydrogen atoms.)

Aromatic hydrocarbons tend to be carcinogenic (some much more than others), and like all chemicals, they should be handled carefully.

Proteins

The three principal classes of foods are proteins, fats, and carbohydrates; all of these consist of complicated organic compounds. A few of the compounds are used in the graphic arts.

Proteins are compounds of carbon, hydrogen, oxygen, nitrogen, and sulfur. Many foods have high protein content. Some common commercial protein products are casein (from milk), soybean protein, gelatin, and glue.

Protein molecules are very large. For example, the molecular weight of egg albumin, a protein material, is about 34,000 (as compared with sucrose or cane sugar, which is 342). A chemist can differentiate between different proteins by decomposing them with acid or base in water. They are split apart to form compounds called ***amino acids.*** All proteins (vegetable or animal) are formed from only 20 amino acids, and any given protein may have from four to 20 different amino acids in its structure. Proteins differ from fats and carbohydrates in having nitrogen and sulfur atoms in their molecules. All amino acids contain an amino group ($-NH_2$) and a carboxyl group ($-COOH$). The amino group is basic, the carboxyl group is acidic so that the individual acid is ***amphoteric,*** that is, it has the properties of both bases and acids. The simplest amino acid is glycine or amino acetic acid.

$$
\begin{array}{c}
H \\
| \\
H-C-COOH \\
| \\
NH_2
\end{array}
$$

Fats and Oils From the chemical viewpoint, fats and oils are the same. If the material is a solid, it is a ***fat.*** If it is a liquid, it is called an ***oil.*** Of course, it is easy to convert one into the other merely by changing the temperature. The term "oil" as used here refers to vegetable and animal oils and not to lubricating oil that comes from petroleum.

Fats and oils are esters, which are formed by the combination of an organic acid and an alcohol. Fats and oils are particular esters in which the organic acid portion is a fatty acid and the alcohol portion is glycerol (or glycerine) $C_3H_5(OH)_3$. A fatty acid is an aliphatic acid with one carboxyl group. Since there are three $-OH$ groups in a glycerol molecule, three molecules of fatty acid react with one molecule of glycerol.

Fats and oils, like proteins, are produced by plants and animals. Common fats and oils are lard, butter, beef tallow, soybean oil, linseed oil, and cottonseed oil. Each contains a mixture of different fat molecules. The type and amount of fat molecules determine the properties of a fat or oil.

The principal fatty acids and the corresponding esters, or fat molecules, are shown in table 3-III.

Stearic and palmitic acids are saturated acids, they contain no carbon-to-carbon double bonds. As you can tell from the relationship between the number of carbon and hydrogen atoms, oleic is an unsaturated acid with one double bond (it is "mono-unsaturated"), linoleic acid has two double bonds,

Table 3-III.
Fatty acids and fat molecules.

Fatty Acid	Number of Double Bonds	Fat Molecule
Stearic acid, $C_{17}H_{35}COOH$	0	$(C_{17}H_{35}COO)_3C_3H_5$
Oleic acid, $C_{17}H_{33}COOH$	1	$(C_{17}H_{33}COO)_3C_3H_5$
Linoleic acid, $C_{17}H_{31}COOH$	2	$(C_{17}H_{31}COO)_3C_3H_5$
Linolenic acid, $C_{17}H_{29}COOH$	3	$(C_{17}H_{29}COO)_3C_3H_5$
Palmitic acid, $C_{15}H_{31}COOH$	1	$(C_{15}H_{31}COO)_3C_3H_5$

and linolenic acid is a highly unsaturated acid, with three double bonds in a molecule (it is "polyunsaturated").

Drying oils, used for printing inks and varnishes, are made from vegetable oils with a high polyunsaturated content. Linseed oil and chinawood oil (tung oil) are good examples. The greater the degree of unsaturation, the faster the varnish dries: linolenic acid groups make a molecule dry faster than linoleic acid groups. Olive oil, which contains oleic acid, is not a drying oil.

Carbo-
hydrates

Carbohydrates are compounds of carbon, hydrogen, and oxygen in which there are twice as many hydrogen atoms as oxygen atoms. Sugars, starches, dextrins, and cellulose are all carbohydrates.

Sugars. Sucrose (cane or beet sugar), lactose (milk sugar), and maltose all have the formula $C_{12}H_{22}O_{11}$. Dextrose (glucose) and fructose (fruit sugar) are simpler sugars; both have the formula $C_6H_{12}O_6$.

Starches. Starches are high-molecular-weight polymers produced in nature from sugars. They have the general formula $(C_6H_{12}O_5)_n$, where "n" is about 250 to several thousand. Antisetoff spray powders consist of special grades of starches.

Dextrins. When starch is heated in the presence of a small amount of HCl, a catalyst, the big starch molecules are broken down into shorter molecules called *dextrins.* They are more soluble in water than starches. They are used as adhesives on postage stamps and envelopes.

Cellulose. Cellulose, like starch, is a natural high-molecular-weight polymer. It has the same formula as starch, $(C_6H_{10}O_5)_n$. Both are polymers of glucose. However, the glucose units within the polymer atoms are arranged differently, and cellulose has a higher molecular weight.

Cellulose is not soluble in water, even though cellulose fibers swell in the presence of water. (Low-molecular-weight polymers, called *hemicellulose,* are soluble in caustic. These naturally occurring polymers strengthen the bonding of cellulose fibers in paper, but they are largely removed in pulping of wood and bleaching of the fibers.) Cellulose is of great interest to all branches of the graphic arts because it is the basis of printing paper. Most cellulose fibers for making paper come from wood pulp, but cellulose fibers can also be obtained from cotton, linen, jute, hemp, kenaf, cornstalks, and straw.

Organic Solvents

Many of the compounds discussed in this chapter are organic solvents. A solvent is a liquid that can dissolve another substance. Thus, methyl alcohol is a solvent for shellac; and ethyl acetate is a solvent for nitrocellulose. A solvent that dissolves one material may have no solvent power for another. Millions of gallons of solvents are used annually in the graphic arts for inks, roller and blanket washes, plate developers, and lithographic dampening additives.

The solvent powers of individual solvents are determined by their physical and chemical properties. These properties include specific gravity, refractive index, kauri butanol (K.B.) number, aniline point, vapor pressure at different temperatures, and heat of vaporization. If needed, this information can be obtained from the manufacturer or distributor of the solvent.

For many people in the graphic arts, the two most important properties of a solvent are its flammability and its relative toxicity. Both of these are directly related to the safety of its use, which is discussed in chapter 11.

Heatset Oils

Many hydrocarbon solvents are used in printing inks. In heatset inks, linear and alicyclic hydrocarbons from petroleum are carefully fractionated to provide *heatset oils.* These products are often referred to as "Magie Oils," the tradename of a familiar brand. Petroleum oils containing unsaturated groups (carbon-to-carbon double bonds) cause smoke and may be carcinogenic, so heatset oils are fully hydrogenated before they are used in inks. (*Hydrogenation* is the direct addition of hydrogen to the double bonds of unsaturated molecules, resulting in saturated molecules.)

Hydrogenation of heatset oils with hydrogen gas, at high temperature and pressure, in the presence of a catalyst, converts the aromatic compounds into alicyclic hydrocarbons (sometimes called naphthenes) and the olefins into paraffins. Inks made with these are the "low odor" or "low smoke" type of heatset inks offered by ink manufacturers. Because these purified heatset oils are poorer solvents for many resins, varnish makers must choose suitable resins.

Mineral Oil

News inks contain *mineral oil,* which contains hydrocarbon solvents with higher molecular weight than those in heatset oils. The higher molecular weight gives mineral oils higher boiling points and higher viscosity.

Polymers and Plastics

Polymers are very large molecules formed from reactive molecules called monomers. The word "mer" means part; so that monomer means one part, dimer means two parts, and polymer means many parts. Proteins, starch, and cellulose are examples of polymers that are vital to life and useful to the printing industry. Polymers are divided into two groups: addition polymers and condensation polymers, depending on the method by which they are formed.

Addition Polymers

Addition polymers are usually formed from vinyl monomers: compounds with the structure $CHR_1=CHR_2$ where "R_1" and "R_2" may be H or an organic group.

Table 3-IV.
Addition polymers.

R_1	R_2	Monomer	Polymer
H	H	Ethylene	Polyethylene
H	CH_3	Propylene	Polypropylene
H	Cl	Vinyl chloride	Poly(vinyl chloride)
H	$OCOCH_3$	Vinyl acetate	Poly(vinyl acetate)
H	C_6H_5	Styrene	Polystyrene
H	CN	Acrylonitrile	Polyacrylonitrile

Properties of the polymer depend not only on the monomer but on the catalyst and the method or conditions of polymerization. Many kinds of catalysts are known. A catalyst starts the polymerization reaction by opening the double bond and creating a free radical (unpaired electron) that is extremely reactive. It attacks another monomer, bonding to it and creating a new free radical:

$$CR_2{=}CR_2 \rightarrow -[CR_2{-}\overset{*}{C}R_2] \rightarrow -CR_2{-}CR_2{-}CR_2{-}\overset{*}{C}R_2$$

$$\rightarrow -CR_2{-}CR_2{-}[CR_2{-}CR_2{-}]_nCR_2{-}\overset{*}{C}R_2$$

where n can be as low as 5 or 10 or up to thousands, and "*" indicates a free radical (an unpaired electron).

Another type of addition polymer is initiated with ionic catalysts that open the double bond, creating a positive or negative ion that attacks the next monomer.

Because the reaction can be repeated with each activated molecule adding another monomer and activating a new carbon atom, the reaction is known as a ***chain reaction.***

The curing of sheetfed litho inks is a type of addition polymerization, but since the oil molecules that are involved contain an allyl group instead of a vinyl group, the reaction proceeds very differently. The reaction is slow, occurring in 2–4 hr. instead of in a matter of seconds, and the molecular weight is much lower than with vinyl polymers such as those listed above.

Polyolefins. The most familiar of these are polyethylene and polypropylene. Depending on the molecular weight and the degree of branching in the polymer, resins with a wide variety of properties are produced. Molecular weight and degree of branching are controlled by the conditions under which the polymerization is carried out.

Vinyls. The most common vinyls are polymers and copolymers of vinyl chloride and vinyl acetate, and they are used in a variety of films and coatings and many other applications. Poly(vinyl acetate) is hydrolyzed to form poly(vinyl alcohol), used in adhesives.

$$\begin{array}{ccccc} & OCOCH_3 & OCOCH_3 & & OH & & OH \\ & | & | & & | & & | \\ -CH_2{-}CH{-}CH_2{-}CH{-} & \rightarrow & -CH_2{-}CH{-}CH_2{-}CH{-} \end{array}$$

Polystyrene. Polystyrene is a hard, rather brittle molding resin, which can be blown to form foamed polystyrene. It is also formed into sheets for printing of backlit point-of-purchase displays. Copolymers of styrene with maleic anhydride are soluble in base and have been used as inexpensive water-based inks for flexography.

Maleic anhydride is a vinyl compound formed from maleic acid, an unsaturated dicarboxylic acid. It will not polymerize with itself, but it will copolymerize with other vinyl monomers.

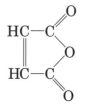

Condensation Polymers

Condensation polymers are produced by formation of an ester or amide with the elimination of water. A diamine reacts with a diacid to form an amide:

$$H_2N-X-NH_2 + HOOC-X-COOH \rightarrow$$
$$H_2O + H_2N[X-NH-CO-X-CONH]_n-$$

where X = $-CH_2CH_2CH_2CH_2-$
or $-CH_2CH_2CH_2CH_2CH_2CH_2-$
or $-CH_2CH_2CH_2CH_2CH_2CH_2CH_2CH_2-$

Ethylene glycol reacts with terephthalic acid to produce a polyester used for films or textiles.

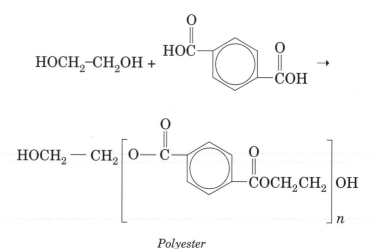

Polyester

Table 3-V lists some familiar condensation polymers, the materials from which they are formed, and their use in graphic arts.

Table 3-V.
Condensation
polymers.

Polymer	Monomer	Application
Proteins	Amino acids	Adhesives
Cellulose	Glucose	Paper and textiles
Starch	Glucose	Adhesives, spray powder, paper coatings
Polyester	Phthalic acid + glycol	Graphic arts film, printing inks
Nylon	Adipic acid + diamine	Packaging films
Glyptal	Glycerol + phthalic acid	Printing inks
Phenolic	Phenol + formaldehyde	Printing inks, varnishes
Urethane	Diisocyanate + glycol	Printing inks, varnishes, adhesives

Polyesters. Polymeric esters are often based on phthalic acid (benzene dicarboxylic acid) and the nature of the polymer depends on the structure of the acid. The para isomer, called terephthalic acid, gives polyesters that are used for graphic arts film, photographic film, and packaging. The meta isomer, called isophthalic acid, gives glyptal resins useful for modifying linseed and other drying oils.

Nylons. Polyamides are often referred to as nylons. Those that are made from straight-chain diamines (such as hexanediamine with six carbon atoms) and straight-chain diacids (adipic acid with six carbon atoms) are designated by numbers such as 6,6 nylon or 6,10 if the acid has ten carbon atoms. Polyamide films are extensively used in packaging. Polyamides are excellent resins for solvent-based flexographic ink. These are made from diamines and dimer acids—unsaturated fatty acids that have been dimerized or polymerized only to the dimer stage.

Plasticizers

Plastic polymers can be molded or drawn. They include most of the synthetic polymers because molding and extruding are major applications for polymers, and great effort has been devoted to their development. However, some materials such as poly(vinyl chloride) are stiff and brittle, and cannot be molded below their decomposition temperature, so they must be softened. ***Plasticizers,*** the materials used to soften them, partially dissolve the polymer. Thousands of different plasticizers are sold commercially. When added to a polymer, such as poly(vinyl chloride), they soften the material so that it can be extruded or blown, and the product remains tough and pliable after formation.

Many plastics such as polyethylene or nylon are sufficiently pliable for forming well below their decomposition temperatures and are not normally treated with plasticizer.

Drying-oil Synthetic Resins

Linseed, tung, tall oil, dehydrated castor, soya, and other drying oils are often heated with alkyd, phenolic, or urethane resins to yield lithographic varnishes with good working properties.

Alkyds. Condensation of glycerol with phthalic anhydride yields a polyester called a *glyptal resin* or an alkyd. The major alkyds used in printing ink are based on isophthalic acid. When linseed oil is heated ("cooked") with one of these resins, it reacts to form a product known as a linseed alkyd. If a little oil is added to the resin, the product is known as a "short-oil" alkyd or resin. If a great deal of oil is added, the product is a "long-oil" resin. The properties of the short-oil alkyd are more like the properties of the alkyd while those of the long-oil resin are more like those of the linseed oil. Reaction with the alkyd improves pigment wetting, film toughness, and adhesion of the dried film.

Alkyds made with non-drying oils make excellent plasticizers for nitrocellulose in gravure inks.

Phenolics. Phenolic resins, made by reacting phenol with formaldehyde, are solids insoluble in any solvent, but the ones used in ink are made from phenol derivatives called alkylphenols. Litho varnishes made by reacting drying oils and phenolic resins show excellent wetting characteristics and produce tough, glossy films with good adhesion and alkali resistance. Gloss varnishes normally contain phenolic resins.

Alcohol-soluble, pure phenolic resins are used in flexo inks and as modifiers for top lacquers.

Urethanes. Polyurethanes, prepared from isocyanates and alcohols, are chemical cousins of the polyamides. They can be combined with drying oils, as can the alkyds and phenolics, to yield products useful in lithographic inks and varnishes. Incorporation of a urethane normally improves adhesion and chemical resistance of the dried film. They also make strong adhesives used in bookbinding.

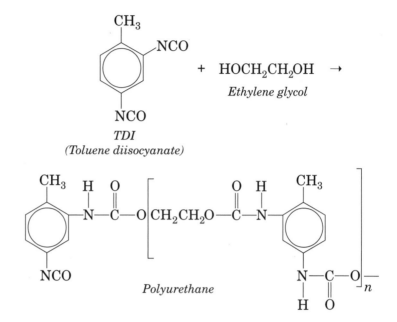

Fessenden, Ralph J. and Joan S. Fessenden. *Organic Chemistry.* 5th ed. (1122 pp) 1994. (Cole Publishing Co.).

Graham Solomons, T. W. *Organic Chemistry.* 6th ed. 1998. (John Wiley & Sons).

McMurry, John and Mary E. Castellion. *Fundamentals of General, Organic and Biological Chemistry.* 2nd ed. (880 pp) 1996 (Prentice Hall).

4 Chemistry of Photography and Proofing

Digital printing and proofing are reducing the use of chemistry in prepress operations. But the prepress area with color separations, layouts, and proofing, still relies heavily on chemistry. Except for digital printing and proofing, the major printing processes use photography and its chemistry in image converting and in platemaking. Understanding the printing processes requires some knowledge of the chemistry of photography.

Photographic Materials

Most photography is accomplished with films or papers coated with an emulsion that contains compounds of silver. These materials are exposed to light that has either been passed through imaged film or reflected from copy (in printing, the original is called "copy"). Graphic arts processes also use lasers and CRT devices to expose film. Following exposure, the film or paper is developed, then fixed, washed, and dried.

Bases for Photographic Images

The light-sensitive photographic emulsion* must be applied to a base. The base of photographic paper is a special grade of paper usually coated with barium sulfate. Most graphic arts films employ a polyester (poly[ethylene terephthalate] or PET) base, polyester having replaced cellulose triacetate that was used for a long time. Polyester, made from dimethyl terephthalate and ethylene glycol (see chapter 3), has better dimensional stability with respect to moisture. For a given

---

*The term "emulsion" usually refers to a dispersion of one liquid in another, like water (or vinegar) dispersed in oil to make mayonnaise. Early photographers used the term "emulsion" to refer to the dispersion of silver and silver compounds in a gel structure. The usage has never changed.

change in relative humidity, cellulose triacetate changes in size about four times as much as polyester.

Some cellulose acetate butyrate is used in duplicating sheet film, but the old cellulose nitrate (nitrocellulose) has been completely replaced because of fires and because of the poor archival qualities (short life) of nitrocellulose photographic films.

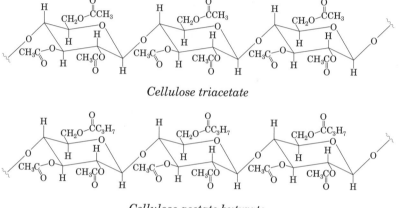

Cellulose triacetate

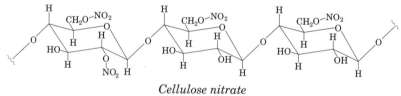

Cellulose acetate butyrate
(the butyrate groups are not distributed uniformly)

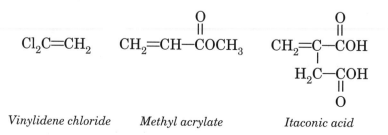

Cellulose nitrate
(the nitrate groups are not distributed uniformly)

Subbing

Gelatin emulsions show poor adhesion to polyester film, and the needed adhesion is ensured by "subbing," that is, by adding a hydrophilic base coat. Subbing is achieved either by oxidizing the surface with a chemical or a flame treatment or by applying a polymeric film that shows good adhesion to both polyester and gelatin. A terpolymer of vinylidene chloride, methyl acrylate, and itaconic acid is one such polymer.

$$Cl_2C\!=\!CH_2 \qquad CH_2\!=\!CH\!-\!\overset{\displaystyle O}{\overset{\|}{C}}OCH_3 \qquad CH_2\!=\!\underset{\displaystyle H_2C-COH}{\overset{\displaystyle O}{\overset{\|}{C}}\!-\!COH}$$

$$\underset{\displaystyle O}{\overset{\|}{}}$$

Vinylidene chloride   Methyl acrylate   Itaconic acid

Silver Halide Film Emulsions

A photographic emulsion consists of tiny particles of a silver halide dispersed in gelatin. The silver halide is sensitive to light. The gelatin serves to hold and protect the silver halide, and acts as an adhesive to hold the emulsion to the film, glass, or paper.

Silver halides (AgX) include silver chloride, silver bromide, and silver iodide. Since the valence of silver is +1 and the valence of the halide ions is –1, the silver halide formulas are AgCl, AgBr, and AgI, respectively. Of these, AgCl is the least light-sensitive and is used in the emulsions of certain photographic papers, particularly papers used to make contact prints. When enlargements are made, a given amount of light passing through the negative spreads out over a greater area on the enlarging paper. Therefore the emulsion for enlarging papers must be faster, or more light-sensitive, if a very long exposure time is to be avoided. Such emulsions usually contain AgBr or a mixture of AgBr and AgCl. When a negative is to be made from the original copy, a fast emulsion is needed. Negative emulsions contain AgBr or a mixture of AgBr and AgI. These mixtures are ionic mixtures and can be expressed chemically as $AgCl_xBr_{1-x}$ or $AgBr_xI_{1-x}$.

Film Speeds

It has been claimed that film speeds have doubled about every ten years since the invention of silver-based photography. This can be verified by an off-the-cuff calculation. Suppose that it took 30 seconds to take a photographic portrait in 1850, fifteen decades ago. The best film can now take that portrait in 1/1000 sec., about 30,000 times as fast. If speed has doubled every decade, in 15 decades it should be 2^{15} (32,768) times as fast, a remarkable agreement. Research continues to increase the speed and quality of film and to reduce the amount of expensive silver required to make satisfactory images.

Preparation of a Silver Halide Emulsion

Film manufacturers use complicated procedures to produce silver halide emulsions. The following is a simplified description of these procedures.

Silver halides, virtually insoluble in water, are formed chemically in an aqueous solution of gelatin. A solution containing from 1 to 5% gelatin is prepared, and enough potassium bromide (KBr) is dissolved in it to give a 10% KBr solution. In a separate container, silver nitrate ($AgNO_3$) is dissolved in water. The solutions are heated to a temperature of 160–195°F (70–90°C), and the $AgNO_3$ solution is poured slowly into the KBr/gelatin solution. The following reaction occurs:

$$AgNO_3 + KBr \rightarrow AgBr + KNO_3$$

Emulsions made this way usually have a wide grain-size distribution and are suitable for continuous-tone film. Graphic arts emulsions usually have narrow grain-size distributions and are made by a double-jet technique where halide and $AgNO_3$ are added to the gelatin solution simultaneously.

In this discussion, KBr is used as an example. If a chloride emulsion is desired, then KCl is used. If a mixture of bromide and iodide is desired, the gelatin solution contains a mixture of KBr and KI. In any case, the chemical reaction is similar to the one given above. AgI is more light-sensitive than AgBr, and AgBr is more light-sensitive than AgCl, but the size of the particles of solid silver halide (AgX) also determines the sensitivity of the emulsion to light. If the $AgNO_3$ is added rapidly, the particles ("grains") are very tiny, and the sensitivity of the emulsion (the dispersion or matrix of AgX in gelatin) is low. If the temperature of the solutions is raised before they are mixed, and the $AgNO_3$ solution is added slowly, the particles are somewhat larger, and the sensitivity to light is greater. Even if the initial size of the grains is very small, they increase in size as the emulsion is allowed to "ripen." After crystal growth is completed, the gelatin is flocculated by adjusting the pH or by adding a salt, and the precipitated gelatin carries the AgX with it, leaving in solution the excess KBr and the potassium nitrate (KNO_3).

A different method is used to obtain larger grains. In this method, precipitation of AgX is made in the presence of ammonia at lower temperatures.

High-speed emulsions used for negatives are prepared by slow addition of the $AgNO_3$ and by long ripening at high temperatures. Lower-speed emulsions used for process work or for positive film are made with a more rapid addition of the $AgNO_3$ at a lower temperature. These lower-speed emulsions have a grain finer and more uniform than high-speed emulsions.

After the emulsion has been ripened, more gelatin is added to bring the total gelatin concentration up to about 10%. Then the mixture is cooled rapidly, whereupon it sets to form a gel. The gel ("emulsion") is cut into small shreds and washed with cold water to remove the excess KBr and the KNO_3. Ultrafiltration is another method used to wash photographic emulsions. It is not desirable to leave these soluble salts in the emulsion (except for emulsions for coating photographic papers, which are usually not washed to remove soluble salts).

After washing, the emulsion is melted, and sometimes more gelatin is added. At this point the emulsion is still low in sensitivity to light. The temperature is held at about 120°F (50°C) for about an hour. This second ripening process increases the sensitivity. How much it increases depends in part on the nature of the gelatin used. The final liquid emulsion is adjusted to contain about 6% gelatin and 4% AgX. It is then coated on the paper or film. Graphic arts films generally contain from 2.5 to 5.0 g/m^2 of silver (0.25 to 0.50 grains/in^2).

One role of gelatin in an emulsion is that of a "protective colloid." If $AgNO_3$ is added to KBr, each of them being in water solution with no gelatin present, the solid AgBr forms clumps and soon settles to the bottom of the container. If one of the solutions contains gelatin, the AgBr is precipitated in such a finely divided state that it is more or less uniformly distributed and suspended in the gelatin solution.

Silver halides are sensitive to light in the ultraviolet (shorter than 400 nm), but not very sensitive to visible light. In order to make them sensitive to visible light, spectral sensitizers (such as organic dyes or sulfur- and gold-containing compounds) must be added.

Ionic Structure of a Silver Halide Grain

The grains of AgX in an emulsion do not consist of individual molecules of AgX. Instead, an ***ionic lattice*** is built up. For AgBr, this lattice consists of a Ag^+ ion positioned next to a Br^- ion, and another Ag^+ ion next to this Br^- ion, and so on in three dimensions, forming an octahedral crystal. A very small two-dimensional part of such a crystal is illustrated in figure 4-1.

Similar layers of Ag^+ and Br^- ions lie in front of and behind the ones shown. In such an ionic lattice, one Ag^+ ion

Figure 4-1.
Two-dimensional projection of a silver bromide grain.

$$Ag^+$$
$$Ag^+ \ Br^- \ Ag^+$$
$$Ag^+ \ Br^- \ Ag^+ \ Br^-$$
$$Ag^+ \ Br^- \ Ag^+ \ Br^-$$
$$Br^- \ Ag^+ \ Br^-$$
$$Br^-$$

does not "belong" to one Br⁻ any more than to any other Br⁻ that is next to it. Instead, every silver ion "belongs" partly to six Br⁻ ions, and each Br⁻ ion in turn "belongs" partly to six Ag⁺ ions.

Bromide Body

Each Ag^+ or Br^- ion in the center of a crystal of AgBr has its electrical forces satisfied. There are oppositely charged ions all around it—left and right, below and above, in front and behind; but at the surface of the crystal, a different situation exists. Here there are no ions on one side of a particular surface ion. Such surface ions still have an electrical force remaining, which could hold something else. If the solution contains an excess of KBr over the amount required to form the AgBr, then the Br⁻ ions of the KBr are attracted to the surface Ag⁺ ions and are held there by the attraction of the opposite electrical charges. Of course, for every Br⁻ ion thus held, a K⁺ ion is left in the solution. These K⁺ ions wander around in the neighborhood of the Br⁻ ion. Thus the final picture of part of an AgBr grain formed in the presence of an excess of KBr is as shown in the accompanying illustration. An AgBr grain formed in this way is called a **bromide body** (figure 4-2).

Figure 4-2.
Bromide body.

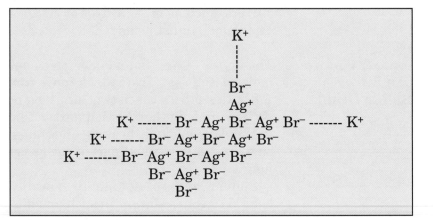

Varying Properties of Photographic Emulsions

Photographic emulsions vary in several properties, such as grain size, speed, contrast, maximum density after development, and sensitivity to the different wavelengths of light. The effect of grain size on the speed of the emulsion is discussed earlier in this chapter. An emulsion with small, uniform-size grains is required for lith (high-contrast) films. Continuous-tone films have a range of grain sizes. This range makes it possible to change contrast with development time. The contrast of lith film is very high, while that of contact film is lower, and that of continuous-tone film, the lowest.

The maximum optical density that can be produced, D_{max}, varies with the type of film. With lith film, it is above 4.0; with contact film, usually more that 3.0; and with continuous-tone film, about 2.0–2.5.

Films also vary in their sensitivity to the different wavelengths of light. All silver halide films are sensitive to ultraviolet light, but it is possible to add certain dyes (sensitizers) that broaden the sensitivity to other wavelengths. Some emulsions are only sensitive to ultraviolet and blue light; orthochromatic emulsions are sensitive to ultraviolet, blue, and green light; panchromatic emulsions are sensitive to ultraviolet and all colors of the visible spectrum (see figure 4-3.)

Reaction of the Emulsion to Light   The developer reacts much more readily with the parts of a photographic emulsion exposed to light (called the *latent image*) than with the unexposed parts. No visible reaction

Figure 4-3.
Sensitivity of the human eye and spectra of various types of film.

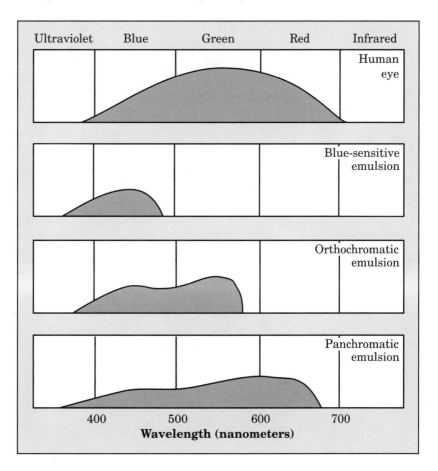

occurs when an emulsion is exposed to light. (Some contact films do show a faint but visible image before processing.) The typical black-and-white portions of a negative do not appear until the emulsion has been placed in the developer bath for a certain length of time.

The actual chemical and physical reactions that occur when a AgBr grain is exposed to light are very complex, and the precise mechanism is still obscure. However, exposed AgX molecules are much more reactive toward developer than are unexposed molecules.

Light Sources for Films and Plates

The most familiar light sources in industry and the home are tungsten filament and fluorescent lights. The most common light sources for cameras and enlargers are quartz-iodine and pulsed-xenon lamps.

The principal light sources for exposing litho plates are metal-halide lamps. Some pulsed xenon and high-pressure mercury lamps are used, but carbon arcs are a thing of the past. Tungsten lights are too weak in ultraviolet to be practical. Figure 4-4 shows the spectral data for various light sources.

The efficiency of a light source is defined as the ratio between actinic light output and total power input. A considerable portion of the light output of most lamps is not in the deep blue and ultraviolet wavelengths to which litho plate coatings are most sensitive. The light that carries out a chemical reaction is called **actinic light.**

Figure 4-4.
Spectral curves for various light sources.

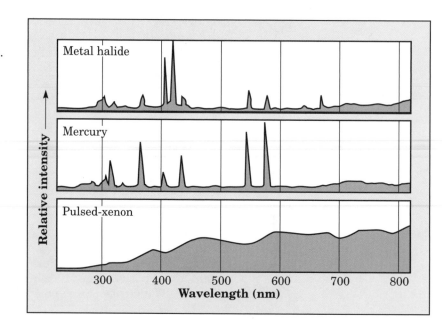

Different light sources produce light with characteristic *color temperatures,* which are commonly measured in "Kelvin" or "degrees Kelvin" (K). Color temperature refers to the spectral distribution emitted by a "perfect radiator" heated to the specified temperature. (Black iron is an almost-perfect radiator.) The higher the temperature, the bluer the light. At low temperatures, infrared and red predominate. The radiation includes a broad spectrum of wavelengths. The numbers chosen in Table 4-I are the peak color radiation.

Table 4-I shows the color temperature of various light sources, and table 4-II compares the Kelvin scale with the Celsius (Centigrade) and Fahrenheit temperature scales. The color temperature of daylight varies, but averages 6,500 K. In

Table 4-I.
Color temperature of various lamps

Light source	Peak color temperature (K)
Daylight (northern exposure)	7,500
Daylight (average)	6,500
Metal halide	6,500
Fluorescent (daylight)	6,500
Pulsed-xenon	5,600
Carbon arc (white flame core)	5,000
Fluorescent (cool white)	4,200
Quartz iodine (high level)	3,400
Tungsten-halogen	3,300
No. 1 Photoflood	3,400
Incandescent (100 watt)	3,000
Quartz iodine (low level)	2,950

Table 4-II.
Comparison of Kelvin, Celsius, and Fahrenheit scales.

K	°C	°F
0	−273	−460
500	227	440
1,000	727	1,340
1,500	1,227	2,240
2,000	1,727	3,140
2,500	2,227	4,040
3,000	2,727	4,940
3,500	3,227	5,840
4,000	3,727	6,740
4,500	4,227	7,640
5,000	4,727	8,540
5,500	5,227	9,440
6,000	5,727	10,340
6,500	6,227	11,240
7,000	6,727	12,140
7,500	7,227	13,040

contrast, ordinary tungsten lamps (bulbs) have a color temperature of about 2,300 K, producing yellow-red light. The color temperature of quartz-iodine lamps is 3,200–3,400 K, producing light that limits their use in color separation work. They are, however, suitable for line and halftone work.

(Color film can be made to give "natural" color either under exposure from tungsten lamps or from quartz-iodine or fluorescent lamps, but the photographer must choose the right film for the illumination to be used.)

The distribution of radiation from a black-body radiator at various temperatures is shown in figure 4-5.

Figure 4-5.
Spectra of black-body radiators.

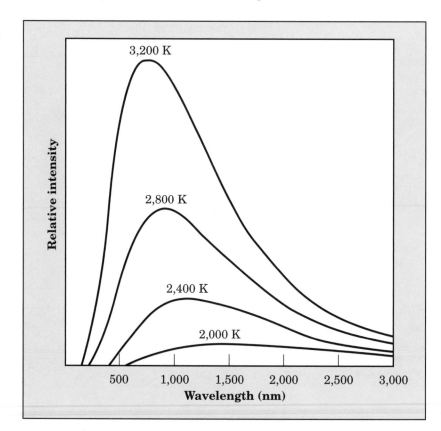

Incandescent lamps. Tungsten has the highest melting point of any metal, and its low vapor pressure makes it resist evaporation at the high temperatures of the incandescent filament. Nevertheless, tungsten reacts with water and oxygen at these temperatures so that they must be excluded. Water is eliminated by the use of "getters" incorporated into the base of the lamp, and air is replaced by a mixture of 93% argon and 7% nitrogen. Nevertheless, tungsten does evaporate slowly, and long-used lamps can be recognized by the gray coating on the inside of the glass bulb.

Photoflood lamps. These tungsten lamps operated at a high temperature, which was achieved by building a lamp to operate at 70 volts and operating it at 110 v. The life of the lamp was very short, but it gave illumination that was brighter and bluer than the conventional tungsten lamp. In use, photoflood lamps experienced a fairly rapid envelope blackening that reduced lamp efficiency and shifted the color. Photoflood lights have been replaced by more efficient lamps.

Halogen incandescent lamps. These lamps are an improvement over photoflood lamps. The inert gas pressure is higher than that of normal incandescent lamps to diminish the evaporation of tungsten. This makes it possible both to operate the lamp at a higher temperature (increase its luminosity) and to use a smaller bulb.

The envelope (bulb) is made of a clear or frosted fused silica (SiO_2) (often called *quartz,* which is a natural mineral of the same composition). The halogen, usually iodine, but sometimes bromine, keeps the fused silica free of tungsten that evaporates from the filament. The blackening of the envelope, which reduces color temperature, is eliminated, and lamp life is greatly extended.

Pulsed-xenon lamps. Pulsed-xenon lamps are made of fused silica tubing filled with xenon gas under low pressure. Xenon is one of the *inert gases,* which are found in very small quantities in air. The light from pulsed-xenon lamps covers the entire spectral range of visible light, with a color temperature of about 5,600 K. In addition, these lamps have the advantage of instant start, constant color temperature, and a life of 300–1,000 hr. For these reasons, they are by far the best lamps for color separation work, either in cameras or enlargers.

Fluorescent lamps. Fluorescent lamps are mercury vapor lights that activate a *phosphor.* Mercury vapor emits radiation in the ultraviolet (185 and 254 nm), and these wavelengths are converted to visible light by the phosphors that are coated on the inside of the lamp. Phosphors are usually finely divided powders of complex inorganic salts or oxides.

One commonly used phosphor bears the formula $Ca_5(PO_4)_3(Cl,F):Sb^{3+}Mn^{2+}$. The Sb^{3+} ions contribute a blue luminescence and the Mn^{2+} ions an orange color. By adjusting the ratio of the two, a white light is obtained. The emission has two peaks, 480 nm and 580 nm, but they are fairly broad. Changing the F/Cl ratio in the phosphor further changes the peak wavelength.

The green lamps (tubes) used in some graphic arts applications are coated with a phosphor that emits only green light. The lamp is suitable for the exposure of line work and halftone prints. It produces only a small amount of heat, has low power consumption, and has a life of about 5,000 hours. The exposure time is approximately 15% longer than is required with four 12-in. (0.35-m) pulsed-xenon lamps.

Metal-halide lamps. Metal halide lamps are high-pressure, mercury-vapor lamps with added halides of iron, barium, titanium, or rare earth elements included in the lamp. By proper choice of additives, the lamp can be made to peak anywhere from 380 to 420 nm. With no additive, the mercury vapor is strong in the 380–400 nm range. This lamp is called a "photopolymer type" because this is the range most useful for exposing photopolymer plates. A typical metal-halide lamp for the graphic arts peaks around 400–410 nm. This metal-halide lamp is sometimes called the "diazo" type, since it is particularly efficient for exposing diazo coatings for proofing systems or printing plates. In practice, lamp manufacturers try to produce lamps that have a strong emission from 380 to 420 nm because these will be useful in exposing any type of plate.

For exposing diazo coatings, a metal-halide lamp gives about twice as much actinic light as a conventional mercury-vapor lamp, and four times as much as a pulsed-xenon lamp, when the particular light sources are compared on an equal-wattage basis.

While a metal-halide lamp is very efficient for exposing plates, it requires about 2–4 min. to warm up before it will operate, and it costs more than other light sources to replace. Also, if turned off after an exposure, it must cool for a few minutes before it can be turned on again. A shutter is used to solve the start-up problem. After the initial warm-up, the lamp is left on low standby power. Then, when an exposure is to be made, the shutter opens while the lamp goes to full power. At the end of the exposure, the shutter closes, and the lamp returns to the standby condition.

Another type of metal-halide lamp requires a warm-up period during which an electric element behind the tube is heating. After warm-up, the tube is in an "instant start" condition; after one exposure, another can be started immediately. With this lamp, a mechanical shutter is not required.

Mercury lamps. Mercury lamps are usually made from tubular, clear fused silica with enough mercury placed in the tube to form an arc between the electrodes at each end. When current flows, it heats the mercury, converting it to a vapor and ionizing it, producing light. The color temperature is determined by the internal pressure, the mercury content, and the current. Mercury lamps are classified as low, medium, or high pressure, depending on the internal pressure.

Mercury lamps are available that peak in the 360–370 nm range, which is in the ultraviolet region. They are more efficient for exposing photopolymer plates than for diazo systems.

Two variations of conventional mercury-vapor lamps are used in the graphic arts. One is a low-pressure mercury lamp with a specially designed ballast that throws power rapidly into the lamp for instant start. The lamp reaches full power in 5–10 sec. With this system, a hot lamp can be turned off, then restarted immediately. No shutter mechanism is required.

The other type of mercury-vapor lamp operates at about 90 atmospheres pressure. It is characteristic that spectra of gases at low pressure (near vacuum) emit sharp lines. As the pressure of the emitting gas rises, the line broadens so that at 90 atmospheres, mercury vapor (gas) emits a broad band of radiation instead of sharp lines. Such a lamp has thick walls to withstand the high pressure, and it must be water-cooled to control the temperature. It has instant start and restart. No shutter is required, but the light does not get up to full intensity for several seconds.

A *light integrator* is a required quality control instrument for exposure by any light. It gives the operator the necessary control information, and it helps prevent unacceptable variation in plate exposure.

Lasers. Lasers are used to expose film and printing plates. Many printing plates carry coatings fast enough to be imaged with a laser beam. On electrostatic plates, the laser discharges a charge carried by a semiconductor on the plate. The image is developed with a toner. Lasers operate differently from other

Table 4-III.
Radiation of various
lasers.

Type of laser	Wavelength (μm)
HeNe	632.8
Argon	351.1, 363.8, 488.0, 514.5
Krypton	647.1
Ruby	694.3
YAG	1060
CO_2	1060

light sources. (See chapter 5 for a more detailed discussion of lasers.)

Most commonly used in graphic arts are the CO_2, argon ion, and HeNe (helium-neon) lasers and the laser diode or light-emitting diode (LED). The CO_2 laser emits in the infrared, the HeNe emits in the red, and there are four main lines in the output spectrum of an argon-ion laser.

Light-emitting diodes. A material that converts electrical energy directly into light is used to make the familiar light-emitting diode (LED). LEDs are used throughout industry; in the graphic arts industry they are used, for example, as indicators on instrument displays on densitometers and scanners, and they can be used to expose film.

The radiation varies from IR to UV, including visible colors, depending on the characteristics of the material used in the light-emitting region of the device. For example, gallium arsenide (GaAs) emits at 850 nm in the IR region, while gallium phosphide (GaP), a mixed gallium phosphide/gallium nitride (GaP:N), and gallium phosphide/gallium arsenide (GaP:As) emit in the red.

Infrared LEDs can be modulated at speeds up to 1000 megabits/sec. and are useful for optical transmission of data.

The familiar red LED is based on a gallium phosphide (GaP), or a blue light is generated by gallium nitride (GaN) and gallium indium nitride (GaInN). Infrared radiation for use in fiber-optic communications is based on gallium (Ga), indium (In), arsenic (As), and phosphorus (P).

The use of red, green, and blue LEDs makes possible flat-screen VDTs and other illumination devices.

Video display terminals. The colors in the familiar video display terminal (VDT) result from the fluorescence of inorganic pigments irradiated by a beam of *cathode rays* (electrons). The red color is produced by an yttrium compound, Y_2O_3 or Y_2O_2S, doped with europium (Eu^{3+}). Green is produced by ZnS + CdS doped with copper and aluminum, and blue is produced with ZnS doped with silver and aluminum. The dopants cause irregularities ("holes") in the crystal lattice that provide different energy levels for the electrons, and it is the transition between energy levels that produces the characteristic visible radiation.

The control of the color emitted by the CRT is of critical importance if the device is to be used in soft proofing.

Reciprocity Law

The intensity of a light source is measured in **lumens.** The reciprocity law states that:

$$E = I \times T$$

where "E" is the amount of the exposure, "I" is the intensity of light, in lumens, at the surface of the film, and T is the exposure time. As long as the reciprocity law holds, the effect on a photographic coating will be the same if the product of light intensity and time remains the same. For example, if a good exposure is obtained in 2 sec. with a light intensity of 2000 lumens at the surface, then exactly the same exposure should be obtained in 4 sec. with a light intensity of 1000 lumens or in 1 sec. with an exposure of 4000 lumens.

Reciprocity is most important in color photography and color printing where films and prints used for each color must behave similarly. If the emulsion used for one color is faster or slower than the others, the color of the print will be distorted.

Reciprocity failure is a function of latent image formation. Two methods of reducing reciprocity failure are the use of dopants in the silver halide grain and the judicious choice of chemical sensitization conditions. While contrast can be affected by reciprocity failure, high contrast does not guarantee its absence.

Measurement of Color in Graphic Arts

Three methods are used to measure or control color in graphic arts processes: densitometry, colorimetry, and spectrophotometry. All can be used for either reflective or transmitted color.

Densitometry. Densitometry, the most basic of measurements, has been used the longest in the printing industry. Densitometry is used in prepress operations to measure the transmittance of light through a film and, in the pressroom, to measure the level of reflectance of light from a page. It is essentially colorblind. It is used in prepress operations to measure film density, proof accuracy, and plate quality. In the pressroom, densitometers are used at the end of presses to make sure the print density is correct and that ink levels are consistent from sheet to sheet. With the use of color filters, densitometers can be used to measure the density of printed cyan, magenta, yellow, and black ink films. Densitometry can be used to determine dot areas.

Colorimetry. Colorimeters measure the color of a print using four filters that are based on the human vision system. The data mathematically describe a color within a specific color space defined by the CIE*, and they can be transformed mathematically into other color spaces, such as Lab, L*a*b*, XYZ, and Luv. Colorimeters cannot be used to estimate dot area, but they can determine color difference, reported by a number called "delta E." Generally a color difference of 1.0 or smaller cannot be seen, but the ability of the eye to detect color differences depends on the color as well as on the observer. Graphic designers use colorimeters to measure and specify colors such as logo colors. Colorimeters can be used to measure the color of proofs and of press sheets, but print density is usually used to control proofs and prints.

Spectrophotometry. Spectrophotometers yield the most accurate and versatile data, but they are more expensive than densitometers or colorimeters. Spectral data can be used to calculate density, dot area, and color as well as to produce spectral curves. Spectrophotometers are powerful for use in ink formulation and (in quality control) to check the consistency of paper and inks. They avoid problems of ***metamerism,*** the process in which a change in the color of the light source causes a visual shift in the hue of an ink. Spectrophotometers are sometimes used at the end of the press where the results can be used to calculate density, dot area, and colorimetric and spectral data.

Optical Density   The light-absorbing strength of a photographic image or a printed image is measured in terms of optical density, abbreviated "D." The optical density is always expressed as a logarithm, and it is defined in the equation:

---

*CIE is the Commision Internationale de l'Éclairage (International Commission on Illumination)

$$D = -\log\left(\frac{\text{Transmitted or reflected light}}{\text{Incident light}}\right)$$

The density of a photographic negative (transmitted light) is defined as the negative of the log of the ratio of transmitted light to incident light. The units used to measure the light are irrelevant since they cancel out in the equation. For a photographic negative, if all of the light is transmitted, then the transmitted light equals the incident light (the light used to illuminate or measure the density of the negative), and density equals $-\log(1/1) = 0$. (The negative logarithm of a number is the log of 1/number.)

If 90% of the light is absorbed, the density equals log 0.1/1 or $-\log 1/10$ or 1. If 99% of the light is absorbed, the density equals $-\log 0.01/1$ or 2.

For reflected density of a print, the mathematics is the same, except that the densitometer measures the light reflected from the print instead of the light transmitted by a negative. In table 4-IV, the reflectance or transmittance, expressed as a decimal, is matched with the optical density or negative logarithm of the number.

Table 4-IV.
Logarithms.

Number	Logarithm	Negative logarithm
0.001	−3	3
0.01	−2	2
0.1	−1	1
1	0	0
10	1	−1

Processing the Image

Development of the Image

When an exposed emulsion is placed in the developer bath, an oxidation-reduction reaction occurs. The photographic developing agent, such as hydroquinone, is oxidized, and the exposed silver ions are reduced to metallic silver. When a substance is reduced, it gains electrons. When the developing agent is oxidized, it loses electrons to the silver ions that were exposed to light. The metallic silver is black in the finely divided state in photographic gelatin, and it forms the black part of the negative or positive.

$$\text{Developing agent} \rightarrow \text{Oxidized developing agent} + 1e^-$$

$$Ag^+ \text{ (exposed to light)} + 1e^- \rightarrow Ag^\circ \text{ (black)}$$

Adding the two equations together yields:

$$\text{Developing agent} + Ag^+ \rightarrow Ag^\circ + \text{Oxidized developing agent}$$

For silver ions to be reduced to black metallic silver, they must react with, or take on, electrons. But electrons are negatively charged and would be repelled from a negatively charged AgBr grain. If the negative "armor" has been broken by the exposure and conversion of some of the Br^- ions to Br°, then the electrons from the developing agent can enter the AgBr grain and reduce the silver ions to black metallic silver.

Thus, the grains that were exposed to light react with the developer and are changed to black metallic silver, while the grains that were not exposed to light still have their negative armor and consequently resist reaction with the developer, at least for a time. Even the unexposed grains are eventually attacked by the developer, causing an overall graying, called *fogging*, of the negative.

It appears that a silver atom on a AgX crystal may act as an electrode, attracting both a silver ion and a molecule of developer. If the reaction is carried out long enough, development can proceed into unexposed areas. The rate of development seems to be controlled by the rate of diffusion of the developer, and stirring or agitation is required to maintain a good development speed.

Developing Agents

A photographic developing agent must be strong enough to reduce the AgX grains that have been exposed to light, but not so strong that it will also reduce the unexposed grains. Most of the materials that meet these requirements are organic compounds that are derivatives of benzene including pyrogallol, hydroquinone, catechol, p-phenylenediamine, p-aminophenol, metol (or elon), amidol, and pyramidol. Most of these compounds contain two $-OH$ groups, two $-NH_2$ (amino) groups, or one of each. The groups are attached to the benzene ring in either the ortho or para position. Some have other groups attached to the benzene ring in addition to the $-OH$ or $-NH_2$ groups.

The two most common developing agents are hydroquinone and metol. The structure of hydroquinone is:

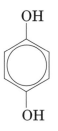

If one of the −OH groups is replaced with an −NH$_2$ group, the result is para-aminophenol, with the structure:

Para-aminophenol is a photographic developing agent. Replacing one of the hydrogen atoms in the −NH$_2$ group with a methyl group (−CH$_3$) produces metol, a more effective developer, with the structure:

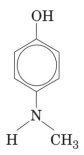

Ingredients in Common Developer Solutions

A photographic developer solution usually contains the following ingredients:

- **One or more developing agents.** These are the basis of the developer solution. The other materials are added to aid the developing agent in some way.
- **Accelerator.** An alkaline material, or accelerator, is added to the developer because the developing agent is much more active in alkaline solution than in a neutral or acid solution. The bases commonly used are sodium hydroxide, NaOH; sodium carbonate, Na$_2$CO$_3$; and sodium tetra-

borate, $Na_2B_4O_7$ (borax). These accelerators also neutralize the acid that is formed during development, preventing the solution from becoming acid. Developers with sodium hydroxide have a pH of about 12.0 and are very active. Those with sodium carbonate have a pH of about 10.2 and are less active. Those with borax have a pH of about 8.5 to 9.0 and are even less active.

- **Preservative.** Sodium sulfite, Na_2SO_3, is usually the preservative. It reduces the effect of oxidation of the developing agent by the oxygen of the air and helps to keep the solution colorless during mixing and storing. It also enters into the chemical reactions of development.

- **Restrainer.** A small amount of KBr is often added to restrain the formation of fog. The Br^- ions of the KBr attach themselves to the unexposed AgX grains. These grains are so completely surrounded by Br^- ions that it is very difficult for the developer to reach the Ag^+ ions and reduce them to $Ag°$ (metallic silver).

A common all-purpose developer contains metol and hydroquinone (two developing agents), sodium carbonate (accelerator), sodium sulfite (preservative), and potassium bromide (restrainer). A low-contrast developer often uses borax as the alkali, instead of sodium carbonate, and contains no KBr.

Infectious Developers

Two types of developer are used in graphic arts photography: the infectious type and the noninfectious type. All rapid-access developers are noninfectious, and they are used in developing camera, scanner, and contact films.

Infectious developers are used for high-contrast halftones and line work exposed in the camera. They are also used to obtain the hardest dots in halftones produced by screening continuous-tone originals. With this type of developer, only hydroquinone is used for the developing agent.

The ***induction period*** is the time interval between immersion of the film in the developer and the time at which the image begins to appear. With infectious developers, there is a long induction period (perhaps half a minute); but when the image begins to appear, its density builds up rapidly. The final result is an image with very high contrast.

The various chemicals in infectious developers are divided into two parts (usually called A and B). How the chemicals are separated varies with the manufacturer. One method is to include the hydroquinone and the preservative in part A, and the alkali and KBr in part B. When solutions A and B are mixed, they begin to react slowly with each other. With infectious developers, operators must be careful about the development time, the temperature of the solution, and the amount and kind of replenishment.

Hydroquinone molecules lose the two hydrogen atoms attached to the oxygen atoms and form ions with a -2 charge. These negatively charged ions are responsible for image development, but they are strongly repelled by the Br$^-$ ions that are adsorbed to the AgX grains. This condition makes the Br$^-$ ion concentration in the developer critical.

When the hydroquinone ions react with exposed AgBr, the AgBr is reduced to Ag$^\circ$ (metallic black silver), and the hydroquinone is oxidized to semiquinone:

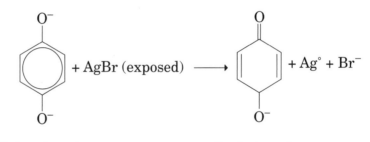

Hydroquinone ion   *Semiquinone ion*

Semiquinone is an even more active developing species than hydroquinone so that reduction of AgX accelerates, as semiquinone is oxidized to quinone:

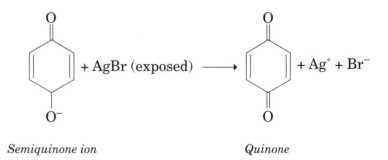

Semiquinone ion   *Quinone*

The quinone reacts with more hydroquinone ions to form more semiquinone:

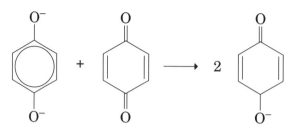

This is a chain reaction that can be repeated many times. The extent of this reaction is limited only by the quantity of hydroquinone and reducible silver halide in a given location. The result is a rapid development rate, once past the induction period, and the production of images with high contrast.

Sodium sulfite (Na_2SO_3) reacts with quinone to form hydroquinone monosulfonate:

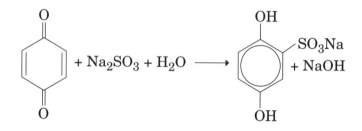

To the extent that this reaction takes place, the chain reaction just described is partially stopped, producing images with somewhat lower contrast. However, sulfite ions are needed in a developer to protect it from oxidation by air. Although a developer must contain some sulfite ions, it should not contain too many. This requirement is often fulfilled by the use of a sulfite buffer to replace part or all of the Na_2SO_3. The buffer most commonly used contains sodium formaldehyde bisulfite (SFB). In an alkaline solution, a small percentage of this compound dissociates to form Na_2SO_3 and formaldehyde:

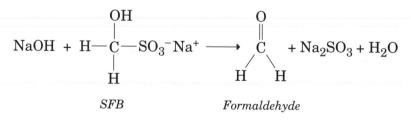

As the Na_2SO_3 is used up in the reactions given above, more of the SFB dissociates to provide more sodium sulfite. This dissociation is typical of materials that serve as buffers.

To summarize, infectious developers can be characterized as follows:
- They contain hydroquinone as the only developing agent.
- The developer has a low sulfite ion concentration.
- A sulfite buffer controls the concentration of sulfite ions.
- The developer is sensitive to bromide ion concentration and to pH.

Automatic Film Processing

Lith developers have functioned satisfactorily for tray processing for many years. Their use in automatic film processors has been another matter and has led to problems of uneven development ("drag lines") and to difficulties in restoring the original developer concentration with a replenisher. Various methods for diminishing these problems have been devised.

The usual infectious developer has a long induction period. In machine processing, development begins when the film enters the developer and ends when it enters the fixer. During the period when the film leaves the developer tank and crosses over into the fixing tank, exhausted developer has no chance to be replenished by fresh developer. As a result, development may be terminated prematurely, at a time when the rate of development is building to its maximum. If premature termination occurs, halftone images will have reduced contrast.

Drag lines and other problems have been reduced by changes in film emulsions and improvements in the design of the development system. These improvements include a blending system that maintains satisfactory levels of hydroquinone, sulfites, and bromide in the development bath.

The developer always loses some strength during shutdowns and over weekends. Loss of strength is nearly eliminated in the blending system by the addition of a special stabilizer.

Noninfectious Developers

Noninfectious developers produce densities in proportion to exposure. They usually contain both metol and hydroquinone for developing agents, and they have a high concentration of sodium sulfite. Such developers are often referred to as "MQ" developers. They have good stability and a considerable latitude in development time and in the amount of replenishment required. Another advantage is longer life, since the rate of oxidation is lower with noninfectious than with infectious developers.

Figure 4-6.
Blending system
for developer
replenishment.

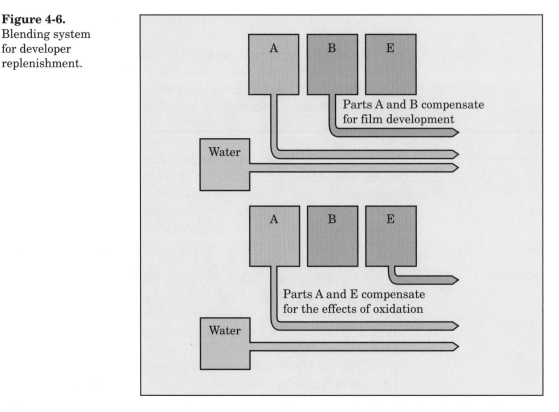

Rapid access developers are noninfectious. Noninfectious developers are used for continuous-tone, contact, and duplicating films. If a scanner produces halftones directly with a laser beam, it is possible to develop the film with noninfectious developers, since the laser gives a "go/no go" type of exposure.

Rapid-Access Processing

Automatic film processors have required variations in the formulas of developers designed for tray development. One of these variations is rapid-access processing. With a modification in developer formulation, it is possible to develop films in about 60–90 sec. by using a developer temperature above 100°F (38°C).

The developer is a low-contrast continuous-tone type, containing metol and hydroquinone (MQ) for the developing agents, and using a special restrainer to hold down fogging of films at the high developer temperature.

The developer also has a high concentration of sodium sulfite. Because of the formulation of the developer solution and the high temperature, the induction period is very short, and a visible image begins to appear almost as soon as the film enters the developer solution.

Rapid-access processing works best if used with films that have a low fogging tendency. Even lith films can be developed by this process if they have a low fogging tendency. The process is used mostly for the development of contact films. It can be used for developing camera line work and for developing laser-exposed halftone color separations produced on a color scanner.

Hybrid Systems

High-contrast or lith films require a special dual developer, which, however, rapidly deteriorates, requiring careful attention during development. The by-product of the developer is a high bromide material that produces drag lines in the film.

Rapid-access processing, on the other hand, gives no drag lines, and minimal attention to the processor is required. The film is not easily overdeveloped. The film, however, does not yield the hard edges to the halftone dots and the high contrast required for much graphic arts work.

A hybrid system combines the advantages of high contrast without critical attention to the processing. These systems are replacing the lith film systems.

Hydrazines serve as nucleating agents for silver halides, and their presence in the system gives a high-contrast film in rapid-access processors. The developer is stable and useful over a long period.

Fixing the Emulsion

The parts of the negative or positive that have not been reduced to black metallic silver during development still contain AgCl or AgBr. These compounds will gradually be reduced to metallic silver if the negative or positive is exposed to strong light, and it is necessary to remove them from the emulsion. This process is called *fixing.*

A chemical that dissolves AgCl and AgBr is needed in the fixing bath. Sodium thiosulfate *("hypo"),* $Na_2S_2O_3$, or sometimes ammonium thiosulfate, $(NH_4)_2S_2O_3$, is used to dissolve them. Either of these compounds forms a complex ion with silver ions, and the sodium or ammonium compound of this complex ion is soluble in water. Two reactions occur with silver halides:

$$AgX + Na_2S_2O_3 \rightarrow Na(AgS_2O_3) + NaX$$

$$AgX + 2Na_2S_2O_3 \rightarrow Na_3[Ag(S_2O_3)_2] + NaX$$

Compounds like this are soluble in water in accordance with the rule (Chapter 1, Table I-VI) that practically all compounds with sodium or ammonium ions are soluble in water.

Purpose of Other Ingredients in a Fixing Bath

While sodium thiosulfate is the main ingredient in a fixing bath, the bath usually contains other chemicals such as acetic acid, sodium sulfite, potassium alum, and boric acid. Potassium alum has the formula $K_2SO_4 \cdot Al_2(SO_4)_3 \cdot 24\ H_2O$. It is a **hardener** in the fixing bath, preventing undue swelling of the gelatin of the emulsion. Potassium alum is not stable in neutral or alkaline solutions because of the precipitation of aluminum hydroxide. So the bath is kept acid with acetic acid. Since the fixing bath is acid, it neutralizes the alkalinity of the developer solution and stops development immediately.

When a solution of sodium thiosulfate is made acid, as with acetic acid, it begins to decompose, forming elementary sulfur as a finely divided white precipitate. The white precipitate forms when the hardener solution is added too rapidly to the sodium thiosulfate solution in the preparation of a fixing bath. If the fixing solution contains Na_2SO_3, this chemical combines with the sulfur to form more sodium thiosulfate, as follows:

$$Na_2SO_3 + S \rightarrow Na_2S_2O_3$$

Thus, the sodium sulfite in the fixing bath stabilizes the solution and prevents formation of a cloudy or milky sulfur suspension.

The boric acid acts as a buffer to limit the change of pH of the fixing bath solution. It helps to prevent the precipitation of aluminum compounds that occurs when the pH changes too much.

It is possible to add other chemicals or to replace some of the above-mentioned chemicals. For example, for processing in tropical regions or at elevated temperatures, the potassium alum may be replaced with potassium chromium sulfate, $K_2SO_4 \cdot Cr_2(SO_4)_3 \cdot 24\ H_2O$ (called chrome alum).

Chrome alum is a more effective hardener than potassium alum, but chrome alum solutions lose their hardening action faster than do potassium alum solutions when the fixing bath is allowed to stand. Also, chrome alum has a greater tendency to form sludge than potassium alum has. Because of these difficulties, chrome alum is sometimes used in a separate hardening bath, preceding the fixing bath. Aluminum chloride, $AlCl_3$, is also used as a hardener.

Useful Life of a Fixing Bath

As more and more film or paper is fixed in a bath, more silver is added to the bath in the form of complex silver compounds. When the concentration of silver reaches a critical amount, some relatively insoluble compounds are formed that cannot be removed from the emulsion during the washing process.

When prints are being fixed, the maximum concentration of silver is about 0.27 oz./gal. (2 g/L) of fixing bath. For fixing films, this maximum is about 0.8 oz./gal. (6 g/L)—although the maximum concentration can be twice this amount when ammonium thiosulfate is used instead of sodium thiosulfate.

The amount of silver present can be checked with silver-estimating test papers that are available from most graphic arts suppliers.

A typical, fresh fixing bath has a pH of about 4.1. As films or prints are transferred to it from the developer solution, they carry some of the alkaline developer with them. This transfer of developer gradually raises the pH of the fixing bath. When the pH reaches about 5.5, the hardener becomes less effective. The bath should then be discarded or the pH lowered by the addition of more acetic acid. The rise in the pH of the fixing bath is largely prevented if an acid stop bath is used before the films or prints reach the fixing bath.

Reclaiming Silver from Fixing Baths

There are several reasons for reclaiming the silver from fixing baths. First of all, governmental environmental agencies require the control of heavy metals being discharged to the sewer and to the environment. Furthermore, silver is expensive and some costs can be recovered by selling it. Since silver ore is an exhaustible resource, silver conservation is important.

Today, the electrolytic method of silver recovery is the one most widely used (figure 4-7). The spent fixing bath is transferred to the silver recovery unit, which contains a carbon anode and a stainless steel cathode. A low-voltage direct current delivers electrons (e^-) to the cathode, and these electrons reduce silver ions, depositing metallic silver on the cathode:

$$Ag^+ + e^- \rightarrow Ag^\circ$$

The treated fixer can be used again, provided pH and sulfite levels are adjusted prior to reuse.

Figure 4-7.
CBS6 computerized
electrolytic silver
recovery system.
*(Courtesy X-Rite Incor-
porated)*

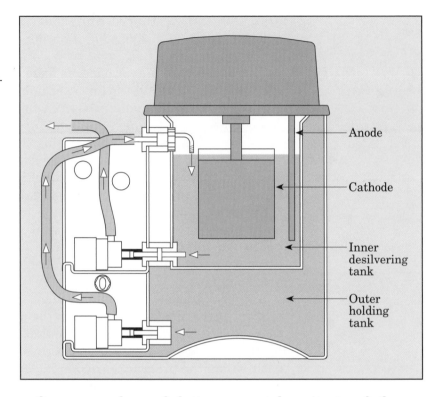

One ampere-hour of plating current deposits 4 g of silver
(0.125 troy ounce), which can usually be sold by a plant at 80%
of its market value. If a recirculation system is employed,
the makeup for the fixing bath can consist of 50% desilvered
solution and 50% fresh solution.

For photographic processing plants or laboratories that
have small volumes, silver is conveniently and economically
recovered by ion exchange using a cartridge such as the one
illustrated in figure 4-8.

The amount of silver recovered from a cartridge depends
on the concentration of silver in the spent bath. The effluent
from a cartridge can be checked with silver-estimating test
papers. When the amount of silver in the effluent begins to
increase, it is time to replace the cartridge. The cartridge is
then shipped to the supplier, who separates the silver and
pays the company for the amount recovered.

**Washing
Negatives and
Positives**

After the negative or positive is fixed, the gelatin emulsion
layer contains the fixing bath chemicals in solution dispersed
throughout the gelatin. If the sodium thiosulfate (hypo) present in this solution is not removed completely from the

Figure 4-8.
Model MR 200 small
photochemical waste
filtration system for
recovery of silver.
*(Courtesy Canadian
Silver Recovery
Service, Inc.)*

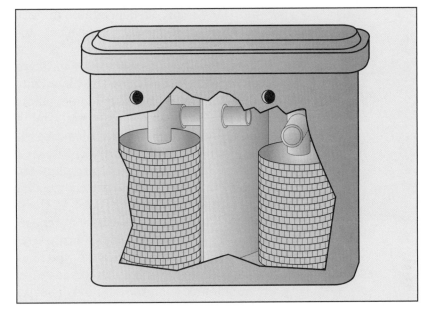

gelatin emulsion layer, it will slowly react with the silver in
the emulsion to form yellowish-brown sulfide:

$$Na_2S_2O_3 + 2Ag° \rightarrow Ag_2S + Na_2SO_3$$

If complex silver salts remain in the emulsion layer along
with hypo, silver sulfide forms causing a discoloration of the
film over the entire surface. To prevent this, the films must
be washed in water to remove hypo and silver salt as completely as possible from the emulsion layer. Complete removal
of these materials is facilitated with a well-designed washing
system. The old method of hand washing in trays caused fre-
quent problems. Raising the temperature of the wash water
from 40–80°F (5–27°C) helps remove the hypo and is estimated
to reduce the washing time by 30%.

The pH of the wash water should be above pH 4.9 to reduce
absorption of hypo by gelatin. Addition of NaCl or other salts
(called *washing aids)* greatly reduces the time required to
remove the hypo.

Disposal of Photographic Processing Solutions

Regulatory agencies require improvements in the quality of
effluents discharged to sewer systems. It is easy to remove
dissolved silver with a silver recovery system, but this leaves
the chemicals in developers, hardeners, and fixing baths.

None of these substances is highly toxic, and most are *biodegradable:* they are gradually consumed and oxidized by
bacteria present in water, lowering the oxygen content in the

water. If too many biodegradable materials enter a stream or lake, the oxygen content may become too low to support the fish.

The amount of biodegradable material in an effluent is determined by measuring the biochemical oxygen demand (BOD). Regulatory agencies make such BOD readings.

If applicable regulations permit, industrial plants that have a small amount of photographic processing waste can combine them in a holding tank and then trickle them into the sewer system. A small or medium-size plant can eliminate up to 95% of the BOD value of photographic waste solutions by installing a waste treatment facility based on the ***activated sludge*** principle. Packaged units can be purchased, or a plant can make its own. The system employs a large steel drum with intake and exit pipes and a perforated pipe leading to the bottom of the drum. Compressed air is forced through the perforations and bubbles through the solution. The oxygen of the air reduces the BOD value of the solution. In the process, a sludge forms on the surface, giving the name "activated sludge system" to this treatment. This system is used by municipal wastewater treatment facilities to reduce BOD.

Special Photographic Films

With conventional films and papers, a negative is produced from a positive, or vice versa. With contact and slow-projection-speed duplicating films, a positive is produced from a positive, or a negative from a negative, by exposure with white light.

Duplicating Films

These duplicating films are chemically "fogged" during manufacture. After exposure, they are developed with an infectious-type developer. The *unexposed* areas develop to D_{max}; that is, they become black, but the exposing light defogs the emulsion, clearing the *exposed* areas. The result is a duplicate of the film through which the exposure is made.

The original duplicating films were very slow and required exposure with quartz-iodine lamps or other powerful light sources. Modern duplicating films have a much faster emulsion speed and can be exposed with a point light source or even in a graphic arts process camera.

Diffusion Transfer Film for Direct-Screen Color Separations

The mechanism involved in the preparation of diffusion transfer lithographic plates is explained in Chapter 5, "Lithographic Plates." A diffusion transfer system has also been developed for panchromatic film.

Two different films are needed. The light from a colored original passes through a contact screen and exposes the panchromatic negative film. This film has a considerably faster speed than regular pan films, reducing the required exposure.

In the processing step, the exposed negative film and a sheet of special positive film are passed simultaneously through a diffusion transfer processor. The positive film has a nucleated coating on both sides, applied to a polyester base. This film is not light-sensitive.

The developer reduces the light-exposed areas of the negative film to metallic silver. Then the two films are tightly mated, and the hypo in the developer causes the unexposed silver compound in the nonexposed areas of the negative film to diffuse to the surface of the positive film. On this surface, the nucleated coating reduces the silver to metallic silver, creating a high-contrast positive halftone on the film. Because of the developing method, there is no problem with adjacency effects. The processing time is about 2 min.

Silverless Films

Some photographic systems do not require the use of silver. Few of them are used because, although silver is expensive, the chemistry of silver has evolved so that less and less is required to give good performance. In addition, recovery of spent and unused silver reduces its cost.

Micrographics, the reproduction of images on microfilm, is closely related to graphic arts. The microfilm may be either a silver content film or a silverless film.

Diazo films are often used in which a light-sensitive dye is coated on a polyester base. These dyes give a grainless image and can be enlarged many fold without the appearance of grain.

Color Photography

All modern color processes use a conventional silver halide emulsion as the light-sensitive medium. The latent image is converted into a silver image as in black-and-white processes, but in most color processes, a dye image is formed simultaneously with the silver image by the process known as ***color development.***

An image is made of the photographic object on each of three different emulsion layers: one records the red light, one green, and the third, blue. Dye images of the complementary color are produced in each emulsion layer: cyan in the layer containing the red record, magenta for green, and yellow for blue. These are negative images.

Color films are based on the tripack, a film in which the three emulsion layers are superimposed as shown in figure 4-9.

The top layer is sensitive to blue light, the bottom layers to green and red. Since the bottom layers are also sensitive to blue, a blue-absorbing (yellow) filter must be placed between the blue- and green-sensitive layers.

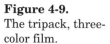

Figure 4-9.
The tripack, three-color film.

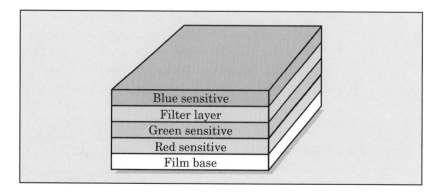

If dye images are developed at the same time as the negative silver images, color negatives are formed. If each emulsion layer is subjected to reversal processing, the dye images will be positive images (as in color slides). When these three dye images are superimposed, a color picture of the original object is formed.

To produce a photographic color negative, the exposed film is developed: the exposed silver is reacted with the color developer and the unexposed silver halide is washed away. To produce a positive, the negative silver images are developed in a black-and-white developer, the residual silver halide in the emulsion layers is re-exposed, and the appropriate dye is developed in each layer. The residual silver is then removed, leaving the fixed dye in each layer.

The key chemistry involved is the process of color development. The reaction is based on development of the exposed AgX with a substituted para-phenylenediamine (such as N,N-diethyl-para-phenylenediamine) and an azomethine or indoaniline dye.

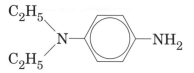

N,N-diethyl-para-phenylenediamine

The para-phenylenediamine reacts with the silver to fix the dye, thereby generating the color. The color coupler contains an active methylene group that reacts with the para-phenylenediamine to produce the color:

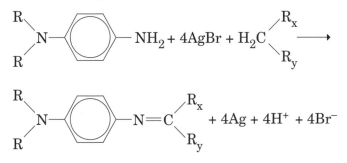

To produce a positive, the negative silver images are developed in a black-and-white developer, the residual silver halide is then re-exposed and coupled with the dye. As only the dye is required, residual silver is bleached out.

Proofing

Proofing is a critical part of the prepress work in graphic arts. It often involves photography, although electronic methods of proofing are popular.

Proofing is carried out to determine whether the text, artwork, and photographic halftones are properly made. Proofing shows the printer and the customer what the job will look like so that any desired changes can be made before the job has been mounted on the press and the printing started.

Soft Proofing and Hard Proofing

Proofing uses two types of images: *soft proofs* and *hard proofs.* Proofs (images) appearing on a CRT monitor are called soft proofs. They are easily manipulated, they can be generated quickly, and they are easily stored on disks. To judge the way a digital file will appear in print requires careful calibration of the equipment, standards for handling digital files, and a highly-trained eye for judging whether an image on the monitor will match reflective (printed) copy. In addition, there is no hard copy on which the purchaser can sign off. Unless stringent quality controls are maintained, a digital color job will not appear the same on different monitors.

Accordingly, for most purposes, hard proofs (proofs produced on film or paper) are still required. They may serve as contracts between customer and printer, they can be carried about or shipped, and their appearance is closer to that of the printed page. Hard proofs may be produced on press or off press. Proofing of digital files is often done with inkjet printing.

On-press proofing involves the preparation of printing plates and printing an image with them. On-press proofing is slow and expensive, but it gives many copies of the proofs, and they closely resemble the final product. Copies are sent to the publisher, the advertising agency, and the advertiser. The printer also retains a set. In package printing, proofs go to the vendor, the package manufacturer, and to the agency handling the work. The chemistry of on-press proofing is identical to the preparation and printing of any artwork, including photography, platemaking, and presswork.

Progressive proofs, called ***progs,*** comprise prints of the individual colors and overlay colors (such as magenta over cyan or yellow over magenta and cyan) that allow a printer to study the appearance of the print as colors are added during printing. These are used mostly to follow the printing of process color on a one-color or two-color press.

Proofs of the film can be prepared off press by using off-press (or prepress) proofing materials. These offer the advantages of lower cost and quicker results, and with modern materials, they give an excellent visual prediction of what the final print will look like. (They do not, however, look exactly like the print.)

Monotone Proofing

Proofs of single-color work show the layout, the absence of marks or dirt, and the overall appearance of the job to be printed. The proof may go to the ad manager, the editor, and/or the author.

One popular single-color (blue or black) proofing system consists of a paper coated one or two sides with a leuco dye (the colorless form of the dye) which is oxidized by a bis-imidazole dimer when irradiated for 30 sec. with UV radiation from a "blacklight" fluorescent tube or a metal-halide lamp from which UV light is masked. The photosensitive coating develops color in UV light (200–400 nm with a peak at 330 nm) and is rendered inactive upon exposure to visible light so that after the image is developed, the proof is stabilized by exposure to room light. When the photosensitized paper is exposed for 30 sec. through a negative, the exposed areas develop color, giving a positive print. A suitable lamp is a blacklight fluorescent tube or a metal-halide lamp with a filter that excludes wavelengths longer than about 400 nm. The unexposed area is cleared by exposure to visible light from which UV has been blocked. The system can be used to proof positive film by exposing the proofing sheet and film to visible light (which inactivates the clear areas) then developing the unexposed areas with UV.

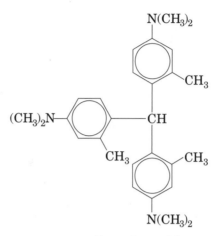

Leuco form of dye

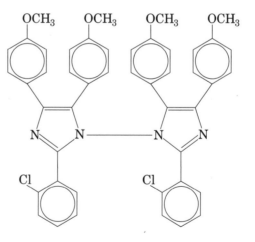

Bis-imidazole

Color Proofing

Prepress proofing is most important in checking color work, especially process color. Prepress color proofs are made by one of three methods: overlay proofs, single layer (or integral) proofs, and digital or electronic proofs.

Overlay proofs consist of (usually) four sheets of transparent film, each containing one of the process colors. Several different systems are on the market, and each involves somewhat different chemistry. One comprises a polyester film that is coated with a diazo, light-sensitive coating that in turn is covered with a pigmented lacquer. Films are available in about 40 different hues. The diazo coating is slow enough to permit handling the films in normal room light. An intense UV light is required to expose it. The light fixes the diazo coating so that it is no longer soluble in the developer. After exposure, the film is wet with a developer and rubbed or brushed to remove the unexposed diazo coating. The developer contains a volatile, organic solvent, and it should be used in a well ventilated area and kept away from heat and open flame. Owing to environmental concerns, we may expect variations that will permit the use of water-based (aqueous) developers.

Another system includes negative- and positive-working products on polyester film. With one product, the diazo coating is hardened by light, and with the other, the diazo coating is made soluble on exposure. The exposed films are developed with a water-base developer that contains no volatile organic solvent.

A third system works on a different principle. It is a peel-apart product, consisting of two sheets of polyester film with a pigmented negative-working photopolymer layer between them. The system requires no chemical processing or development. The films are treated so that they have different adhesive characteristics. Upon exposure to UV radiation, the photopolymer changes its adhesive characteristics so that the exposed, pigmented photopolymer adheres to one of the sheets of film, and the unexposed layer adheres to the other. The exposed films are peeled apart, the positive print becomes one of four films required to make a progressive proof.

Single-sheet (integral) proofs. Single-sheet proofs usually give a better simulation of the printed sheet than do overlay proofs. Six basic methods are used to prepare a single-layer proof:

- Adhesive polymers to which pigment powders are applied
- Pigmented sensitized coatings transferred to a single base
- Superposed sensitized coatings of inks on a plastic base
- Diffusion transfer of layers of dyes to photographic paper
- Electrostatic imaging
- Inkjet imaging

The chemistry of each method is described briefly.

Several adhesive polymer systems are used. In one positive-working system, the adhesive contains acrylic monomers formulated with other image-forming and stabilizing components. The product is furnished in a sandwich between a polyester sheet and a polypropylene sheet (figure 4-10A). The proofing sheet is prepared by removing the polypropylene sheet and laminating the adhesive and polyester with heat and pressure to a specially treated, glossy paper that is thick, smooth, and heat resistant (figure 4-10B). Exposure to UV radiation hardens the tacky polymer, and the polyester film is removed. Exposure through a positive film converts the adhesive layer to a hardened acrylic polymer. The tacky (unexposed) image area is developed by brushing with a fine, micron-sized pigment (e.g., cyan). Another adhesive layer is

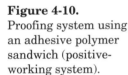

Figure 4-10.
Proofing system using an adhesive polymer sandwich (positive-working system).

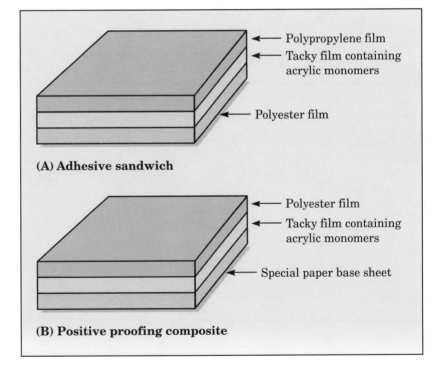

applied over the first color, and the process is repeated, using a pigment with a different color (e.g., magenta). The process is repeated two more times to give a four-color proof.

For negative-working proofing systems (figure 4-11), a hardened photopolymer is laminated to a paper base. Exposure to light makes the photopolymer tacky in the image areas (the clear areas of the negative). The tacky areas retain a colored toner (e.g., cyan) when it is brushed over the exposed film.

Figure 4-11.
Proofing system using an adhesive polymer sandwich (negative-working system).

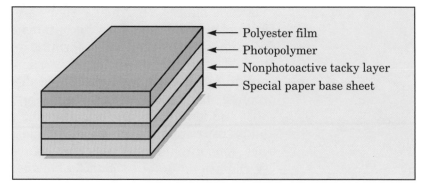

Where the opaque nonimage areas of film block UV light, the photopolymer remains hard and will not accept toner. The procedure is repeated for magenta, yellow, and black.

Another system consists of a polyester cover sheet to which is laminated a 1-micron layer of colored photopolymer (e.g., yellow poly[vinyl acetate]) (figure 4-12). To make a proof, the product is laminated under heat and pressure to a paper substrate, the cover sheet is removed, the separations

Figure 4-12.
Proofing system using an adhesive polymer and pigmented layer (positive- or negative-working).

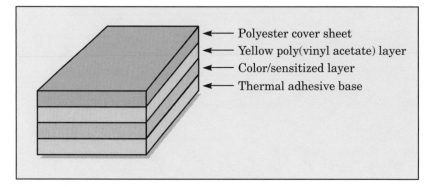

are mounted on top, and an exposure is made with UV radiation. (The proofing laminator is illustrated in figure 4-13.) In the positive mode, where UV radiation passes through the clear, nonimage areas of the separation film, the colorant is rendered soluble in the alcohol-water processing solution and washes off. The dark image areas protect the colorant from exposure; it remains insoluble in the developer and stays on the base sheet to form the image. In the negative mode, the opposite occurs: unexposed colorant is soluble in the alcohol-based processing solution, while UV-exposed colorant in the image areas is rendered insoluble. After development, the process is repeated until the four process colors have been applied to the base sheet.

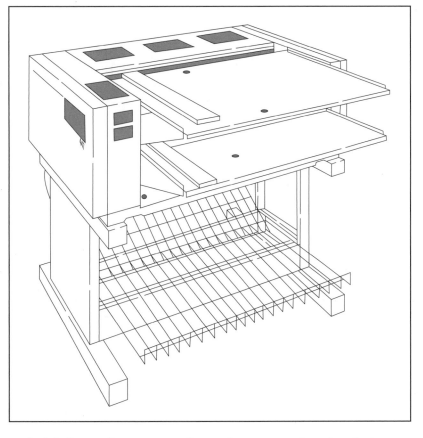

Figure 4-13.
447L laminator used for Matchprint™ digital halftone color proofs. *(Courtesy Imation Enterprises Corporation)*

A third proofing system depends on transfer of a photo-sensitive, pigmented, gelatin layer to the base sheet or to the actual printing substrate from a carrier sheet or film. After the carrier sheet is removed, the layer is exposed, and the unexposed gelatin is washed away to develop the color. After the process is carried out four times, a final proof is obtained that consists of the four colors on a single sheet.

Printing ink pigments can be dispersed in photosensitive coatings and coated one over the other on a polyester base. The first coating is exposed to the appropriate color separation negative, developed with an aqueous developer, and dried. The next color is then applied and the process is repeated. Several proofing products use variations of this process.

Diffusion transfer, the photographic process used in instant cameras, is based on the transfer of a soluble silver complex and reduction of the transferred silver to form a positive image on a receiver sheet. The materials include a negative-working color film coated on an opaque polyester base. Processing is done automatically with a premixed, single-batch chemical without any need for temperature control, fixing, washing, or even running water. To make a color proof, four exposures are made on a negative proofing film. The exposed film is placed in emulsion-to-emulsion contact with a receiver paper. Ten minutes after this film/receiver sandwich leaves the processor, it is peeled apart. A color proof is now on the paper.

Electrostatic Proofing

One commercial proofing process uses a photoconductive, sandwich on a polyester base to prepare a process-color proof that is then laminated onto a coated sheet of paper. The system is based on a proofing sheet comprised of a polyester base, a conductive layer, a photoconductive layer, and a thermoplastic release overcoat.

The emulsion side of the photographic negative from which the proof is to be made is placed against the release overcoat, and they are exposed under a corona discharge to develop an electric charge or potential. Exposure through the negative (using a UV/blue lamp) now discharges the photoconductive layer and develops a latent image because where light (photons) hit the charged layer, the charge leaks into the conductive layer.

The first toner is applied in an electrostatic field, and it adheres to the unexposed areas of the film, the areas that retain a charge. The sandwich, which now carries one color, is dried, and the process is repeated until all four process colors have been applied.

The proofing sheet now contains all four colors, and it is laminated onto a coated litho paper.

Digital Electrostatic Proofing

Electrostatic proofing or xerography (figure 4-14) requires a semiconductor and a toner. The semiconductor is charged in the dark. The charge dissipates when it is illuminated (either by exposure to a visual image or to a laser beam). The electrostatic image is treated with a proofing ink or liquid toner to develop a visual image. One system uses an organic photoconductor on a transparent polyester to prepare a four-color overlay. Another, based on a paper coated with zinc oxide, develops the image on the paper. Another system that uses a cadmium sulfide semiconductor coated on a stainless steel base develops the visual image on the plate with a liquid toner and transfers it to an offset sheet and then to the proofing paper.

Figure 4-14.
Diagram of electro-
static proofer.

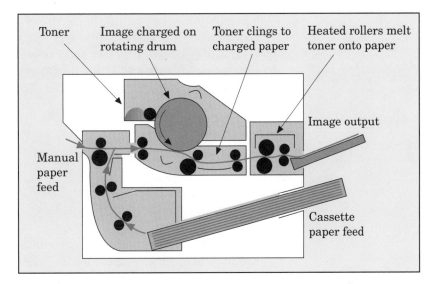

Toner   Image charged on
rotating drum   Toner clings to
charged paper   Heated rollers melt
toner onto paper

Image output

Manual
paper
feed

Cassette
paper feed

Inkjet Proofing

Inkjet proofing (figure 4-15) is a refinement of inkjet printing discussed in Chapter 8. The inkjet colors are made to match the printing ink colors. Inkjet has the advantage that the image can be produced on any printing or proofing stock directly from digital data.

Digital and electronic proofing are growing in importance together with electronic handling of digitized images. Most of the proofing systems used for digital or electronic printing use chemistry described above. Some 50 proofing systems

Figure 4-15.
Diagram of a drop-on-
demand inkjet color
proofing system.
*(Courtesy Imation
Enterprises Corporation)*

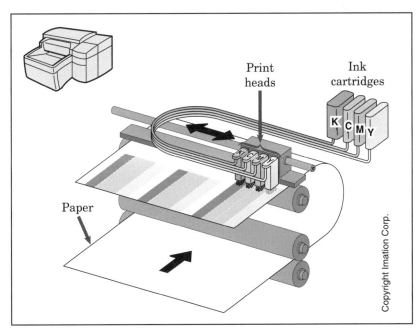

Print
heads

Ink
cartridges

K C M Y

Paper

were available in 1990, and many variations of these systems are in use. Although the technology is that of electrical engineering and physics, there is a great amount of chemistry involved in building both the equipment and the materials used.

Further Reading

Adams, Richard M. II, and Joshua B. Weisberg. *GATF Guide to Practical Color Management.* 2nd ed. 2000 (Graphic Arts Technical Foundation, Sewickley, Pa.).

Bruno, M. H. *Principles of Color Proofing.* 1986 (Gama Communications, Salem, New Hampshire).

Field, Gary G., *Color and Its Reproduction.* 2nd ed. 1999 (Graphic Arts Technical Foundation, Sewickley, Pa.).

Mason, L. F. A. *Photographic Processing Chemistry.* 2nd ed. 1975 (Halstead Press, Division of John Wiley, New York).

Mees, C. E. K. *The Theory of the Photographic Process.* 3rd ed., 1966 (Macmillan, New York).

Othmer, Kirk. *Encyclopedia of Chemical Technology.* 4th ed. 1996 (John Wiley, New York), vol. 18, pp. 905–963.

Thompson, Bob. *Printing Materials.* 1998 (Pira International, Leatherhead, Surrey, UK).

Ullmann's Encyclopedia of Industrial Chemistry. 5th ed. 1992 (VCH Verlagsgesellschaft), vol. A20, pp. 1–159.

5   Lithographic Plates

Introduction

Printing processes are distinguished by the method of forming the image. The major printing processes can be divided into five groups as shown in table 5-I and figure 5-1.

Table 5-I.
The printing methods and processes.

Method	Process
Planographic	Lithography (litho or offset)
	Collotype
Relief	Flexography (flexo)
	Letterpress
Intaglio (engraving)	Rotogravure (gravure)
	Steel die
Stencil	Screen
	Mimeograph
Nonimpact	Inkjet
	Electrostatic

(The terms litho, offset, flexo, and gravure are used even in formal writing.)

Lithography is the most popular method of publication and commercial printing. A lithographic plate is planographic: the image areas are level with the nonimage areas. (Actually, except for electrostatic litho plates, the image area consists of a very thin coating applied to the base.) Owing to the chemical nature of the surface of the plate, water forms a film on the nonimage areas but not on the image areas. When a properly made plate is run on the press, the water rollers keep the nonimage areas of the plate moist so that they do not accept ink. The ink rollers then transfer ink only to the image areas on which there is no film of water.

Most lithographic plates have a photosensitive coating that develops an image after exposing through either a negative or

Figure 5-1.
Printing surfaces of
different processes.

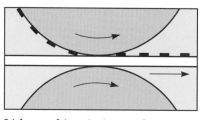

Lithographic printing surface

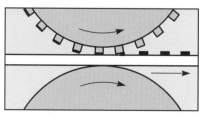

Relief printing surface

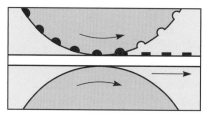

Gravure printing surface

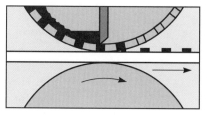

Screen printing surface

a positive film, depending on the type of plate. In the early days of lithography, grained metal plates were coated with a bichromated albumin solution, dried, and exposed through a negative. This process and the deep-etch platemaking process (which used a coating of bichromated gum arabic that was exposed through a positive) are now both obsolete.

As the lithographic process continued to gain popularity for commercial work, printing of newspapers, packaging, and offset duplicating, other methods for making plates were developed. These methods use anodized aluminum, paper, or polyester as base materials. Multimetal plates use stainless steel, mild steel, aluminum, and brass for base metals, which are electroplated with an ink- or oil-receptive metal, usually copper.

It is convenient to divide the types of lithographic plates into those that are exposed through photographic negatives (negative-working) and those exposed through photographic positives (positive-working) or from positive copy. The negative-working plates include:

- Additive diazo presensitized
- Subtractive diazo presensitized
- Photopolymer presensitized
- Wipe-on
- Waterless
- Multimetal
- Projection speed
- Photodirect
- Thermal

The positive-working plates include:
- Presensitized
- Diffusion transfer
- Multimetal
- Electrostatic
- Projection speed
- Photodirect

In addition, electrospark and laser plates are designed to work from digital copy rather than from positive or negative film.

The choice of a suitable plate includes consideration of the base material, the sensitivity of light-sensitive coatings to specific wavelengths, dark reaction of coatings, light sources for exposure, use of a sensitivity guide to check exposures, and making nonimage areas water-receptive. These topics are discussed in this chapter.

The Plate Base

Most duplicator-type plates use paper, paper laminates, or polyester plastic as a base. For larger plates, anodized aluminum or polyester is now used, although some multimetal plates are still in use (see figures 5-2 and 5-3).

Aluminum is a good lithographic plate material for several reasons. It is reasonable in cost, available in uniform thicknesses, and strong enough for the purpose. It is stiff enough to resist significant stretching when it is properly mounted on the press, and it maintains good register of halftone dots,

Figure 5-2.
Cross section of an aluminum plate.

Figure 5-3.
Cross section of a plastic plate.

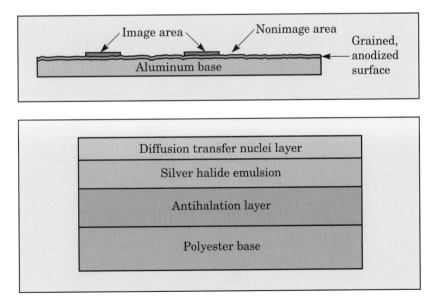

even on large plates. Aluminum is lightweight, weighing only about 38% as much as zinc, 35% as much as steel, and 32% as much as brass (metals that have also been used). Although pure aluminum is highly reactive, under normal pressroom conditions, it is protected by an oxide film that clings tenaciously to the metal surface. The oxide film is self-healing when damaged, and it is easily made water-receptive.

Several aluminum alloys are used for lithographic plates including 1100, which is 99.0% aluminum, and 3003, which contains 1.0–1.5% manganese and up to 1.8% of other metals. Generally, 1100 alloy is used when plates are to be grained mechanically, and 3003 alloy is used for chemical graining. Some European manufacturers who use electro-chemical graining prefer the high-purity types of aluminum, although they are mechanically weaker.

The reason that aluminum plates are chemically stable is that aluminum oxidizes spontaneously in air to produce a thin, tenacious, transparent film of aluminum oxide or alumina (Al_2O_3). The film is generally impervious to water, further oxidation, and mild base or mild acid such as dampening solution. Aluminum is attacked by strong bases and acids that dissolve the aluminum oxide and allow the underlying metal to react with air or water. By contrast, iron, which is less reactive toward air and water than aluminum, forms a colored oxide (rust) that does not form a glassy film to protect the metal from further attack. Thus, uncoated iron or steel corrodes rapidly and is unsuited for litho plates.

Press operators often use the term "plate oxidation" to refer to a printing problem in which the nonimage area of the plate becomes sensitive to ink. Since the surface of aluminum is always oxidized, the term is chemically incorrect. Sensitivity of the nonimage area is often overcome in the pressroom by wiping the plate with an acid-gum solution.

The water-holding properties of aluminum plates are achieved by a combination of increasing the surface area by graining and of building a porous structure by anodizing. It is essential that the dampening solution be carried in the grain pits below the mean surface plane of the metal. Consistency and uniformity of the grain and the honeycombed structure achieved by anodizing are therefore important.

Uniformity of the caliper of the plate, its surface, the pore size, and the depth of surface treatment are critically important to assure uniformity of imaging and wear resistance of the plate.

Graining of Plates

The term "graining" means the roughening of the surface of a plate so that it has a matte finish that consists of many thousands of tiny hills and valleys. This process adds to the real surface area. The area after graining is between two and four times the area of a smooth plate. Graining can be done mechanically, electrochemically, chemically, or by combinations of these.

Many presensitized litho plates and certain multimetal plates run satisfactorily with practically no grain on them, but grained plates have more latitude on press, and they run at a higher water setting without looking wet or shiny. They are said to "hold water" better on the press. More importantly, graining improves adhesion of image coatings and other films, giving them longer life. Grained plates are more durable, they draw down quicker in the vacuum frame, and they cause fewer problems with halation, hickeys, dirt, and piling.

An ungrained plate appears to be wetter on press, but it does not have the water reserve of a grained plate. Thus, controlling dampening with smooth plates is more difficult. Smooth plates do not have the latitude of grained plates, that is, ink emulsification and plate dry-up occur more readily.

Mechanical graining. Litho plates are grained mechanically or electrochemically or both in combination. The mechanical graining process in common use is brush graining. Rolls of smooth aluminum are conducted through a precleaning process, then into a brush-graining machine. The brushes are wet with a continuously recycled abrasive, often a mixture of pumice and sand (SiO_2). The rotating, oscillating brushes support the abrasive particles as they impact the surface, giving a pattern of fine scratches or dents. Three or four brushes produce a more uniform surface than do a pair of brushes—especially at high web speeds. As the aluminum web continues through the graining machine, it is often lightly etched with a solution of strong alkali to remove embedded grit, and then carefully rinsed. Drying is optional. The aluminum may or may not be given a finish grain chemically or electrolytically, and then it is usually anodized. The grained and anodized web is usually treated with a sodium silicate (Na_2SiO_3) or other desensitizing solution and coated with a photosensitive layer or layers and cut to size for the press. Almost all aluminum plates (except bimetal plates) are anodized. Wipe-on plates are usually brush-grained and anodized. Some smaller plates are chemically grained and relatively smooth.

Electrochemical graining. Electrochemical graining is carried out at 68–104°F (20–40°C) in dilute hydrochloric acid or nitric acid. Graining is commonly performed by imposing an alternating current on a moving web, using stationary electrodes and very high current density for short process times. After 30–150 sec., a uniform, intensely pitted matte surface is produced. Combining chemical or electrochemical graining with brush graining produces a broad range of surface textures.

Electrochemical graining, preferred in Europe, is more versatile, producing a rougher surface. Brush graining is a dirty process, brushes are expensive, and they wear unevenly. If the process is not carefully controlled, directionality of brush graining is apparent, as shown in figure 5-4. Despite its disadvantages, brush graining is still widely used because it is fast and relatively inexpensive. Brush grains are often improved by chemically graining the brush-grained surface. Inexpensive wipe-on plates are almost always brush-grained.

Figure 5-4.
Photomicrograph of a mechanically grained plate showing directionality and nonuniformity of structure.
(Courtesy Agfa Corporation)

Good printing performance requires that the grain in the aluminum plate be consistent and uniform. Figure 5-5 shows examples of uniform and nonuniform electrochemical graining.

Chemical graining. Some lithographic plates are grained chemically, using ammonium bifluoride (NH_4HF_2) or trisodium phosphate (Na_3PO_4). These chemicals dissolve the surface oxide allowing the raw metal to react to produce the grain.

Figure 5-5.
Photomicrographs showing uniform electrochemical graining *(left)* and poor, non-uniform structure *(right)*.
(Courtesy Agfa Corporation)

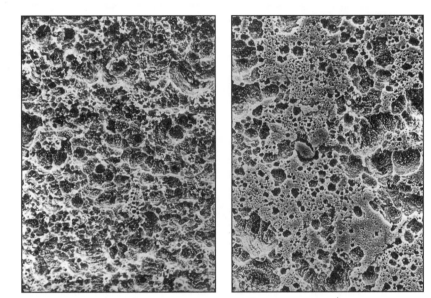

Anodized Aluminum

Almost all presensitized and wipe-on aluminum plates are anodized. Anodized aluminum has an excellent surface for litho plates, providing ceramic qualities that make it behave much like the old lithographic stone. It is hard and scratch-resistant, characteristics that add greatly to the plate's press life. On the original Mohs' scale for relative hardness, with numbers varying from 1 (soft talc) to 10 (diamond), the alumina (Al_2O_3) on the surface of anodized aluminum rates between 5 and 6, about the hardness of fully-hardened steel. The surface alumina is a glass film, plus some hydrated, crystalline alumina, and it is softer than the alumina of emery or corundum (9 on Mohs' scale).

Although aluminum oxidizes immediately when exposed to air, the natural alumina film is extremely thin and does not confer wear resistance. Anodizing protects the plate surface against chemical and mechanical damage and greatly increases its press life by producing a film with a porous structure that is much thicker (a million times thicker) than the film that forms naturally. Anodizing does not change the "hills and valleys" of a grained plate surface significantly, so plates must be chemically, electrochemically, or mechanically grained before they are anodized.

An aluminum plate is anodized by making it the anode (positive pole) of a cell that commonly contains sulfuric acid, although phosphoric acid or other chemicals can be used. A heavy, direct current passed through this cell, converts the surface of the aluminum into aluminum oxide. The thickness of the oxide coating formed depends directly on the quantity of electricity (coulombs or faradays) used per unit of surface. Time, temperature, purity, and concentration of the anodizing solution are very important. The anodized film thickness can vary from 0.0001–0.0012 in. (0.1–1.2 mil or 0.0025–0.030 mm).

The porosity, composition, and thickness and texture of the anodic layer strongly affect plate life. Other factors, particularly press factors, are also very important, and the optimum surface condition for one printing plant may not be best for another.

The anodized coating of litho plates is porous, consisting of millions of very tiny hexagonal-shaped cells. In the center of each is a star-shaped "pore" or "well" (figure 5-6).

These cells are so small that they can be seen only with an electron microscope. Figure 5-7 shows an electron micrograph of a cross section of an anodized plate.

The size of these tiny cells depends primarily on the makeup of the anodizing electrolyte. Anodizing conditions that give a particularly wear-resistant anodic film often increase the resistance and require a higher voltage, especially for thicker anodic layers. Pulsed direct current is often used, but alternating current may also be used, especially at high current density and low anodizing times.

After being anodized, the plates are treated to increase the hydrophilic nature of the surface and to improve shelf stability of presensitized plates. This process is often (incorrectly) called "sealing." Solutions of sodium silicate are very effective but not generally compatible with positive-working coatings. For these, a polymeric acid with phosphonate groups may be used.

Anodized aluminum is not sufficiently water-receptive to fully reject ink or organic contaminants. It is easy, however, to make it water-receptive where desired (in the nonprinting areas) by treatment with an acidified solution of gum arabic or other hydrophilic polymer. The aluminum oxide that forms the anodized film resists mild acids and bases, but strong ones attack it. It is pitted by salt solutions and acids of medium strength.

Plate Washes

Negative-working aluminum plates are treated at 190–210°F (88–99°C) for at least 30 sec. with sodium silicate ($Na_2O:SiO_2$ = 1:2.5) or other chemical after they are anodized and before the photosensitive coating is applied. The treatment forms a

Figure 5-6.
Photomicrographs of a grained, anodized aluminum surface at three magnifications: 200× *(top),* 1000× *(middle),* and 4500× *(bottom).*
(Courtesy Imation)

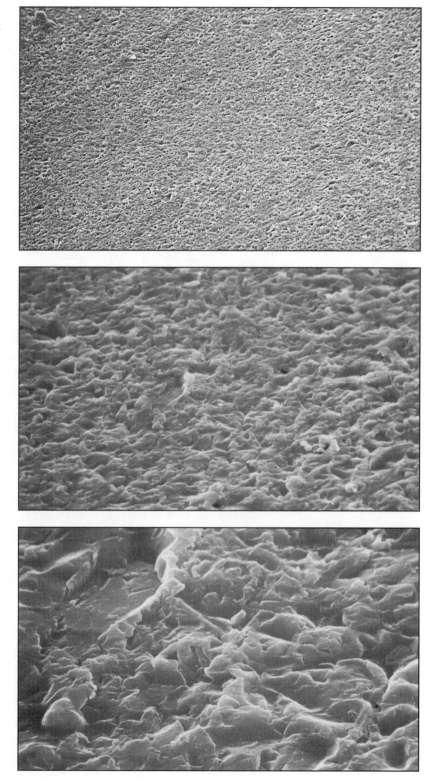

Figure 5-7.
Electron micrograph of a cross section of an anodized aluminum surface.
(Courtesy Agfa Corporation)

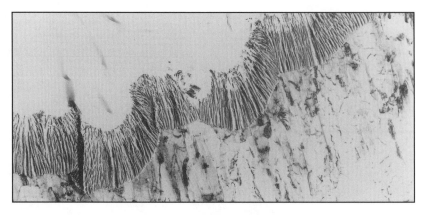

barrier layer that protects aqueous diazo coatings from interaction with the aluminum and leaves the nonimage area highly water-receptive. The silicate layer favors the adhesion of exposed diazo coatings and the release of unexposed diazos. The silicate treatment reduces the adhesion of solvent-based coatings and is never used on positive plates.

Paper Plates

Paper plates are used on offset duplicators. It is no longer less expensive to produce a dozen copies on a duplicator than on a copier. Paper plates must produce short runs inexpensively. When the run gets to a few thousand, a larger commercial press is often less expensive (although the best paper plates are capable of running 10,000 copies or more). A few companies produce paper plates in the United States, a few in Japan, and a few in Europe. Even fewer paper companies produce the paper base.

Several types of paper plates are in common use: electrophotographic, direct image, photodirect, diazo-coated, and diffusion transfer. Electrophotographic plates carry a zinc oxide (ZnO) coating on the paper. Zinc oxide is a photoconductor: it can carry a charge in the dark, and when it is exposed, light discharges the nonimage area. The remaining electrostatic image attracts toner to develop a visible image. Image transfer plates can be typed on directly, but they usually are imaged from a copier or laser printer. Diazo-coated paper plates have a light-sensitive coating on them, like presensitized aluminum plates. Diffusion transfer is a silver halide, photographic process. Photodirect plates are usually based on silver halide chemistry.

Regardless of the process, the paper base contains long fiber for strength and resistance to cracking at the clamps and short fiber for smoothness and uniformity. The paper is treated with wet-strength resins, sizing (either acidic or basic), and a waterproofing material to give a sheet that retains its strength and dimensions when wet with dampening solution. A major requirement of the paper is uniformity: uniformity from sheet to sheet as well as uniformity over each sheet. Cost is also important because the printer would use aluminum plates if costs were comparable. On long runs, cost of the plate is less important than on the short runs typical for paper plates. Caliper—commonly 5, 8, or 12 mils (0.13, 0.20, and 0.30 mm)—is controlled to suit the reproduction process.

After manufacture, the paper is converted to plates by applying an appropriate coating. It is sold in rolls or sheets.

Paper-Aluminum Laminates

With presensitized paper-aluminum laminates, the shiny side of the aluminum foil is adhered to the paper, leaving a naturally matte face exposed. It is unanodized and ungrained. The plates carry a subtractive coating that develops with aqueous developer. The smooth aluminum produces a sharp halftone dot. With a nominal run length of 10,000 impressions, they are useful for duplicators and even for short-run newspapers.

Plastic Plates

Plastic plates are made to run with most printing inks. They last longer on press than most paper plates. Because polyester is not as stiff as aluminum and is more easily stretched, tension must be carefully controlled when tightening the plates on press. Since these plates are rarely used for process color, dimensional stability is not a major issue.

Plastic plates are made on polyester film (see figure 5-3). Polyester is ink-receptive and water-repellent, so it must be coated to accept water. The photosensitive layer is applied over this coating. The coating, light exposure, and development of the plate leave the nonimage areas water-receptive. If this thin, water-receptive layer is damaged on press, the plates will scum. The polyester is of optical quality, so the plates can be exposed through the back.

Like paper plates, most plastic plates use special dampening solutions but are compatible with most printing inks. Run lengths of 25,000 to 50,000 are reported, which is longer than paper plates. Presensitized laser-speed (diffusion transfer, silver halide chemistry) polyester printing plates are supplied in rolls up to 32¼ in. (820 mm) in width. They are supplied in 0.004- and 0.007-in. (0.10- and 0.18-mm) thicknesses.

Light-Sensitive Coatings

Relief and engraved plates and cylinders can be prepared mechanically or photographically, but lithographic plates are imaged photographically. This usually requires that the base material be coated with a photosensitive material.

The coating material depends on the type of plate: the variety of plates available makes it difficult to generalize. No light-sensitive coating is required on the base material for diffusion transfer plates even though light-sensitive coatings are used to create images on the master sheet. The image is transferred from the master to the carrier sheet, which can be paper, metal, or plastic.

Sensitivity of Coatings to Light of Different Wavelengths

Energy travels through space in the form of electromagnetic waves as shown in figure 5-8. Light sources used in the graphic arts consist of visible, infrared, and ultraviolet light of many wavelengths.

Figure 5-8.
The electromagnetic spectrum.

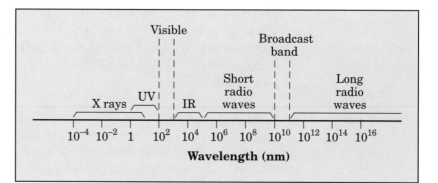

The various coatings used on lithographic plates have a peak sensitivity over a narrow range of wavelengths; the sensitivity drops off rapidly at higher and lower wavelengths. The peak sensitivity varies from one coating to another: photopolymer coatings are most sensitive around 365 nm (nanometers) while diazo coatings are most reactive at about 400 nm. This region extends from the ultraviolet area into the deep blue area of visible light.

Most lithographic plates are insensitive to yellow, green, and red light. In other words, when white light is used, much of the light employed to expose plates is wasted. Many coatings are sensitive to lower wavelengths in the ultraviolet region, but these wavelengths are absorbed by the film base and the glass of the vacuum frame. A polyester film base and the plate glass of the vacuum frame absorb most of the ultraviolet radiation shorter than 330 nm.

If a reaction is to occur when a plate is exposed to light, the light must be absorbed by the coating. While light consists of waves of different wavelengths, it can also be considered to consist of bundles of energy, called **photons.** The amount of change in the coating of an exposed litho plate is related to the number of photons that are absorbed.

The effect of exposure to light on plate coatings follows an S-shaped curve, similar to that for photographic films. For the first few seconds of exposure (the **threshold time),** there is no measurable effect. The effect increases slowly in the portion of the curve called the **toe.** Next, there is a rapid increase in almost a straight line. On this portion of the curve, the effect produced in the coating is proportional to the number of photons of light absorbed. The proportionality is represented by the slope—the steepness—of the straight line. Finally, the effect levels off; very little further change occurs with increasing exposure. This part is called the **shoulder** of the curve (see figure 5-9).

Figure 5-9.
Effect of light exposure on a plate coating.

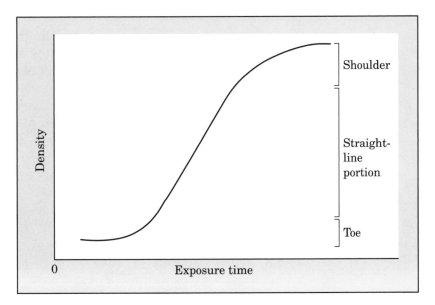

Dye-Sensitized Photopolymerization

Most photopolymer systems in graphic arts and other industrial systems are sensitive only to ultraviolet light. For many processes, such as proofing and laser platemaking, a reaction that is initiated by visible light is highly desirable.

Photopolymerization initiated by visible light is made possible by a complex process called **dye-sensitized photopolymerization.** Certain dyes can be activated by visible light (each dye is sensitive to a specific color) so that they react with another molecule to generate a material that initiates polymerization. If S represents the sensitizer dye and D represents an electron donor, the following equations can be written:

$$S \xrightarrow{h\nu} S*$$

$$S* + D \rightarrow S^- + D^+$$

$$D+ \rightarrow \text{polymerization initiator}$$

This is called ***photoreducible dye sensitization*** since the dye, upon activation by light and reaction with material D becomes reduced (it gains an electron). The $D+$ becomes the polymerization initiator. Other processes are possible, but the photoreducible dye sensitization appears to be the most commonly used.

Sensitizer dyes mentioned in the chemical literature include xanthenes, thiazenes, and acriflavine. Donor materials mentioned include tertiary amines, sulfinates, and enolates.

Dark Reaction

Storing plates in the dark reduces the threshold time because some reaction occurs before the plates are exposed.

This ***dark reaction*** has the effect of an overall preexposure, and plate coatings that have undergone some dark reaction require less exposure. Dark reaction is significant for wipe-on plates, but other modern plates have long shelf life. Nevertheless, plates (like film) should be used before the expiration date. Printers should store presensitized plates in a cool, dry place and use a sensitivity guide on all plates.

Post-Exposure of Developed Plates

The image area of some negative-working plates can be toughened by exposing the plate after it has been developed (for most plates, the developer destroys the photosensitivity). By further cross-linking the polymeric coating, this technique sometimes gives a significant increase in run length. Most positive-working plates and a few negative-working plates can be heated to 460–490°F (240–255°C) after development for better durability on press.

Safelights

Since most lithographic plate coatings are insensitive to yellow, green, and red wavelengths of light, it is usually safe to use a red, orange, or yellow light in the platemaking room. Some yellow fluorescent lamps are satisfactory. Red or yellow plastic tubes, placed over fluorescent lights, eliminate the actinic light (light that induces a chemical reaction) and make fluorescent lights safe. The plate manufacturer should be consulted for the recommended safelights.

Reciprocity Law   The reciprocity law, discussed in chapter 4, also applies to the exposure of most printing plates, but it does not apply to diazo coatings or to the thermal plates that are imaged by infrared lasers.

Sensitivity Guide   The best practical way to measure the plate exposure is to include, along with other copy on the plate, a stepped continuous-tone transmission gray scale. This film is called a platemaker's sensitivity guide (see figure 5-10). The most common increment of density is 0.15 although sensitivity guides are available in other increments of density such as 0.05 or 0.10.

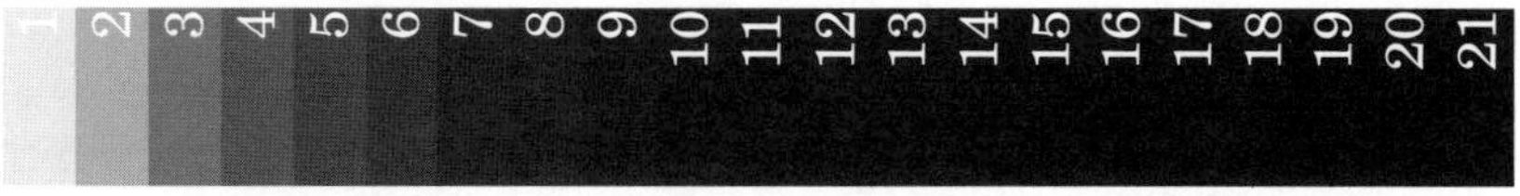

Figure 5-10.
Platemaker's
sensitivity guide.

In general, moving up two steps on a sensitivity guide divided into density increments of 0.15 means the transmission of light is half as great; step 5 lets through only half as much light as step 3, and step 7 lets through only half as much light as step 5. In other words, the plate under step 5 has received half as much exposure as under step 3 and twice the exposure of step 7.

Lithography is a "go-no-go" process. That is, any area on a plate either takes ink or does not take ink. This is true even in the reproduction of pictures that have many tone values. Each small halftone dot takes ink while the spaces between dots do not take ink. The color of the print is controlled largely by the area of the dots: ink-film thickness or color density has less effect on the color of the print.

Since the light-sensitive coatings on most lithographic plates are high-contrast, a continuous-tone gray scale does not produce a continuous-tone image when a plate is developed. If the gray scale is exposed onto a negative-working plate, the first few steps (perhaps five or six) will hold ink completely. The next step will appear as a dark gray, and the next step or two as lighter grays. These last few steps are referred to as the "gray steps." Beyond these, the coating is exposed so little that the remaining steps will not hold ink at all. The highest numbered step on a negative-working plate that still holds a solid film of ink is called the ***critical step.***

On a positive-working plate, a sensitivity guide works the opposite way. Negative plates are exposed to a solid step number; positive plates to a clean (or clear or open) step number. On positive plates, the higher numbered steps print solid. Two or three steps appear gray, and the steps lower than the gray steps are clear and do not hold any ink. With such a plate, the critical step is the lowest numbered step that will remain clean or free of ink.

A sensitivity guide shows the extent of exposure to the platemaker. If "step 5" gives good press performance, one that shows a "step 6 or 7" will have more dot gain on negative-working plates or sharpening on positive-working plates when the plates are run on the press. The critical step is the result of all the factors that affect the reaction of a coating, including dark reaction, exposure to light, coating thickness, coating sensitivity, temperature, and processing conditions. With this information, the platemaker can be sure that every plate is properly exposed and processed to produce good-quality prints on the press.

With the diazo-type presensitized plates, a sensitivity guide's critical step can be varied no more than one step without producing a perceptible change in the tone values of halftones.

When plates are exposed, halftone dots are always subject to some light undercutting that causes dot gain on negative-working plates and sharpening on positive-working plates. This effect is especially important with negative-working plates that have a heavy coating, especially with photopolymer plates where cross-linking tends to continue after exposure. Tone or halftone values vary slightly, depending on the intensity and length of exposure. For this reason, it is important to expose these plates to a constant step on a sensitivity guide. The manufacturer supplies a recommended number in the literature supplied with the plates. The recommended level, usually 5, 6, or 7, varies, depending on the on-press wear-resistance of the coatings at different exposures.

Dot Gain

Plates never produce, on paper, a halftone dot that is exactly the same size as on the film. While the digital production of printing plates differs in details, variability between original and the printed product still exists. The entire system is variable including exposure, developing, manufacture and type of plate as well as film, paper, ink, and press. The best that management can do is to keep the variations under control so that the negative or digital copy carries dots of a size that will produce the desired printed dot. Owing to dot undercutting during exposure, negative plates tend to produce dots that are larger than on the film while positive plates tend to produce smaller dots (tend to sharpen the dot). In other words, negative working plates gain and positive plates sharpen with increasing exposure.

A gain of 2–4% on a negative-working plate is generally accepted to be satisfactory, but gains up to 10% are frequently found, and even greater dot gains from film to plate occur under extreme conditions. Positive-working plates may sharpen as much as 10–15%. (The actual numbers depend on the instrument used to measure them.) When the plate is used to make a print, more dot gain occurs on press. It is far greater on uncoated than on coated stock. It should therefore be apparent that the same plates cannot be used to produce good pictorial work on both coated and uncoated paper. Color separations must be corrected for press conditions, including the ink and the paper to be printed.

With more than 100 different plates available from over 20 manufacturers, great variations in dot gain between plates are to be expected.

Treatment of Nonimage Areas

After a plate is exposed, it must be developed to differentiate the image from the nonimage areas. Then additional treatment is used to increase the ink-receptivity of the image areas and the water-receptivity of the nonimage areas. Nonimage areas so treated are said to be *desensitized*—that is, they are not receptive to ink as long as they are kept wet with water. Four terms are used to describe the surface areas of lithographic plates:

Technical term	Derivation	Meaning
Hydrophilic	Water-loving	Water-receptive
Hydrophobic	Water-hating	Water-repellent
Oleophilic	Oil-loving	Ink-receptive
Oleophobic	Oil-hating	Ink-repellent

The objective in litho platemaking is to make the image areas as oleophilic and hydrophobic as possible and to make the nonimage areas as hydrophilic and oleophobic as possible. The best-known method to desensitize the nonimage areas on aluminum plates is to use an acidified solution of gum arabic.

A desensitizing gum must fulfill two functions: it must be water-receptive, and it must adhere tightly to the surface of the metal. On the press, a plate is wet continuously by the dampening system. If the gum is dissolved from the plate, bare metal is exposed, and it becomes "sensitive" (it begins to take ink).

Desensitizing materials do not react chemically with the metal but are adsorbed on the surface. A good desensitizing gum such as gum arabic is soluble in water. When "washing off the gum," the press operator removes a considerable amount that has been dried on the nonimage areas of the plate, but a thin, adsorbed film of gum arabic remains, not removed by water. This film desensitizes the nonimage areas.

Why some materials desensitize better than others.
A desensitizing material must be hydrophilic. Most suitable substances are water-soluble organic materials that contain hydroxyl groups (–OH) in their molecules. The hydroxyl groups are partly responsible for the hydrophilic nature of these materials.

Most good desensitizing agents are weak organic acids of high molecular weight. Most organic acids can be expressed with the general formula R–COOH, where R stands for an organic group and –COOH is the carboxyl group. The potassium compounds of gum arabic can be written R–COOK where K represents potassium. When an acid, such as phosphoric acid, is added to a water solution of gum arabic, the calcium, potassium, and magnesium salts of arabic acid are converted into free arabic acid. One ionic equation is:

$$\text{R–COOK} + \text{H}^+ \rightarrow \text{R–COOH} + \text{K}^+$$

Gum arabic   *Arabic acid*

It is well known that gum arabic does a better job of desensitizing a plate if its solutions are acidified.

Chemistry of gum arabic. The dried exudation from the gum acacia tree, gum arabic, was first used as a plate desensitizer by Alois Senefelder, the inventor of lithography, and 200 years later, it still does more things and does them better than any other plate desensitizing material that has been tested.

Gum arabic, as it occurs in nature, is a mannogalactan gum, a high-molecular-weight carbohydrate consisting primarily of two sugar monomers named mannose and galactose that are closely related to glucose.

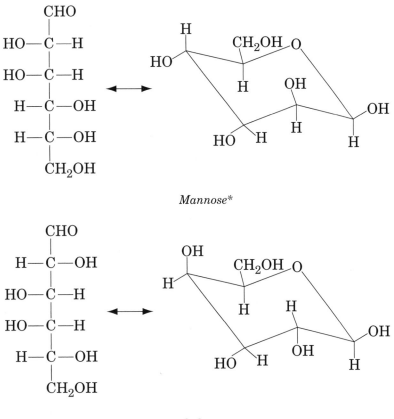

*Mannose**

Galactose

(The formulae show an equilibrium between the aldehyde and hemiacetal forms of each of these two sugars as well as two common methods of representing the structure of sugars.)

Gum arabic also contains several carboxyl groups that are partially neutralized with calcium, potassium, and magnesium. An average sample of gum arabic contains about 0.7% of calcium and 0.6% of magnesium, plus much smaller amounts of many other metals. It has a molecular weight of about 42,000.

The highly polar carboxyl (–COOH) groups of arabic acid are responsible for its adsorption to a metal surface. As an acid is added to a solution of gum arabic, lowering the pH, the solution becomes a better desensitizing agent down to a pH of about 3.0. At this point, most of the gum has been con-

---

*These two sugars are represented in their two forms: the open-chained aldehyde form and the closed-chain acetal form. Both methods of showing the structures of sugars are commonly used.

verted to free arabic acid. Such a solution desensitizes a metal plate very well; if more acid is added, there is no improvement. Converting the salt to the free acid decreases its solubility.

Steel offers an exception. A solution containing nothing but phosphoric acid and water can desensitize a steel ink roller so that it will repel ink and accept water. To a limited extent, phosphoric acid will also desensitize aluminum and chromium, but it will sensitize copper so that it will accept ink. Measurements at the GATF laboratory with phosphoric acid containing a tracer amount of radioactive phosphorus showed that a considerable amount of phosphoric acid remained adsorbed to aluminum and chromium after washing, while only a small amount remained adsorbed to copper. A film of phosphoric acid is adsorbed on hydrophilic steel, aluminum, and chromium but not on oleophilic copper.

Other natural and synthetic gums. Many materials have been used as replacements for gum arabic. The more successful replacements often have carboxyl groups in their molecules. Some, the alginates, are compounds of sodium, potassium, or ammonium ions with a weak organic acid called alginic acid. On the other hand, hydrophilic materials such as starch, dextrin, and methyl cellulose are poor desensitizing agents; these do not have any carboxyl groups in their molecules.

Many natural and synthetic materials are hydrophilic. Among these are gum tragacanth, gum arabic, cherry gum, larch gum, mesquite gum, methyl cellulose, hydroxyethyl cellulose, carboxymethyl cellulose, arabogalactans, manno-galactans, dextrins, alginates, oxidized starches, poly(vinyl pyrrolidone), and poly(vinyl alcohol). However, these materials vary widely in their ability to adhere to a metal surface.

Natural or synthetic gums are commonly used in automatic plate processors where gum arabic often causes a problem of gum blinding by coating the image area as well as the non-image area. This is especially true with aqueous plates.

Negative-Working Lithographic Plates

Lithographic plates exposed through negatives are called *negative-working plates.* Several types were listed at the beginning of the chapter.

When a negative-working plate is exposed, the light-sensitive coating is hardened so that the developer does not attack the exposed area. The developer dissolves the coating in the nonimage areas, exposing the metal. Finally, a gum solution desensitizes the metal in the nonimage area, so that on the press, it will hold water instead of ink.

Automatic plate processors are usually used. Some machines process only plates of a particular manufacturer; others process certain types—diazo plates, presensitized, or wipe-on, or certain multimetal plates, for example. Most processors develop a plate, rinse it, apply a desensitizing film of gum arabic, and dry the plate. Thus processed, the plate comes out of the machine ready for the press. Figure 5-11 shows a computer-to-plate system that integrates plate handling, imaging, and processing in one unit.

Figure 5-11.
The Galileo platesetter, which integrates plate handling, imaging, and processing.
(Courtesy Agfa Corporation)

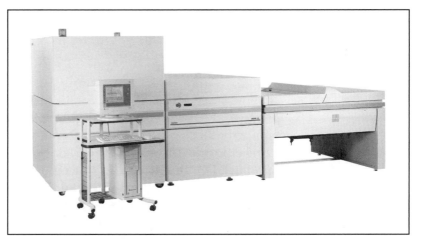

Automatic plate processors have many advantages over the old, hand processing. They process the plate more uniformly, saving time and chemicals. The chemical saving, in turn, reduces the amount of chemicals added to the plant effluent.

Additive Presensitized Plates

Except for some newspapers and specialty printers, who use wipe-on plates, most lithographers use *presensitized* plates. This term means that the plate manufacturer applies the light-sensitive coating to the metal. The term *additive* is used to indicate that a wear-resistant lacquer or pigmented resin is added to the image areas when it is developed. Very thin diazo coatings can be used for additive presensitized plates. Without the additive lacquer, the poor wear resistance of the thin diazo coating would cause poor plate life.

Development is accomplished in a plate processor made expressly for use with additive plates. Additive developer removes the unexposed diazo sensitizer and adds a pigmented resin to the exposed diazo, producing a visible image and reinforcing the fragile image area.

To improve the adhesion of the diazo coating (and the running life of the plate), it is necessary to give plates a surface treatment. One method is to immerse the plates in a hot solution of sodium silicate (Na_2SiO_3), then wash them with water, and seal the layer with a weak acid.

The diazo compound commonly employed is produced from a complicated organic compound called 4-diazo-1,1′-diphenyl-ammonium chloride. Its structure is:

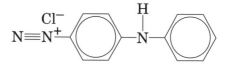

To make a water-soluble light-sensitive diazo resin, this material is polymerized with formaldehyde in the presence of zinc chloride and sulfuric acid. Solvent-soluble resins are prepared by replacing the zinc chloride with another salt.

Other organic compounds can be used as light-sensitive materials. These include certain azide compounds, hydrazine derivatives, quinone diazides, and esters of quinone. These compounds have very complicated molecular structures.

When a diazo coating is exposed to light, the light splits nitrogen from the diazo molecule, and it escapes as a gas. The remainder of the molecule undergoes a variety of reactions. The important thing is that exposure makes the coating insoluble so that the nonimage area is removed during developing. On negative plates, the diazo coating becomes insoluble upon exposure to light. On positive plates, an insoluble diazo film is made soluble upon exposure.

The dark reaction of diazo coatings is very slow if packages of presensitized plates are stored at room temperature. (Higher temperatures accelerate aging; lower temperatures retard it.) Aging studies on typical negative-working presensitized plates showed that, on the average, plates aged for six months gave about one more solid step on a 21-step platemaker's sensitivity guide than did freshly coated plates given the same exposure. In general, modern presensitized plates can still be processed successfully after storage periods of six months to two years.

Additive plates are usually developed with a one-step emulsion, which consists of a water phase containing gum arabic and an acid, and a solvent phase containing an oleophilic resin and a pigment. When the emulsion is rubbed over a plate, it breaks, allowing the resin and pigment to be deposited on the hardened image areas. The water phase dissolves the unexposed coating and deposits gum arabic on the nonimage areas.

Subtractive Diazo Presensitized Plates

Subtractive presensitized plates have a thicker coating than additive plates. The coating incorporates an oleophilic resin. When it is exposed to light, the diazo hardens the resin and makes a durable, ink-receptive image. The developer removes ("subtracts") the coating from the nonimage areas. The developer is usually an aqueous mixture of higher alcohols and surfactants. When the plate is gummed, it is ready for the press.

It is obvious that different developers are required for additive and subtractive plates. Because of the complex formulations involved in making plates, it is important that the proper developer be used for each plate. No one single developer is best for all plates.

Photopolymer Presensitized Plates

As pressruns grew longer, printers needed presensitized plates that would run as long as those made by the deep-etch copperized-aluminum process. This demand has been met by photopolymer presensitized plates that have replaced the expensive and time-consuming deep-etch plates. Improvements in diazo coatings have also produced presensitized plates with excellent run lengths.

The term "photopolymer" has become popular among the manufacturers of litho plates, and it has sometimes been applied indiscriminately to any plate on which exposure through a negative makes the image areas very hard and wear-resistant.

Two principal mechanisms are involved in the production of a photopolymer. One is linear polymerization. Under the influence of ultraviolet or deep-blue light, comparatively low-molecular-weight molecules called ***monomers*** combine with each other to form long, straight chains as explained in chapter 3. A simple illustration is shown below.

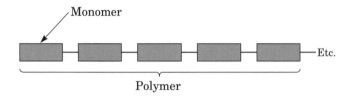

When long chains are formed, the coating becomes hard-ened and insoluble in the developer used to remove the coating from the unexposed area.

The second mechanism is called ***cross-linking.*** It may occur along with linear polymerization. When this happens, one linear chain hooks onto another by bonding between chains until the whole area exposed to light is one giant "molecule." This linkage is illustrated below.

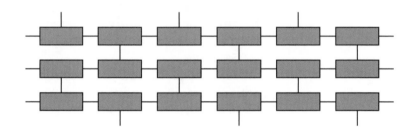

There are many types of bonds between chains. For exam-ple, carbon-to-carbon double bonds in adjacent polymer chains may form a new carbon-to-carbon cross-link:

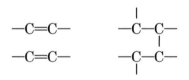

Linear polymers that melt when heated sufficiently are called ***thermoplastic resins.*** But if extensive cross-linking is involved, the polymers are harder and do not melt when heated. They are called ***thermoset resins.***

Since such plate coatings are proprietary, exact formula-tions are unavailable. However, one type uses a combination of light-sensitive acrylate and modified styrene monomers. A ***photoinitiator*** is also in the coating. The light first affects the photoinitiator, which, in turn, brings about the polymerization of the monomers. This involves a series of reactions that are presented in more detail in chapter 3.

In any case, the image areas are hard and wear-resistant; plates of this kind give pressruns as great as 2,000,000 impressions. The press life is extended even more if the light-sensitive coating is applied to an anodized aluminum surface.

It is usually recommended that photopolymer plates be exposed to give a solid "step 5 or 6" on a 21-step platemaker's sensitivity guide. It is important to expose these plates to some constant step on a sensitivity guide because the coating thickness is usually greater than on diazo-type plates and is therefore subject to more undercutting by light. If successive plates are exposed to give the same step on a sensitivity guide, the amount of undercutting will be the same, with the same amount of dot gain.

Thermal Plates

Computer-to-plate technology often uses plates that undergo a cross-linking reaction upon exposure to infrared lasers. The chemistry involves phenolic or melamine prepolymer resins that are cross-linked by the infrared laser radiation.

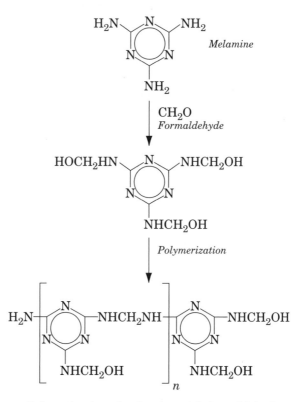

Polymerization of melamine with formaldehyde

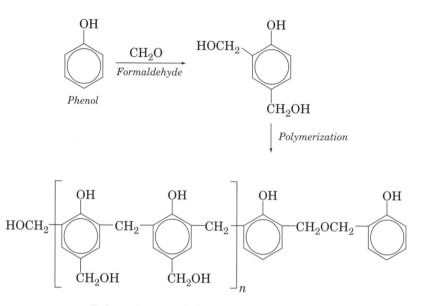

Polymerization of phenol with formaldehyde

A phenolic coating has peak sensitivity at 750–830 nm in the infrared and between 380–400 nm—near UV, producing a plate that can be used both in the infrared and ultraviolet. It can be exposed either with a laser or in a conventional exposure frame.

Initiation of the cross-linking requires a beam with a high power density attainable with a laser beam. Power sources include YAG lasers or laser diodes. Lower power densities have no effect on the polymer, and the plate is completely stable under ordinary illumination so that it can be handled without a darkroom.

Exposure to an infrared laser causes partial cross-linking, which forms a latent image. The latent image is developed by heating for 1 min. at 250°F (120°C) to set the image area. The nonimage area is removed with an aqueous surfactant. After washing and rinsing, the plate is baked at 485–575°F (250–300°C) to increase run length and solvent resistance of the image. The plate is then gummed and treated like any litho plate.

The cross-linking reaction does not occur below a specific temperature, and above that temperature, it occurs rapidly. Thermal coatings give a sharper ("harder") dot because of the nature of the system, and there is no "shoulder" on the dot. Back scattering from the surface of photoreactive litho plates

Figure 5-12.
Difference in sharpness of dots produced from optical and thermal radiation.

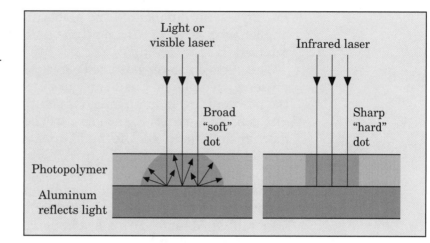

contributes to dot gain. Thermal conductivity of the aluminum base removes any excess thermal energy, reducing dot gain as illustrated in figure 5-12.

Aqueous-Developable Plates

Because of the problems of disposing of the solvents required to develop the common types of diazo and photopolymer plates, companies making plates have invented new coating systems. These plates (or coatings) have the awkward name of "aqueous-developable coatings" or, often, just "aqueous plates." After the plate is exposed, the nonimage area can be washed away with chemicals that are diluted with a large amount of water. Even some of these developers contain hazardous solvents, and attempts are continuing to find coatings that can be processed in pure water after exposing and fixing the image. These are referred to as ***"zero solvent"*** systems.

These aqueous plates have less contrast than the older plates requiring developers in organic solvents—that is, the gray scale has more steps between the clear step and the full step. The entire printing system (notably inks and pressroom chemicals) is being adjusted to suit the difference in the image on these ecologically desirable plates.

Positive-working plates (those that create a nonimage area where they are exposed) are developed with dilute alkali that dissolves the indene carboxylic acid that is generated on exposure. These developers contain no organic solvent.

Wipe-on Plates

The first diazo plates were presensitized. Later, it was found that a solution of a diazo compound could be wiped over a plate, rubbed down smooth, and fanned dry. Wipe-on plates are popular with newspapers and other printing plants that handle a great number of litho plates. The coatings are now applied with a roll coating machine, but the plates are still referred to as "wipe-on."

The same diazo resin is used as with many of the additive presensitized plates. Customary practice is to use 3–5% of the powdered resin in an aqueous solution. The solution may also contain a material, such as citric acid, to stabilize the diazo resin for a longer life, and a wetting agent to help spread the coating more evenly with a roll coater.

The diazo powder used for wipe-on plates should be stored in the refrigerator. Heat and humidity accelerate its deterioration; the material remains useful for a year if it is stored at 50°F (10°C). Manufacturing has improved over the years, and the modern product has greater purity and solubility than that produced in the 1970s.

Wipe-on plates are usually brush-grained and anodized. They must, of course, be treated with a solution of sodium silicate to provide a barrier between the aluminum and the diazo coating, as already explained. Most wipe-on plates will produce 200,000 impressions, although some are claimed to produce 500,000. Plates still printing acceptably but starting to show early signs of wear can have their life extended by applying the additive developer to the plate image while the plate is still on press.

The diazo content of the coating influences exposure speed. One coating that is used has less diazo content than the typical wipe-on coating. As a result, the exposure time is about 65% of that required by the usual wipe-on plates. A thermoset resin is incorporated in the developer. After development, the plate is washed, dried, and baked at 450°F (230°C) for 4 min. Baking the developed plate extends the polymerization making the photosensitive material hard and insoluble. The effect on structure is illustrated in figure 5-13. As a general rule, the longer and more cross-linked the polymer molecule, the more insoluble it becomes.

After cooling, the plate is gummed to desensitize the non-printing areas. The press life of this plate is much longer than that of conventional wipe-on plates.

Waterless Plates

Conventional lithographic plates are desensitized in the non-printing areas with a film of gum arabic or other hydrophilic gum in the dampening solution so that these areas accept water instead of ink during printing. Waterless plates (also called driographic or dry plates) are planographic plates so

Figure 5-13.
Structure of thermoset
resin.

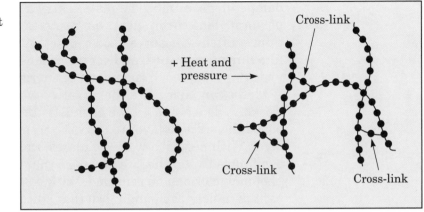

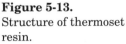
Figure 5-13.
Structure of thermoset
resin.

prepared that they can be printed without the use of water.
The background or nonimage area of these plates consists of
a low-energy material, usually a silicone rubber or polymer
that is not wet with printing ink.

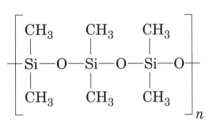

Segment of a silicone polymer

Figure 5-14 is a diagram of a waterless offset plate. The
plate comprises a silicone layer adhered to a photopolymer
layer on an aluminum base. The ink-repelling silicone layer
keeps the ink off the nonimage area, serving in place of the

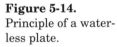
Figure 5-14.
Principle of a water-
less plate.

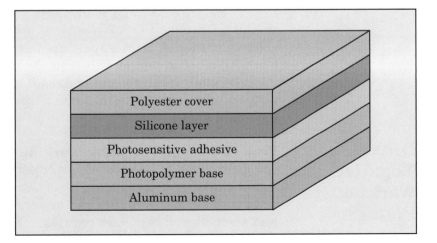

dampening solution. The plates carry a polyethylene or polypropylene cover sheet that protects the surface silicone from scratches, assures good contact with the photographic film during exposure, and serves as a barrier to oxygen, which would quench the photopolymerization process.

Plates can be made either positive-working or negative-working, depending on the chemistry of the adhesive that bonds the silicone layer to the ink-receptive photopolymer layer. With negative-working plates, radiation breaks the bonding between the two layers so that the silicone over the photopolymer can be removed. With positive-working plates, light enhances the bonding so that silicone over unexposed areas is removed during development.

The plates offer many advantages. By eliminating water, they eliminate the need to constantly monitor and adjust the ink/water balance during printing, and they eliminate color variation and washed-out color caused by water. Waterless plates reduce paper curl and paper waste during press start-up. Waterless plates print solid, hard halftone dots, especially in dot areas from 50% to 99%. This is an important advantage in printing stochastic screens, screenless lithography, and halftones made at screen rulings as fine as 300–500 lines/in. (120–200 lines/cm).

Waterless plates also present some problems. Because there is no water to cool the printing nips, refrigerated cooling of the oscillators is necessary. The coating on the nonimage areas is soft and easily scratched by dirt from any source, making long-run work difficult. Since there is no water to wash away dirt, the paper must be exceptionally clean. The tack of the specially formulated inks tends to be higher than that of usual offset inks, and this tack sometimes causes the paper to pick. Because no water is involved, static electricity is troublesome when the relative humidity is low. Furthermore, the tack of the ink is greatly reduced as the press warms up, and without good temperature control, toning of background areas becomes troublesome. The relatively high cost of these plates and the expensive inks and paper limit their use to top-quality jobs.

Other Negative-Working Plates

Projection-speed negative-working plates. Plates with sufficient speed to be exposed by projection must contain a fast photopolymer. They usually contain a mixture of diazo and photopolymer, and they can be very fast (which means poor shelf life and high dark reaction). There are many patents on other sensitizers. Other types of projection-speed plates use silver halide and/or diffusion transfer technology.

Manufacturers vary the speed of the plate by varying the diazo content of the coating. The diazo in the coating absorbs the light in the film. Decreasing the diazo coating improves the adhesion of the coating to the plate. Since less light is absorbed by the diazo, more light reacts with the polymer, improving the cure at the metal interface and yielding better adhesion. The plate manufacturer's problem is to find a formulation that gives both high speed and good adhesion. The solubility of the unexposed coating depends on the diazo content. Decreasing the diazo content reduces the solubility of the unexposed coating, and the plate tends to scum.

Projection-speed plates are used for automated imaging, the reproduction of engineering drawings, financial printing, short-run books, and for step-and-repeat machines. The projection systems can be used in place of step-and-repeat machines to directly expose the plates for process color with screen rulings up to 175 lines/in. (69 lines/cm).

Positive-Working Lithographic Plates

Some positive-working plates are exposed through positives. Others do not require film. The practice followed here is to call any process "positive-working" if it starts with positive copy, either film or paper, and ends with an image on the printing plate, regardless of the intermediates steps that are involved. The various types of plates based on this definition, listed earlier in the chapter, are treated in detail here.

Presensitized Positive-Working Plates

The method of producing an image on positive-working presensitized plates is different from the method used on negative-working plates. With negative-working plates, the light makes the coating in the image areas insoluble in the developer, and this hardened coating remains on the plate as the ink-receptive image area.

With positive-working presensitized plates, the light has the opposite effect. The plates are exposed through a positive, and the light photo-degrades or solubilizes coating in the nonprinting areas. Unexposed coating in the image areas is insoluble in the developer. Light-exposed coating, however, is changed chemically so that it becomes soluble in an alkaline solution. When an alkaline developer is applied, the coating is removed from the nonimage areas, leaving the unexposed coating as the ink-receptive image areas.

The base metal for these plates is anodized aluminum. The coating of a positive plate consists of a support resin, a diazo sensitizer, and a pH-sensitive dye that is applied to the grained and anodized metal substrate as a uniform, thin film.

The compound usually used for the light-sensitive coating is a naphthoquinone diazide (a derivative of naphthalene), which has the diazide group ($=N_2$) and an oxygen atom ($=O$) attached to the adjoining, or ortho, position. The material is also called a diazo-oxide. When this compound is exposed to light in the presence of trace amounts of water or other polar materials, nitrogen gas is liberated and the original compound is converted into indenecarboxylic acid. A sulfonic ester ($-SO_2R$) derivative is often used.

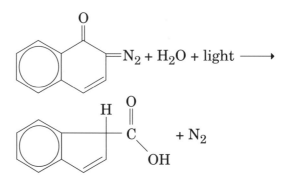

The indenecarboxylic acid is soluble in an alkaline solution, while the unexposed coating is not. A plate can be produced by exposure through a positive because the unexposed coating is oleophilic, or ink-receptive. After development to remove exposed coating, it is necessary only to gum the plate to desensitize the nonimage areas.

These plates are exposed through a positive, with exposure times about 1.5–2 times longer than with most negative-working plates. They are developed with an alkaline developer. This is followed with a finisher solution to remove the last traces of exposed coating and to deposit gum so that the plates will not scum on press.

Baking of positive plates. For long runs, after a plate has been dried, a prebake solution is applied and the plate is baked for about 5 min. at approximately 450°F (230°C). Any plate to be "baked" cannot be gummed since the high baking temperature chars both synthetic and natural gums and causes scumming. The baking polymerizes the coating in the image areas and makes it very hard and abrasion-resistant. Thus, before baking all plates should be carefully inspected, and any dirt or unwanted image areas should be carefully removed. The prebake solution contains synthetic detergents that prevent oils and other materials from baking into the plate and sensitizing it. Baking also makes the image areas much less soluble in solvents such as blanket washes, increasing the life of the plate from 500,000 up to 2,000,000 impressions. After baking, the plate is given a light rinse with water, and a synthetic gum is applied in the plate processor.

Screenless lithography. Screenless lithography depends on special positive-working presensitized plates that are exposed through a continuous-tone positive and then developed.

In screenless lithography, the grain of the plate itself provides the halftone screen, and the coating thickness affects the tone response. Screenless litho requires very close control of both the press and the plate.

Screenless lithography has the ability to resolve fine detail, eliminate moiré patterns in multicolor printing, give a higher saturation of pastel colors, and avoid the problem of "50% jump" (an increase of dot gain as halftone dots approach the 50% dot) that is experienced with halftone reproduction.

Dot gain is of maximum significance with fine-screen patterns, and a small gain quickly plugs the shadow areas of a print.

Exposing plates for screenless lithography and control of the press are critical, and they severely limit use of the method, but its advantages over the use of a halftone screen have intrigued many people.

A 300- or 500-line/in. (120–200 lines/cm) screen gives similar results and is less critical.

Screenless lithography requires a special, uniform yet relatively rough grained plate with the proper coating thickness to print ten or more specific gray steps when exposed through a 21-step platemaker's sensitivity guide (gray scale). Under magnification, these steps do not print a truly continuous tone, but rather a random, very fine "scum" that increases progressively in amount from highlight steps through midtones and shadows to solid areas. Similarly, the ***collotype process*** (see chapter 6), which is also referred to as a continuous-tone process, is not truly continuous-tone, but consists of a finely reticulated pattern.

Studies have shown that the grain and coating thickness on the plate, proper exposure, and precise development are all extremely important for good continuous-tone reproduction. If the coating completely covers the grain and remains after development, unexposed areas print solid. With decreasing density in the gray scales or films, more light gets through progressively increasing exposure of the coating. During development, the coating is progressively removed from the valleys of the grain in relation to the amount of light penetrating the photographic film. The remaining coating takes less ink as the tone values become lighter with increasing exposure.

The critical control that must be maintained throughout plate manufacture, platemaking, and presswork has severely limited the commercialization of screenless lithography.

Positive-/ Negative- Working Plates

The chemistry of some commercial plates makes it possible for them to be processed with either negative or positive copy or both. Their popularity depends on sharpness of the half-tone dots as well. The plate, coated with a naphthoquinone diazide, is exposed to a negative. Illumination degrades the naphthoquinone diazide to indenecarboxylic acid, which is soluble in an alkaline developer. Heating it to 275°F (135°C), however, gives off carbon dioxide and converts the indenecarboxylic acid to a compound that is insoluble in the developer.

Suppose that the plate was exposed through a negative film to UV light, which converts the exposed areas to indenecarboxylic acid, and that the plate is then heated for 2.5 min. at 275°F (135°C) in the plate processor to generate an insoluble image area. After cooling to 105°F (40°C), the entire plate is exposed to UV light to convert the unexposed naphthoquinone diazide to indenecarboxylic acid, which is removed by the developer in the usual fashion. The plate now prints the image from the negative film.

A modification of these steps makes a plate that properly produces images from both positive and negative films. One part of the plate is exposed to UV light through a negative film. After heating, the other part is covered by a positive film, and the entire plate is exposed to UV light and developed in the usual manner.

Multimetal Plates

Despite the long run length and excellent lithographic properties of modern aluminum plates, multimetal plates are still used for very long runs. Both negative-working and positive-working multimetal plates are available.

Multimetal plates are frequently preferred by large publication printers, package printers, metal decorators, and newspaper insert printers who like the high ink density and brilliance that result from the relatively low amounts of dampening solution required. The plates are unaffected by electron-beam (EB) and ultraviolet (UV) inks but are more expensive and difficult to make than surface plates.

If properly made and if properly handled on the press, a multimetal plate is capable of printing several million impressions. The reason for the long life of most bimetal plates is that the image and nonimage areas are established by two different metals; the holding of image areas is not dependent entirely on a hardened coating or lacquer.

A bimetal plate is made by electroplating a metal onto the base, almost always aluminum. If a thin layer of a third metal is electroplated over the second metal, the result is a trimetal plate.

Choice of metals. The use of contact angles helps to determine what metals are best for the image and nonimage areas respectively. The contact angle can be determined by placing a small drop of water on a well-cleaned piece of metal. If a tangent line is drawn to the surface of the drop at the point where it touches the surface of the metal, the angle between this tangent line and the surface of the metal is called the ***contact angle.*** The accompanying diagram shows that a small contact angle indicates that the metal is easily wet by water. A large contact angle indicates the opposite. Contact angles vary from about 15° for a metal that is easily wet with water, to about 130–140° for a metal that is poorly wet (see figure 5-15).

Contact angles can be measured for liquids other than water. A study made in a Pira research laboratory in Eng-

Figure 5-15.
Contact angles.

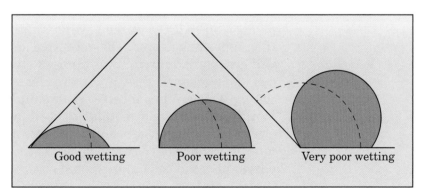

Good wetting Poor wetting Very poor wetting

land determined the contact angles of small drops of oleic acid applied to several metals previously immersed in water. The study showed the tendency of an oily material to displace water from the surface of the metal.

Metal	Contact Angle of Oleic Acid
Copper	18°
Aluminum	60°
Chromium	77°
Iron	178°

With oleic acid, a small contact angle means the metal is easily wet by oil—not easily wet by water. These results show that copper should be a good metal for the image areas of a plate and that aluminum, chromium, and iron (or stainless steel) should be good metals for the nonimage areas. The larger contact angles show that oil does not easily wet their surfaces.

A word of caution is appropriate. Contact angles measured on smooth, clean metal do show which metals are most easily wetted by any liquid. Plates, however, are grained and have rough surfaces, they often have grease or coatings on them, and they operate under dynamic rather than static conditions. The simple experiments described here are insufficient to predict much about the performance of a finished printing plate.

Multimetal plates use the metals predicted by contact angles. Copper or brass is used for the image areas; stainless steel, chromium, or aluminum is used for the nonimage areas.

This combination of metals is interesting in another way. It is desirable to be able to repair plates that are either scumming or going blind on the press by the use of one chemical that makes the image areas ink-receptive and nonimage areas water-receptive. A 2% solution of nitric acid (HNO_3), a 2.5% solution of sulfuric acid (H_2SO_4), or a 10% solution of phosphoric acid (H_3PO_4) worked over a multimetal plate will make the copper image areas ink-receptive and at the same time will desensitize chromium, stainless steel, or aluminum. Special "copper sensitizers" and plate cleaners formulated for these plates are much better than simple acid solutions.

Although different metals are used for the base and the electroplated layer, certain procedures are common to all multimetal plates. Thus, if copper is the electroplated metal, development of the light-exposed coating must uncover the copper in the nonimage areas so that it can be etched away to bare the stainless steel or aluminum underneath. The principal multimetal plates in use today are copper-coated aluminum and copper-coated stainless steel, but because of the lighter weight, aluminum is replacing both steel and brass as the base metal. Copper is too easily drawn or stretched to make a good base sheet.

Producing a multimetal plate. Two methods are used to expose and develop a multimetal plate. With one method, a negative-working plate is exposed through a photographic negative. The exposure hardens the coating in the image areas, and the developer removes it from the nonimage areas so that the exposed metal can be etched away.

In the other method, a positive-working plate is exposed through a photographic positive. In this case, the developer removes the coating from the light-exposed nonimage areas so that the exposed metal can be etched away. The unexposed coating over the image areas acts as a stencil to protect the image during etching.

The above methods uncover the copper in the nonimage areas, so that the copper etch that is applied next can remove the copper and expose the metal underneath. For high-quality work the coating or "stencil" that protected the copper during etching is removed with a stencil remover that leaves the copper very ink-receptive. A plate finisher is used to gum up and protect the nonimage areas of the plate.

Etches for multimetal plates must be formulated so they will dissolve (oxidize) the top metal but will not attack the metal underneath. An etch containing ferric chloride, $FeCl_3$, is often used to etch copper when the metal underneath is stainless steel. Ferric nitrate, $Fe(NO_3)_3$, may be used to etch copper when the metal underneath is aluminum. With either etch, a copper atom loses two electrons, to form a cupric ion, Cu^{2+}, and these two electrons reduce two ferric ions, Fe^{3+}, to ferrous ions, Fe^{2+}. The ionic equation is:

$$2Fe^{3+} + Cu^{\circ} \rightarrow 2Fe^{2+} + Cu^{2+}$$

Although chromium is an excellent metal for nonimage areas, chrome plating is expensive and requires extensive measures to protect the workers and prevent pollution. Thus, chrome-plated plates are essentially obsolete in the United States.

Other Positive-Working Plates

Deep-etch plates. Once popular for long-run printing, copperized aluminum deep-etch plates have been replaced by negative-working photopolymer plates and by baked, positive-working plates.

Electrostatic plates. Many small plates for offset duplicators are made by an electrostatic process. Electrostatic printing and electrostatic printing plates depend on the use of photoconductors (semiconductors), materials that are non-conductive in the dark but conductive in the light. Zinc oxide, selenium, and some organic compounds have this property. The semiconductor is charged in the dark and exposed by laser or by light reflected from the image to be copied. The charge dissipates from the exposed areas, leaving an electrostatic charge that attracts toners to the image areas.

Two general methods are used to prepare electrostatic plates. In xerography, the semiconductor consists of selenium coated permanently on a drum or plate. A corona discharge is used to charge the selenium surface. Some electrostatic copiers use an organic semiconductor on the drum instead of selenium. When exposed to camera-ready positive copy, the charge is removed from the exposed areas (nonimage), but remains on the unexposed areas (image). This invisible, electrostatic charge must be toned to create an image. This is done by applying a toner, a dry powder that contains a pigment (usually carbon) and a thermoplastic resin such as poly(vinyl butyral). The charged areas attract and hold the finely divided toner particles. The toned image is then transferred to a heavy paper plate, which, after the toner is fused with heat, becomes a lithographic printing plate.

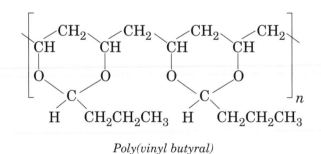

Poly(vinyl butyral)

Most of the paper plates used for offset duplicators use a different electrostatic process. The paper plate is coated with zinc oxide (ZnO) in a suitable binder so that the plate will accept a charge of static electricity. The plate is charged in the dark, exposed to the image, which discharges the background area, and toner is applied to the plate. If dry toner is used, it must be fused. This forms a visual image that also attracts ink, making the printing plate. These paper plates can be used for runs up to 10,000 impressions. Since paper is not dimensionally stable, they are not suitable for close register or process-color work.

Static electricity will not remain on aluminum. Therefore, to prepare an aluminum electrostatic plate, a coating must be applied over the aluminum. In one process, the plates are coated with a layer of zinc oxide, much the same as for paper plates.

Zinc oxide is not water-receptive, and after exposure to camera-ready copy, a "conversion solution" containing potassium ferrocyanide, $K_4Fe(CN)_6$, is applied to zinc-oxide-coated plates. This solution converts the zinc oxide to water-receptive zinc ferrocyanide, $Zn_2Fe(CN)_6$. The ferrocyanide solution is also used in the dampening solution.

A major disadvantage of all electrostatic plates using dry toners is their relatively poor image fidelity. They are satisfactory for line work, but halftone screens in excess of 100–120 lines/in. (39–47 lines/cm) are too fine for quality printing using electrostatic plates.

Photodirect plates. To achieve the speed required for laser plates or for camera speed, three methods are available: electrostatic process with semiconductors (discussed above), silver halide chemistry, and photopolymer chemistry. A great variety of these plates are used in camera platemakers.

One common camera platemaker exposes a positive-working camera-speed litho plate to camera-ready copy. This plate has three layers of gelatin emulsion on a paper base: an emulsion layer containing a photographic developer next to the paper, an unexposed silver halide emulsion in the middle, and a pre-fogged silver halide emulsion on top.

Light from the copy penetrates the top layer and exposes the silver halide in the middle layer. Following exposure, the plate is moved into a processor that contains an activator that dissolves the developer in the bottom layer. The developer reduces the light-exposed areas in the middle layer to metallic silver. All of the developer is exhausted in these areas, so none can reach the top layer.

In the image areas, the developer migrates through the middle layer and reduces the prefogged emulsion in the top layer to metallic silver. Thus a positive image is formed in this top layer. Since a tanning type of developer is used, the gelatin in the image areas of the top layer is hardened, and these areas become ink-receptive. The gelatin in the nonimage areas of the top layer is not hardened, and these areas become water-receptive. Like other paper plates, these are relatively short-run plates for use when close register is not needed.

Two types of photodirect plates are based on chemistry similar to that used in diffusion transfer, but neither requires a separate photographic paper. One photodirect plate consists of two layers mounted on a polyester base. The top layer is a silver halide emulsion, the bottom layer is "nucleated"—that is, it contains a material that can reduce unexposed silver halide to metallic silver.

Upon exposure and development, a silver image develops in the bottom layer. The top layer is stripped off with warm (115°F, 46°C) water, and the plate is treated with an activator to make the metallic silver image ink-receptive. The plate is washed with water, and it is ready for printing.

The second type of photodirect plate based on diffusion transfer has four layers on a paper base. (The fourth layer simply prevents undesirable light reflections from the base. This is called an antihalation film.) After exposure to positive copy, the plate passes through a diffusion transfer developer. The developer penetrates the top two layers and reduces the light-exposed (nonimage) areas of the third layer to metallic silver. The unexposed silver halide in this layer diffuses through the second layer and is reduced to metallic silver by the thin nucleated coating in the surface layer. (This process is explained in the description of the first type of diffusion transfer plate.) The final treatment is with a solution that makes the silver image on the top layer ink-receptive. This type of plate resembles the structure and chemistry of the polyester plates discussed at the beginning of the chapter and shown in figure 5-3.

Diffusion transfer plates. In accord with the earlier definition of positive-working plates as those that start with positive copy, diffusion transfer plates are discussed in this section.

Lithographic diffusion transfer plates are available with aluminum or paper backing. A photographic paper containing a silver halide is exposed, either by contact or in the camera.

This paper negative, as it is called, is placed in close contact with the aluminum or paper plate. The assembly is then passed slowly through a special developer in a diffusion transfer processor. When the paper negative is peeled from the plate, a positive image remains on the plate. After a treatment to make the image areas ink-receptive, the plate is ready for printing.

The method by which a positive image can be formed on the printing plate involves some interesting chemistry. If the exposed photographic paper were developed in the usual way, a negative of the original copy would be formed: the image areas of the original copy still contain unreacted silver halide in the photographic paper.

In the diffusion transfer process, the processing fluid contains, among other things, a developing agent such as hydroquinone and a fixer such as sodium thiosulfate ("hypo," $Na_2S_2O_3$). The developing agent reacts with the light-exposed areas of the paper, changing them to black, metallic silver. The sodium thiosulfate acts much as it does in any fixing bath, dissolving the unexposed silver halide and causing it to diffuse from the surface of the paper. Since the paper and the lithographic plate are in close contact at this point, the dissolved silver halide transfers or diffuses to the surface of the plate.

Something must now happen to reduce the silver halide to metallic silver. With one process, the plate is grained aluminum. Aluminum is a more active metal than silver: aluminum atoms can lose electrons to silver ions, causing the silver ions to be deposited as metallic silver. The ionic equation is:

$$Al^\circ + 3Ag^+ \rightarrow Al^{3+} + 3Ag^\circ$$

If the plate surface is anodized aluminum or paper, a different process must be used. In this case, the surface of the plate is treated with a "nucleated coating." This is gelatin containing small amounts of a metal in a very finely divided (colloidal) state. The metal used could be any metal, such as aluminum or magnesium, that is more active than silver.

The nucleated coating reduces the transferred silver halide to metallic silver, by a process similar to the one described above for aluminum. The reaction involved is an oxidation-reduction reaction. The finely divided metal is oxidized (losing electrons), and the silver ions of the silver halide are reduced to metallic silver (gaining electrons).

By this process, a positive image of metallic silver forms on the plate. After a short wait, the paper negative is peeled from the plate. The silver image on the plate looks good, but it is not ink-receptive. To make it ink-receptive, an "etch" solution is applied. With one process, this etch contains an oxidizing agent and an organic compound with an oleophilic group. The oxidizing agent converts the silver atoms to silver ions. Then the silver ions combine with the oleophilic group of the organic compound to form an insoluble ink-receptive material. Since the gelatin in the nucleated coating on the nonprinting areas is hydrophilic (water-loving), the plate is ready for printing.

Computer-to-Plate Systems

Computer-to-plate (CTP) systems offer many advantages. Exposing a plate directly from digitally controlled illumination increases the speed, efficiency, and convenience of the plate-making process, and it eliminates the use of expensive film. The development was delayed for decades after the invention of the laser because of multiple problems—problems of emulsion speed, proofing, and process control. Now that these problems have been largely resolved, most printing copy is in digital form, and the generation of a film represents an additional step in the process.

One advantage of computer-to-plate laser exposing is that digital data can be used to make a plate and can, at the same time, be sent to some distant location to make a duplicate plate, or by changing the signals, a negative can be produced on film. When this negative is developed, it can be used to make a conventional litho or relief plate.

The machine illustrated in figure 5-16 is a digital contract proofing and platemaking system designed to process digital halftone proofs or thermal litho plates.

Computer-to-plate has replaced much film-to-plate imaging by direct exposure of printing plates. (As a matter of fact, the Hell Helio-Klischograph is a computer-to-plate system used for years to prepare publication gravure cylinders.)

Laser-Exposed Plates

With the usual method of making litho plates, a light source exposes all of the plate at once, for times varying from about 20 to 60 sec., depending on the strength of the light source, its dominant wavelengths, distance from the plate, and the light sensitivity of the plate. When a plate is exposed with a laser beam, the light scans the plate very rapidly, while the plate moves slowly under the scanning beam.

Figure 5-16
Heidelberg/Creo Trend-
setter Spectrum™
digital halftone proof-
ing and platemaking
system.
*(Courtesy Creo Products,
Inc.)*

With a laser, each tiny printing spot (pixel) may receive an exposure of only microseconds duration. Thus the exposure time is a million times less than that used with ordinary light sources. To get sufficient exposure, the intensity of the laser beam must be a million times as great, the plate coating must be a million times as fast, or there must be some combination of the two. (Actually, the paths of the beams overlap slightly to provide some additional exposure.)

A number of high-speed lithographic plates are available that can be imaged by lasers directly from digital data without the need for intermediate films. These are of six primary types:
• Plates with silver-halide coating on film and metal bases
• High-speed photopolymer plates with dye-sensitized coating on aluminum
• Hybrid plates
• Plates with a thermal-based coating
• Plates imaged by inkjet technology
• Electrophotographic plates on an aluminum base

Silver halide. Silver-halide plates for computer-to-plate are positive-working plates. They are exposed in "write white" mode on the platesetter, which means the laser writes the nonimage area of the plate. Light-sensitive silver halide (e.g.,

AgCl) is suspended in the emulsion layer, which lies on top of a positive or "nuclei" layer. The two layers are separated by a "barrier" layer. During exposure of the nonimage areas, laser light catalyzes the reduction of AgCl to molecular silver, which remains in the emulsion layer and does not affect the hydrophilic printing surface. During the first processing step, unexposed AgCl from the emulsion layer diffuses through the barrier layer into the nuclei layer, where it is reduced to molecular silver. During the second processing step, the emulsion layer, barrier layer, and unexposed areas of the nuclei layer are removed. The image areas on the processed plate are formed by oleophilic molecular silver on the plate surface.

Silver halide plate processing involves two stages: activation and stabilization. In the activation stage, a strongly alkaline solution dissolves the unexposed silver halide in the emulsion layer, after which it diffuses into the positive layer. In the positive layer, centers of development cause silver to precipitate, leaving metallic silver on the plate surface. After the development is complete, the plate enters the stabilizing tank where the weakly acidic solution neutralizes the alkali in the emulsion layer of the plate and stabilizes the precipitated silver.

Photopolymer. Photopolymer plates are negative-working plates exposed in "write black" mode. During processing, unexposed emulsion is dissolved by the developer, leaving exposed photopolymer behind to form the image areas of the plate. One disadvantage of photopolymer plates is that they require light-sensitive developing solutions that have a tendency to foam. In addition, the plate has to be heated after exposure and yet does not offer the best light sensitivity. On the other hand it has very good run lengths and print characteristics.

Hybrid. Hybrid plates are positive-working plates that are exposed in "write white" mode on the platesetter. Hybrid plate technology is unique in that it uses two separate and distinct photosensitive coatings on the metal plates. The top coating is a silver-halide emulsion whose light sensitivity can be varied to handle a full range of speeds from contact to film speed and a full gamut of spectral responses from UV to visible blue, green, red, and infrared lasers. The bottom coating is a photopolymer known for good performance on the press.

The plate is imaged by a low-power argon-ion or yttrium-aluminum-garnet (YAG) laser, creating an image on the top emulsion. Then the plate is processed through two stages. The first stage uses a silver developing process like silver-halide film, but without the clear film substrate. Instead of that, this silver-halide image puts the photopolymer emulsion directly on the top of the printing plate. The second stage is to image the photopolymer emulsion with a standard UV light source, using the silver image as a mask to image the emulsion on the plate conventionally.

Exposure of the silver halide (nonimage areas) causes it to precipitate as molecular silver in the first stage of processing. Black silver forms a mask over the photopolymer layer. A secondary exposure, made inside the processor, hardens the photopolymer exposed through the silver halide mask. Subsequently the silver halide and unexposed photopolymer are dissolved in processing, leaving the insoluble, exposed photopolymer behind to form a positive image.

Thermal. Thermal plates (see pages 167–169) are true digital plates. Once imaged, the dot size conforms to the size of the laser's imaging spot, and it doesn't change if the plate is processed. They can hold very fine dot structures; they are environmentally clean with no chemical processing and can be handled in either yellow safelight or controlled-daylight conditions; and they have the same handling characteristics as conventional presensitized plates.

As their name implies, thermal plates are imaged by heat rather than light. Thermal plates require higher laser energy to create images than the energy used to image conventional plates. Most suppliers have created plates sensitive to either wavelengths of 830 nm or 1064 nm or both to satisfy the needs of internal and external drum users.

Inkjet. A new approach to CTP is the use of inkjet compounds deposited on grained aluminum. One approach uses the inkjet ink to create a mask for light exposure and a processing wash. Another approach creates a positive image on each of the four plates for process-color printing.

Electrophotographic plates on metal bases. The first of these plates was the Kalle Elfasol plate, used with argon-ion lasers in EOCOM laser platemakers. This and other electrophotographic plates were developed mainly for the newspaper market. One brand of electrophotographic plate uses an organic photoconductor (OPC) and liquid toners to obtain quality suitable for the commercial printing market.

Lasers for CTP

Starting from the original conventional high-powered yttrium-aluminum-garnet (YAG) water-cooled laser the industry has seen the adoption of argon ion (Ar), through to helium-neon (HeNe), frequency-doubled YAG (Fd:YAG), and more recently the availability of solid-state devices such as laser diodes (LD) and light-emitting diodes (LEDs). Lasers are either single beam, using a high-speed rotating mirror to reflect the beam at great speed onto the plate, or are used in an array so that coverage is effectively increased by a factor of 6, 8, or 10, depending on the array format, or a single beam split into 32 arrays, which in turn can be used in parallel to provide coverage.

Imagesetter and plate manufacturers have an increasing range of laser sources to choose from, including Nd:YAG. Operating at 1064 nm in the high-infrared range, Nd:YAG lasers offer a great deal of power, making them suitable for thermal-reacting materials and especially in the development of infrared (IR) plates. Internal-drum platesetters, for example, use Nd:YAG.

Table 5-II shows the primary laser sources used for laser exposure of plates. A basic requirement for all laser imaging systems, whether imaging to film, paper, or plates, is the relationship between laser power (in watts), scanning efficiency (the percentage of the laser power actually delivered for imaging purposes), coating sensitivity (in joules/cm^2), and scanning speed (ft^2/min).

Table 5-II.
The primary laser sources used for laser exposure of plates.

	Wavelength (nm)	Power (milliwatts)	Life (hours)
Nd:YAG	1064	5,000–20,000	9,000–12,000+
Infrared laser diode	830	5,000–20,000	20,000+
Red laser diode	670, 680	3–7	50,000+
Helium neon (red)	633	3–7	10,000–23,000+
Frequency-doubled YAG (green)	532	5–300	10,000–25,000+
Argon ion (blue)	488	5–300	5,000–12,000+

Spark-Discharge Plates

New technologies permit direct-to-press imaging without the use of light. The spark-discharge imaging plate (figure 5-17) makes it possible to image a plate after mounting it on the press. Electrospark imaged plates have a nonprinting surface of silicone rubber, which is ablated or evaporated to open the image area that lies beneath. The silicone layer is impregnated with inorganic oxide particles that improve sparking.

Figure 5-17.
Principle of an electro-spark ablative plate.

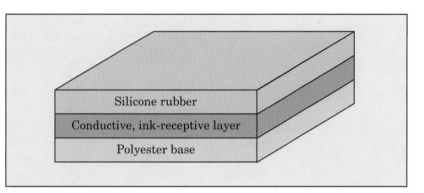

The coating is applied over an ink-receptive, electrically conducting layer that accepts the electrical surge or spark. Where the spark discharge strikes the plate, the layer releases the silicone rubber and cross-links the ink-receptive layer, forming an ink-receptive image area. This provides "computer-to-press" or "direct-to-press" imaging of the plate from digital input or copy.

Spark erosion technology can also produce a conventional litho plate if the silicone rubber is replaced with a hydrophilic layer that accepts water to form the nonimage area.

Digital Plate-making Processes using Laser Printers

Techniques for producing plates directly from the computer have been available for years. One convenient method used a zinc-oxide-coated plate in a laser printer. Silver chemistry is commonly used to produce paper or polyester plates by exposing photosensitized paper or polyester plates that can be exposed by computer-driven lasers. Although this has been limited to line work and simple halftones, it is important in some printed products. A prime example is check overprinting where the information comprises personalization indicias, the bank logo, and various bank codes. It is a comparatively simple task to combine this information into groups of twelve or eighteen different checks (the number printed at one time), imaged in position with a variety of laser imagesetters or exposure systems and printed in run lengths of 200 impressions or so. Most checks in North America are overprinted in this manner.

Further Reading

Adams, Richard M., and Frank J. Romano. *Computer-to-Plate: Automating the Printing Industry.* 2nd ed. (264 pp) 1999 (Graphic Arts Technical Foundation, Sewickley, Pa.).

Adams, Richard M., and Frank J. Romano. *Computer-to-Plate Primer.* (126 pp) 1999 (Graphic Arts Technical Foundation, Sewickley, Pa.).

Destree, Thomas M. *Lithographers Manual.* 9th ed. (428 pp) 1994 (Graphic Arts Technical Foundation, Sewickley, Pa.).

MacPhee, John. *Fundamentals of Lithographic Printing, Volume I: Mechanics of Printing.* (384 pp) 1998 (Graphic Arts Technical Foundation, Sewickley, Pa.).

Wilson, Daniel G. *Lithography Primer.* 2nd ed. (192 pp) 1997 (Graphic Arts Technical Foundation, Sewickley, Pa.).

6 Image Carriers for Flexography, Screen, Gravure, Inkjet, and Other Printing Processes

Flexography prints from photopolymer or rubber plates; gravure uses chrome-plated copper cylinders; screen printing uses a screen of steel wire or synthetic fibers to support a stencil; and letterpress prints from photopolymer plates. The old, metal letterpress plates are rarely used, except as pattern plates for flexography.

Although flexography and letterpress are both relief (raised-image) printing processes, the chemistry for preparing plates differs significantly. Gravure cylinders have an intaglio or recessed image; most are now engraved by mechanical processes, although some chemical engraving is used. Screen printing plates, which include a stencil attached to a screen, are flat, but they can be used to print on objects of various shapes.

Printing images usually come from a computer, although visual copy (hard copy) is still widely used. Digital copy drives a laser that exposes or engraves the image. Visual images may be captured on a film used to expose a photopolymer plate or to etch a metal plate.

Collotype, heat transfer printing, thermography, and inkjet printing are discussed briefly at the end of the chapter.

Flexographic Plates

The term "flexography" was adopted in 1952 after the Packaging Institute polled converters and suppliers. The earlier term, "aniline printing," described the original coal-tar-derivative inks first used to print products such as paper bags.

Flexo photopolymer plates have a hardness of 40–60 Shore A (figure 6-1); flexo rubber plates have a hardness of about 50. In contrast, letterpress plastic plates are very much harder, with a Shore D hardness of about 70–80.

Figure 6-1.
Durometer: Shore A
hardness.
*(Courtesy Pacific
Transducer Corp.)*

Ink transfer from flexographic plates depends on the chemical nature, the surface free energy, and the roughness of the surface of the plate. These properties depend on UV exposure, solvent washout, and chemical finishing of the plates.

Thickness of Flexographic Plates

Thin plates carry a shallow relief that best reproduces fine detail. Shallow relief plates, however, require very clean surfaces because they tend to pick up dust and dirt from paper and board and to clog, destroying the detail. For printing on corrugated, a deep relief is used.

The thickness of the plate is important because it is to be stretched around a cylinder. After attaching to the cylinder, the outside (or top) of the plate is longer than the inside—or bottom (figure 6-2). The thicker the plate, the greater is the difference in these two lengths, and the greater is the distortion of both the plate and the printing.

Figure 6-2.
Distortion in a
mounted flexo plate.

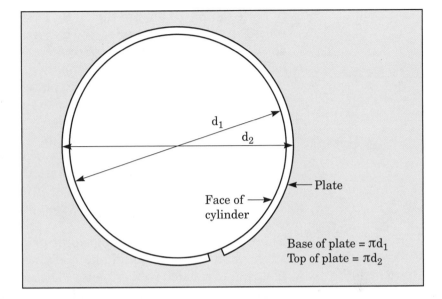

The figure shows that the base or inside of the mounted plate has a length of $\pi(d_1)$, where d_1 is the diameter of the plate cylinder. The outside of the mounted plate has a length of $\pi(d_2)$, where d_2 is the diameter of the cylinder plus twice the thickness of the plate. Thus, if the plate completely wraps the cylinder, the difference in length is $\pi(d_2 - d_1)$ or $2\pi t$, where t is the thickness of the plate. If the plate does not completely wrap the cylinder, then the difference in length, or the distortion of the plate, is $2\pi t \times l/c$, where l is the length of the plate and c is the circumference of the cylinder. The percent distortion is $2\pi t/c \times 100$ or $2t/d \times 100$. Percent distortion is thus independent of the plate length and depends only on the plate thickness and the diameter of the cylinder.

Table 6-I.
Thicknesses of flexo
plates (mils).

	Backing	Photopolymer	Total	Relief
Thin	4, 5, 7	38–41	45	20
		60–63	67	30–35
Medium	4, 5, 7	78–81	85	25–35
		100–103	107	25–35
		105–108	112	25–35
		118–120	125	20–35
Thick	5 ,7	148–150	155	80–100
		180–182	187	80–100
		243–245	250	100--140

(1 mil = 0.001 in. = 0.0254 mm)

For example, if a plate that is 0.117 in. thick is mounted on a 6-in.-diameter cylinder, the distortion of the surface of the plate is $2 \times 0.117/6 \times 100\%$ or 4%.

This view is somewhat oversimplified, because other factors also affect the distortion of a flexo plate when it is wrapped around a cylinder. However, it is true that the thinner the plate, the less the distortion.

Sleeves are printing plates that are first wrapped around a steel cylinder and then engraved, typically with a CO_2 laser, so that the image is formed on a plate that is already cylindrical. The image is not distorted as it is when a flat plate is wrapped around the cylinder after engraving (figure 6-2).

Types of Flexographic Printing Plates

There are four types of flexographic printing plates: solid-sheet photopolymer, liquid photopolymer, laser-etched rubber plates, and molded rubber plates. These plates are prepared in one of three methods: by crosslinking a low-molecular-weight elastomer with UV light, by molding and cross-linking an elastomer with heat, or by engraving a rubber or other elastomer that already has the hardness desired in the plate.

Photopolymer Flexographic Plates

The development of photopolymer plates has promoted the growth of flexography so that flexo now plays a major role, not only in package printing but in publication printing as well. Photopolymer flexo plates make it possible to print a halftone screen of 150 lines/in. (60 lines/cm) or even more. Several companies manufacture photopolymer plates, each using a chemistry that is somewhat different, but the plates can be placed in two groups: those made from a solid base stock and those made from a liquid.

Solid-Sheet Photopolymer Plates

Solid-sheet photopolymer plates are made from a solid elastomer adhered to a polyester film. The plate is hardened by exposure to light through a negative film and developed by washing away the unexposed material, which remains soluble. The formulation contains a block copolymer or an elastomer (often an acrylate) that can be cross-linked, a photoinitiator, and other additives that promote adhesion, dimensional stability, plasticity, and other properties. Acrylates and methacrylates have a low activation energy, but even these require the use of a photoinitiator if curing is to be accomplished in reasonable time with the usual UV lamps.

In a typical solid plate (figure 6-3), the photopolymer layer is sandwiched between a protective, strippable polyester

Figure 6-3.
Structure of a solid
photopolymer flexo
plate.

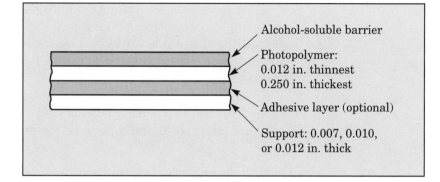

sheet or alcohol-soluble protective sheet and a polyester backing or stainless steel base that provides dimensional stability.
The photopolymer layer contains a thermoplastic elastomer, a
polyfunctional acrylate monomer, photoinitiator, and other
additives. An adhesive layer to promote adhesion of the photoreactive layer to the backing is used in some plates.

Photoplatemaking typically begins with exposure of the back
of the plate, which consumes stabilizers in the formulation and
determines the relief height by hardening the base of the plate.
The polyester cover sheet is removed, and the front of the plate
is exposed to form the image. The plate is developed by removing material from unexposed areas with a solvent. The plate is
dried to remove all solvent and exposed again to reduce tack
and toughen the plate with further cross-linking.

The hardness of the plate is governed by the amount of
cross-linking, which, in turn, is determined by the amount of
exposure and by the choice of materials used to make the plate.

Many different elastomers have been used; even natural
rubber can be cross-linked with acrylates. One popular
photopolymer plate is the product of cross-linking neoprene,
a polymer of chloroprene (2-chloro-1,3-butadiene) and trimethylolpropane triacrylate.

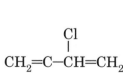

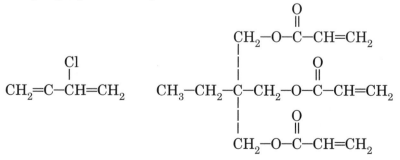

Chloroprene
(2-chloro-1,3-butadiene)

Trimethylolpropane triacrylate

Upon exposure, the trimethylolpropane triacrylate cross-links and hardens the polymer, making it insoluble. The degree of hardness is controlled by the amount of triacrylate added to the formulation—the more cross-links, the harder the resin.

Another popular flexo plate is based on styrene-isoprene rubber, formulated with a polyacrylate, such as trimethylolpropane triacrylate and a photoinitiator.

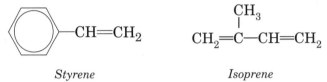

Styrene

Isoprene
(2-Methyl-1,3-butadiene)

Irradiation with UV light cross-links the elastomers so that they are insoluble in the washout solvent or developer.

Many polyacrylates are mentioned in the chemical literature as useful cross-linkers, including:
- 1,4-Butanediol diacrylate
- Triethylene glycol diacrylate
- Ethylene glycol dimethacrylate
- Polypropylene glycol dimethacrylate
- Polyethylene glycol dimethacrylate
- 1,6-Hexanediol dimethacrylate
- Neopentyl glycol dimethacrylate
- Dipropylene glycol dimethacrylate

A wide variety of formulations is possible, and details of these formulations include confidential, proprietary information.

Photosensitizers and Photo-initiators

A ***photosensitizer*** is a material (often a dye) that becomes activated by absorbing visible light and transfers that energy to a photoinitiator.

A ***photoinitiator*** accelerates the polymerization of the monomer by ultraviolet light. When the photoinitiator absorbs ultraviolet light, it splits apart to form ***free radicals.*** Typical photoinitiators are benzophenone, benzoin, benzoin ethers, and benzil.

Benzophenone

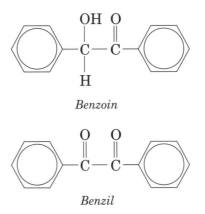

Under illumination by UV light, benzoin dissociates to form two free radicals:

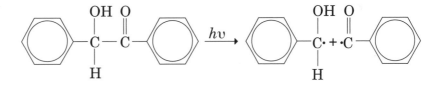

(*hυ* is the chemist's shorthand to indicate radiation energy)

These free radicals open the double bonds of the monomer, starting a chain reaction by which polymerization occurs.

$$R\cdot + R'CH{=}CH_2 \rightarrow RR'CH{-}CH_2\cdot \rightarrow$$

$$RR'CH{-}CH_2{-}R'CH{-}CH_2\cdot \rightarrow \text{etc.}$$

When the free radical opens one of the carbon-to-carbon double bonds of the monomer, it becomes part of the polymer chain and generates a new free radical. Because each benzoin molecule can initiate two very long polymer chains, it requires only a little initiator to complete the reaction, and a small amount of photoinitiator can bring about the polymerization of a large amount of resin.

Washout Solvents   Washout solvents are frequently chlorinated solvents, such as perchloroethylene ($Cl_2C{=}CCl_2$). This solvent, a derivative of ethylene, has a boiling point of 250°F (121°C). It must be used in an enclosed environment to prevent its escape to the environment.

Because of environmental hazards associated with chlorinated solvents, other materials are used as washout solvents. These include a mixture of heptyl acetate and heptyl alcohol, a mixture of nonyl acetate and benzyl alcohol, and a mixture of petroleum hydrocarbons with aliphatic alcohols. Another washout solvent contains terpene mixed esters, terpene alcohol, benzyl alcohol, and a synthetic isoparaffin. Compositions are disclosed in the MSDS (Material Safety Data Sheets; see chapter 10) that accompany the products.

Plates based on polyamides can be washed out with aqueous solvents or alcohol, depending on the chemical structure of the polyamide. After development, they resist swelling in water-based inks.

Copolymerizing with acrylic acid (CH_2=CHCOOH) introduces carboxyl groups into the elastomer and improves the solubility of the unreacted (uncross-linked) elastomer in washout solvents. For example, if they are unexposed and not cross-linked, carboxylated nitrile rubbers can be washed out with glycol ethers.

Liquid Photopolymer Plates

Liquid plates are (solid) printing plates prepared from a liquid containing partially polymerized acrylates, rubber oligomers, or other chemically unsaturated materials. With liquid plates, the light-sensitive material used to create the plate has a consistency of heavy honey. Light hardens this material in the image areas but leaves a viscous liquid in the nonimage areas. After exposure, the uncross-linked material may be washed out, but some liquid plates are developed by using a stream of air to blow away the unexposed liquid, which can be recovered and reused. Equipment for processing liquid plates is illustrated in figure 6-4.

The plates are typically adhered to a polyester foil, although stainless steel may also be used. The developed plates are still flexible and can be bent around a steel cylinder.

When the printing surface of the plate is hard, with a softer layer underneath, distortion caused by printing on an uneven surface is reduced (figure 6-5). One way to achieve this is to put on top of the plate material a thin layer of a resin composition that cross-links more extensively than the main layer beneath. Plates made in this way are referred to as ***capped plates.*** The hard layer reduces "squeeze out" (which produces a halo around letters and solids) and holes in the center of halftone dots, typical problems in flexo printing. The softer layer gives some cushion or compressibility to the plate, enabling it to print well on surfaces that are not very smooth. (Letterpress and gravure do poorly on rough surfaces because the hard plates cannot conform to the surface. Offset does much better because the blanket, like a flexo plate, is resilient.)

Figure 6-4.
Merigraph® 5280 platemaking system. *(Courtesy MacDermid Inc.)*

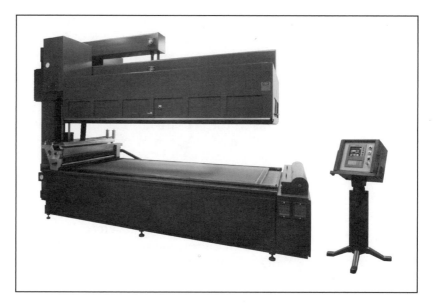

Figure 6-5.
Capped plate improves print fidelity.

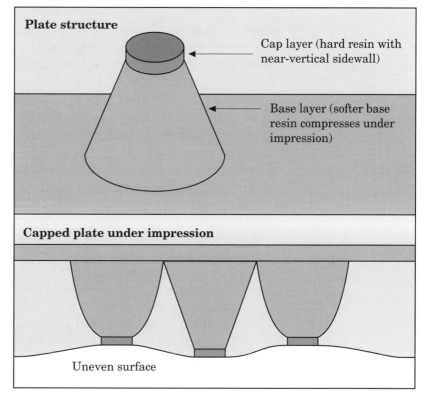

Photopolymer flexo plates are considered better than rubber plates for jobs requiring close register, since the highly cross-linked photopolymer does not stretch as much as rubber. In fact, commercial flexographic work rivaling the best four-color lithography, gravure, or letterpress has been produced with photopolymer plates. They run as long as—or longer—than rubber flexo plates. They run well with alcohol-based or water-based flexo inks, although some solvents, such as ketones and aromatic hydrocarbons, attack photopolymer plates. Such solvents should be eliminated from the inks and from the reducer used to adjust the viscosity of the inks in the pressroom. The printer should specify the type of plate to be used when ordering the inks.

Advantages of photopolymer plates. They produce accurate, multicolor register and hold fine screens and small halftone dots. They have uniform thickness. Few production steps are required (expose, develop, dry) with no matrix (molding) or metal engraving required. The original (called "copy") is a digital file or a film, and it is easily stored. The cost of a single plate is less than the cost of preparing a rubber plate. Liquid systems can produce a capped plate polymer.

Disadvantages of photopolymer plates. The stiff polymer conforms to the cylinder less easily than does a rubber plate. Multiple plates each cost as much as the original. Old plates may curl, delaminate, or tear when they are removed from the press. For most plates, solvents, which may present an environmental hazard, are required to wash out the unexposed photopolymer. Used plates must be stored in rolls.

Rubber Flexo Plates

If two or more photopolymer plates are needed from the same negative, the entire exposure and development process must be repeated for each plate. When ***duplicate plates*** are required, they can be prepared from a matrix molded from an original photoengraving on magnesium or copper. Economics dictates the use of duplicate plates for more than about four or five repeat images because the preparation of the pattern plate is expensive, but the duplicates are less costly than the photopolymer.

Flexographic duplicate plates are made from natural rubber, nitrile rubber, compounded vinyl resins, or EPDM (ethylene-propylene diene monomer). Incoming work is mostly in digital form, and rubbers are conveniently patterned with a CO_2 laser either in flat plates or on sleeves.

Nitrile rubber is often used as the synthetic rubber for flexo plates. It is a copolymer of butadiene and acrylonitrile:

$$CH_2{=}CH{-}CH{=}CH_2 \qquad CH_2{=}CH{-}C{\equiv}N$$

Butadiene   *Acrylonitrile*

$$-CH_2{-}CH{=}CH{-}CH_2\left[\begin{array}{c} CN \\ | \\ CH{-}CH_2{-}CH{=}CH{-}CH_2 \end{array}\right]_n$$

Polymer

The amount of acrylonitrile may vary from 10 to 50%. Increasing the nitrile content improves oil resistance but hurts the low-temperature performance of the rubber. The molecular weight of these amorphous (noncrystalline) polymers is reported to vary between 250,000 and 600,000. The structure can be largely linear or highly branched depending on the polymerization conditions. Nitrile rubber resists oxidation and softening with aliphatic hydrocarbons that are often found in inks and solvents.

Vinyl rubber is a mixture of poly(vinyl chloride) with nitrile rubber. These two polymers are not compatible—that is, they do not dissolve in each other—and the structure of the compounded product contains regions in which one polymer or the other predominates as shown in figure 6-6.

EPDM (ethylene propylene diene monomer) is a terpolymer of ethylene and propylene with a small amount of a diene such as ENB or DCPD.

Figure 6-6.
The nature of a vinyl-nitrile rubber blend.

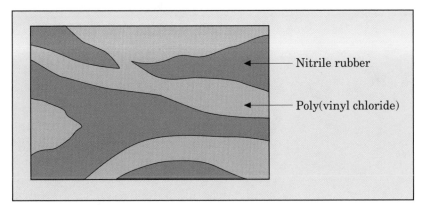

Nitrile rubber

Poly(vinyl chloride)

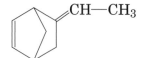

5-Ethylidene-2-norbornene (ENB)   *Dicyclopentadiene (DCPD)*

The structure of the terpolymer can be represented as

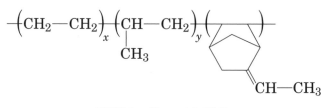

EPDM rubber with ENB

Natural rubber is a polymer of isoprene produced naturally by the rubber tree (Hevea brasiliensis) that is grown in Asia. The rubber exists as a **latex,** a dispersion of tiny droplets of rubber in the aqueous sap, which is collected and treated with acid to coagulate the rubber particles.

Natural rubber is a polymer of isoprene [cis-poly(2-methyl-1,3-butadiene)]. ("Cis" means that all of the methyl groups are on the same side of the chain.) Perhaps the rubber tree carries out the following reaction:

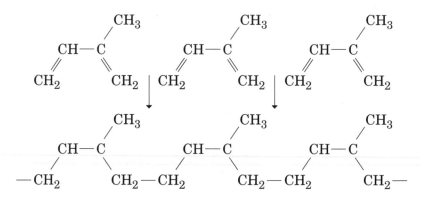

In any case, the resulting isoprene polymer molecule is tightly kinked and stretches easily. During vulcanization, disulfide cross-links are generated between these raw rubber molecules so that they will not break when they stretch. Adding 1–3% sulfur makes the rubber soft and stretchy like a rubber band; 3–10% makes it harder like automobile tires.

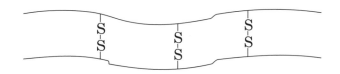

Cross-linked rubber chains

These rubbers can be vulcanized with sulfur, sulfur compounds, or with acrylates, all of which react with the pendant double bonds.

The properties of these rubbers depend on the type, amount, and distribution of the diene monomer, and they are greatly affected by the catalyst and the polymerization conditions. EPDM rubbers have excellent resistance to ozone in comparison to natural rubber or other synthetic rubbers. Furthermore, the EPDM rubbers can be extended with inexpensive fillers and plasticizers to give good performance at low cost.

Natural and synthetic rubbers are compounded with powdered sulfur and other materials (such as plasticizers and pigments). When this soft mixture is molded and cured under pressure for about 10 minutes at 300°F (150°C) in contact with a matrix, the rubber is **vulcanized,** becoming much harder. This is a polymerization reaction that involves cross-linking by sulfur. The rubber takes on the image from the matrix, in reverse, of course.

Butyl rubber is a copolymer of isoprene and isobutylene.

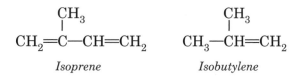

Rubber plates often lack the required uniformity of thickness, and they must be ground to give a variance no greater than 1.5 to 2.0 mils (thousandths of an inch) thus adding to the cost and time to prepare a rubber plate. Although duplicate plates are highly reproducible, they do not carry the fine detail that a photopolymer plate can carry. Consistency is most important if several copies of the same image are to be produced from a single plate. Molded rubber flexo plates are usually about 0.100–0.125 in. (2.5–3.2 mm) thick.

Advantages of rubber plates. Duplicate plates made from a mold or matrix are inexpensive. Rubber plates are easily joined to other plates on the cylinder. They provide better ink coverage of solids. They also conform more easily to the plate cylinder, and they are easily demounted and remounted. Corrections can be made by cutting and adhering corrected plate segments to the cylinder. These plates are often thicker than photopolymer plates, and they print better on rough surfaces such as corrugated board.

Disadvantages of rubber plates. Laser engraving of sleeves has overcome many of the problems of rubber plates—the complicated production procedure and cost of a metal engraving, followed by molding and curing and the long production time. Rubber plates are less stable than photopolymer plates, making four-color register difficult.

Mounting Flexo Plates

Flexo plates, photopolymer or rubber, are commonly mounted on the printing cylinders with "stickyback" tape, a two-sided pressure-sensitive adhesive tape that covers the entire back of the plate. The term ***stickyback*** generally applies to the thinner adhesive plate backings. Stickyback is a foamed polyurethane or polyester sheet coated with a polyurethane, acrylic, or synthetic rubber adhesive on both sides. A typical plate-mounting tape may contain many layers, as shown in figure 6-7.

The foam layer is a foamed polyurethane or polyester cushion, sold in rolls. On the roll, the release paper is on the top and outside. The adhesive, facing the center of the roll, is applied to the plate cylinder. The release paper is removed

Figure 6-7.
Cushion tape for mounting flexographic plates.

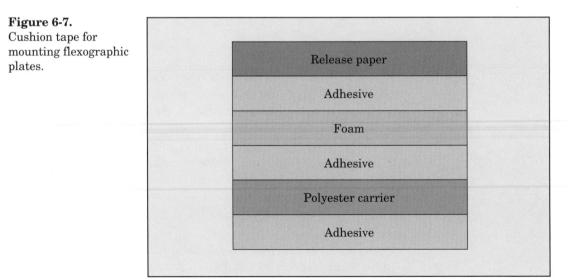

and the plate is attached to the cylinder. The thinnest sticky-back is 2–3 mils (0.050–0.075 mm) thick, and the plate cushions may be as much as 15–20 mils (0.38–0.50 mm) thick.

Pin-register systems are used with rubber and photopolymer plates.

Plates are sometimes mounted on a steel backing and held to a magnetized printing cylinder.

For publication flexo, the plates are clamped onto the press much like litho printing plates. For these applications, a steel or heavy polyester plate backing is required to avoid distortion by the clamps.

Photoengravings and Matrices (Matrixes)

The metal plates once used for letterpress printing are now used as engravings (or "pattern plates") for flexography. The chemistry of these plates is therefore discussed under flexography. Magnesium is preferred for line work, copper for halftone reproductions.

A mold or matrix is prepared from the engraving. Matrix material contains a filler and a binder. The filler is usually cellulose fiber plus a coloring material. The binder is a thermosetting phenolic resin. (Polypropylene is also used for matrices.)

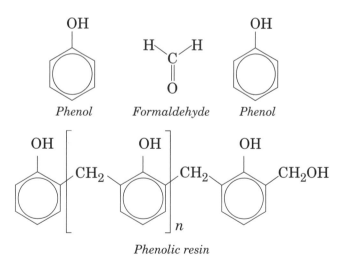

The matrix is formed by pressing the molding compound containing low-molecular-weight polymer against the engraving and baking at 300°F (150°C) for about 6–10 minutes to increase its molecular weight and to produce a hard mold or matrix.

Magnesium Photoengraving

Metal plates are supplied with a photosensitive coating applied by the metal fabricator, often a poly(vinyl cinnamate) coating.

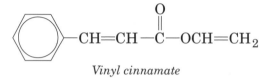

Vinyl cinnamate

Exposure to light through a negative hardens this coating in the image areas, and the unexposed part is removed by inserting the plate in a special trichloroethylene vapor degreaser.

Trichloroethylene

Magnesium plates are etched in machines equipped with paddles that throw the etching solution onto the plate. Etching solutions contain 14–22% by volume of 42° Baumé nitric acid, and the bath is maintained at 85–130°F (30–55°C).

Powderless etching is also used to engrave magnesium matrices. In this process, banking materials, added to the bath, form a film on the shoulders of the image areas to prevent undercutting as the metal is etched deeper (figure 6-8).

Several improvements have been made in additives for this process. Some of them contain five or more ingredients. The materials used are complicated organic compounds, such as

Figure 6-8.
Mechanism of powderless etching.

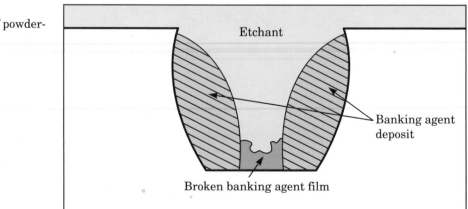

saturated fatty acids with eight to sixteen carbon atoms in their molecules, coupling agents, such as glycol ethers, and anionic surface-active agents.

The chemical reactions of magnesium etching are:

$$4Mg° + 2NO_3^- + 10H^+ \rightarrow 4Mg^{2+} + N_2O + 5H_2O$$

$$Mg° + 2NO_3^- + 4H^+ \rightarrow Mg^{2+} + 2NO_2 + 2H_2O$$

The solid magnesium is oxidized by the nitrate ions, NO_3^- to form magnesium ions, which are soluble in the etching solution. The nitrogen atoms of the NO_3^- ions are reduced to N_2O or NO_2. These reactions also consume the hydrogen ions present in HNO_3. Nitrogen dioxide (NO_2) is highly toxic and operators must not be exposed for extended periods of time to quantities over five parts per million. Therefore, etching machines are equipped with hoods and blowers to remove these vapors.

The question arises as to why the additives in the etching bath can prevent etching at the sidewall while allowing it to occur at the bottom of the etched cavity. One theory is that the force of the splashing etch removes the banking film from the bottom areas. This may be part of the answer, but the following differential heat theory is the explanation most widely accepted today.

Heat is evolved when magnesium is etched with nitric acid, and the additives do not adhere well to a hot surface. But the additives adhere to the cooler sidewalls and prevent further etching.

As magnesium plates are etched, nitric acid is consumed, and more must be added. At the same time, the concentration of magnesium ions in the bath keeps increasing. This increasing concentration creates a ***mass action effect,*** resulting in a decrease in the rate of etching. (As the reactants of a chemical reaction build up in the reaction vessel, they often retard the reaction.) Finally, the etching bath must be dumped. The proper disposal of spent etching baths is important because of local and federal regulations. The magnesium in the bath is not harmful—all it does is make water harder—but the acid must be neutralized by adding sodium hydroxide or sodium carbonate (soda ash or washing soda). Most sewer codes will accept effluents with a pH of 5.5–6.0. The magnesium precipitates from solution if the pH reaches 9.0.

Copper Photo-engraving and Etching

Copper surfaces are etched to produce matrices for forming flexo plates and to produce gravure cylinders. The light-sensitive coating for copper photoengravings is called a "hot top" enamel. Fish glue is still used for this coating. After exposure through a negative, the photoengravings are developed with water. This removes the coating from the unexposed areas. The remaining coating, the *resist,* is then set by heating to temperatures of 550–650°F (285–345°C).

Copper plates can also be presensitized with a positive-working coating. This coating works similarly to that of positive-working presensitized litho plates. The alkaline developer removes the exposed coating but does not remove the unexposed coating.

With the powderless etching process for copper, it is common to give a plate two etches—one for more-deeply-etched line work and another for the flat-etched halftone work. This etching is easy to accomplish with a presensitized positive-working coating.

Copper plates are descummed with dilute hydrochloric acid (HCl) prior to etching. This removes any cupric oxide (CuO), which is soluble in HCl. The molecular reaction is:

$$CuO + 2HCl \rightarrow CuCl_2 + H_2O$$

The plates are etched with a 28–45% solution of ferric chloride, $FeCl_3$. The ionic reactions are:

$$Fe^{3+} + Cu^\circ \rightarrow Fe^{2+} + Cu^+ \text{ (cuprous ion)}$$

$$Fe^{3+} + Cu^+ \rightarrow Fe^{2+} + Cu^{2+} \text{ (cupric ion)}$$

These are called *redox reactions* because oxidation and reduction are involved. In the first reaction, ferric ions are reduced to ferrous ions, and the copper is oxidized from metallic copper, Cu°, to cuprous ions. In the second reaction, the ferric ions oxidize the cuprous ions to cupric ions.

A *powderless etching* process for copper plates was developed by the International Prepress Association. The mechanism for the protection of sidewalls is somewhat different from that for magnesium plates. The additives to the etching bath are derivatives of thiourea. The prefix "thio" indicates that the molecules contain sulfur as a substitute for oxygen. The structural formulas for urea and thiourea are:

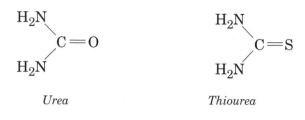

Urea   *Thiourea*

The etching bath additives react with cuprous ions, Cu^+, to form an insoluble film on the sidewalls. This film protects the sidewalls from further etching.

The disposal of spent etching baths presents problems. The baths can be neutralized with sodium hydroxide to precipitate out the copper, but this also precipitates out the iron as a mixture of hydrated iron oxides, which are gelatinous and very difficult to filter out. They rapidly plug most filter screens. The same problem is encountered in the treatment of spent etching baths from the production of copper gravure cylinders. The best answer at present is to arrange for a disposal company to collect such spent baths.

Methods of Preparing Gravure Cylinders

Gravure is an ***intaglio*** process. The image areas are lower than the surface of the chrome-coated copper cylinder. The old, complicated carbon tissue method of etching gravure cylinders is no longer used. A great deal of chemistry is involved in the chemical etching of the image, but most gravure cylinders are now imaged mechanically. Two of the most popular machines are illustrated in figures 6-9 and 6-10.

The term "etching" usually refers to the chemical generation of cells in the copper surface; "engraving" usually refers to mechanical generation of the cells. Ferric chloride etches copper as shown in the chemical equations shown on the facing page; diamond styli and laser beams engrave the surface.

Because most infrared light is reflected by copper, the powerful CO_2 laser cannot be used to engrave a copper cylinder, but laser-beam engraving of a gravure cylinders covered with a zinc alloy is under development.

Figure 6-9.
The K406 Helio-Klischograph gravure engraving machine. *(Courtesy of Hell Gravure Systems)*

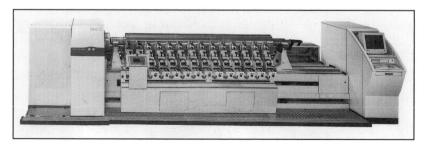

Figure 6-10.
Ohio M820 electronic
engraver.
*(Courtesy of Ohio
Electronic Engravers)*

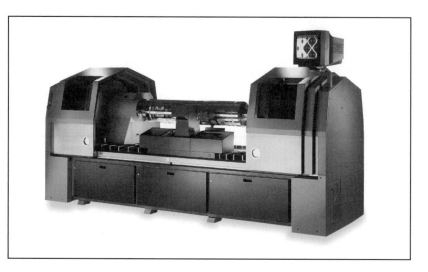

After etching and washing, or engraving, the cylinder is chrome-plated to give it a hard surface that resists wear. After use, when the cylinder is no longer needed, the cylinder must be dechromed, stripped of copper, and replated before a new image is applied. This costly process of cylinder preparation makes gravure uneconomical on short runs, unless gap-free images are required for items such as decorative wood grains or floor coverings.

The cost of preparing a gravure cylinder can be reduced significantly with the use of a ***Ballard shell,*** a layer of copper thick enough to carry the image. The shell is electroplated non-adhesively to the copper of the base cylinder, and engraved. When the cylinder is no longer needed, the Ballard shell is easily removed.

The Electronic Engraving Machine

The gravure cylinder is normally engraved with a diamond stylus controlled with digital information. The old photographic halftone prints (called "bromides" or "opaques"), once used to control mechanical engraving have been replaced with the computer. Two of the most popular electronic engraving machines are illustrated in figures 6-9 and 6-10.

Electronic engravers eliminate the need for bromide prints. The digital output of a color scanner is fed into a computerized imaging system for storage. In this step, originals can be enlarged or reduced and cropped as desired. A page layout is made, and information about where each page is to be placed on the printing cylinder is also fed into the computer, and pages are assembled. All of this information then flows electronically to the unit that operates the diamond stylus.

A good gravure cylinder requires that the copper have a uniform hardness across and around the entire cylinder. For the diamond stylus to cut properly, it is necessary to harden the copper by the use of proper electroplating equipment and procedures and certain additives in the plating bath. If these requirements are not met, the cells may have burrs, or the copper may be torn instead of cut.

Chemical Methods

Etching produces images with subtle differences from engraved images. Etched type is said to have a "smoother" edge. Etching processes in current use require coating the cylinder with a material that resists etching and removing the image area with a laser. Two methods are in use. In one method, a photo-resist is applied to the gravure cylinder and exposed with a digitally driven laser. The unexposed coating is removed and the cylinder is etched. In the other system, a black coating is applied, and the image area is removed with a laser.

Electroplating the Gravure Cylinder

To assure long life, the soft copper of the etched or engraved cylinder is electroplated with chromium, which provides a hard, wear-resistant finish. The first step in etching is to clean and carefully rinse the copper cylinder. Good adhesion of the coating requires that the copper surface be clean and free from grease, oil, and copper oxides. The cylinder is made the cathode (the electrode that bears a negative charge) in a plating bath. The coating is 0.2–0.3 mils (0.0002–0.0003 in., 0.005–0.0075 mm) thick for publication printing, twice as thick for package printing. The cylinder is rotated during coating to assure a uniform coating, and a heavy current deposits chromium atoms on the surface (figure 6-11.)

Figure 6-11.
Electroplating a gravure cylinder.

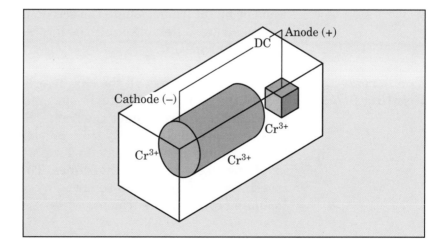

The reaction at the cathode is:

$$Cr^{6+} + 6e^- \rightarrow Cr^\circ$$

The reactions at the anode can be summarized as:

$$3H_2O \rightarrow 6H^+ + 3O_2 + 6e^-$$

The electrolytes are sulfuric acid and chromic acid. Chromium trioxide (chromic anhydride) is readily soluble in water, forming chromic acid.

$$CrO_3 + H_2O \leftrightarrow H_2CrO_4 \leftrightarrow H^+ + CrO_4^-$$

To achieve a useful chrome coating, it is important to control the temperature (110–115°F, 43–46°C) and the current density.

Numerous proprietary additives are used to control the reaction and to provide a useful coating. The anode is an inert metal, usually lead or titanium.

Smoothness of the coating is a complicated subject, but the chrome surface must be neither too smooth nor too rough. Hydrides are always present in the coating. They produce microcracks, which provide "lubrication" and prevent streaking of the print.

When the chrome begins to wear, the cylinder can be reclaimed by dechroming in concentrated hydrochloric acid. Chrome is soluble in hydrochloric acid; copper is not.

Screen Printing

Screen printing, as practiced for hundreds of years, depends on a stencil attached to a piece of fabric stretched over a wood or metal frame. While the screen is held in contact with the substrate, a flexible squeegee forces the ink through the stencil (figure 6-12).

The Screen Printing Process

The screen frame has all the characteristics of a printing press—the plate controls the image on the substrate, which is fed and registered under the frame. Screen printing presses range in complexity from an elementary homemade, hand-operated wooden frame to a sophisticated system involving steel or aluminum frames. These modern machines insert, register, print, and remove any printed products, including cylindrical surfaces.

Figure 6-12.
Ink being poured into a screen printing frame *(left)* and the printed result *(right)*.

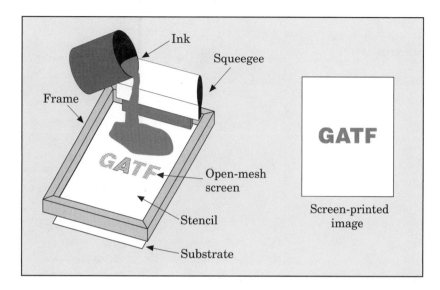

Screen printing was long a hand-operated process. Although it is still often done this way, automation has made screen printing into an important production process. Automatic screen printing presses print on flat stock, and rotary screen presses print webs of paper or plastic.

The rotary screen press (figure 6-13) uses a cylindrical screen. The stencil is applied to the surface of the cylinder, and the ink is forced through the metal screen, from within the cylinder, by a rubber or metal squeegee. Large rolls of fabric (yard goods) as well as wallpaper type products are printed with water-based and solvent-based systems by the ***rotary screen*** process.

Screen printing is the most versatile of all printing processes. It is used for printing paper, paperboard, wood, glass,

Figure 6-13.
Basic principle of a rotary screen printing press.

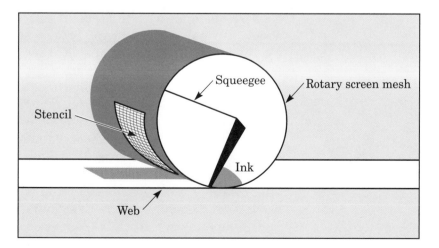

Figure 6-14.
Semiautomatic screen
printing press.
*(Courtesy A.W.T.
World Trade, Inc.)*

metals, plastics, circuit boards, textiles, ceramic products,
and leather. The printing is done not only on flat surfaces
but also on round, convex, concave, and irregular shapes.
One of the most popular uses of screen printing, the decoration of textiles, is done on all types of fabrics, woven, knitted,
and felt, because the surface of the substrate is not critical to
the generation of a good image. Screen printing offers the
widest range of ink deposit thicknesses of any printing
process (from films not much greater than lithographic ink to
films several mils in thickness).

Thick films of ink are particularly important in providing
weather resistance for outdoor advertising, and in producing
good permanence with fluorescent colors. On the other hand,
screen is capable of producing process color printing that
requires thin films of ink capable of reproducing halftone
line counts up to 150 lines per inch (59 lines/cm).

Screen printing was originally called ***silk screen printing,***
because the early screens consisted of silk bolting cloth.

The screen is usually made of stainless steel wire, nylon
monofilament, or polyester monofilament attached to a steel,
aluminum, or wooden frame. Multifilament is difficult to
clean, and it tends to swell, retaining ink and stencil material
after cleaning. This is a major reason that silk is no longer
used. With monofilament, each strand is a single fiber. With
multifilament, each strand is composed of many small fibers.

Figure 6-15.
Advantage in-line,
multicolor, UV
print/cure system for
screen printing.
*(Courtesy M&R Sales
and Service, Inc.)*

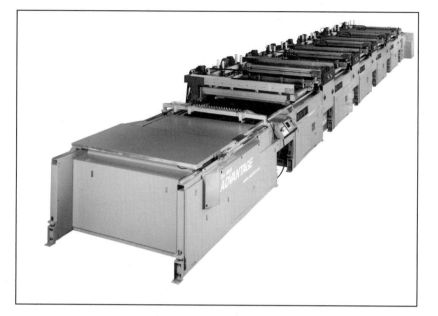

A *squeegee* (figure 6-12) forces ink through the stencil, which is supported on the screen. The squeegee has a blade of neoprene (polychloroprene) or polyurethane rubber supported in a handle. The screen may have a mesh as fine as 120 openings per centimeter (300/in.). Polyurethane rubber is prepared from a diisocyanate such as TDI and a glycol such as ethylene glycol.

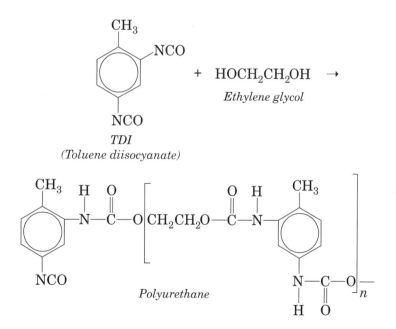

The most common form of nylon is "6,6 nylon" but others are also manufactured.

$$H_2N{-}(CH_2)_x{-}NH_2 + HOOC(CH_2)_yCOOH \rightarrow$$

1,6-Hexane diamine
(x = 6)

Adipic acid
(y = 4)

$$-NH{-}(CH_2)_x{-}NHCO(CH_2)_yCO-$$

Nylon

The most common polyester is poly(ethylene terephthalate) (PET):

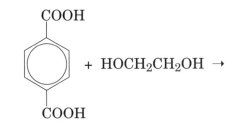

Terephthalic acid   *Ethylene glycol*

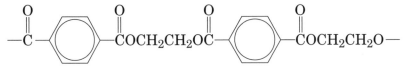

Poly(ethylene terephthalate)

Creating the Stencil

Images in the stencil are made in a variety of methods. The old methods are still used by some artists. In one, the design is cut into a film supported on a base sheet. The film and base sheet are adhered to the bottom of the screen, and the base sheet is removed.

Hand-generated stencils are also produced from a tusche-and-glue process. The ***tusche,*** which is a pigmented drying oil, much like a lithographic ink*, is painted onto the screen by hand to form the image. The nonimage areas are then filled in with glue, and the tusche is washed out with a solvent.

Today, most stencils are made photographically using one of four methods: direct, direct-indirect, indirect, and capillary

---

*The *Merriam Webster Unabridged Dictionary* gives the following formula for tusche: melt together yellow wax, 2 parts; mutton tallow, 2 parts; Marseilles soap, 6 parts; shellac, 3 parts; and lamp black 1–2 parts.

direct film (CDF). Similar chemistry is involved in each method. The main difference is how the coatings are applied to the screen.

In the ***direct method*** (used as the primary method by a large percentage of screen printers) a light-sensitive emulsion is applied to both sides of the screen, first onto the print side, then onto the squeegee side. The best quality direct-emulsion stencils require multiple coatings, applied with horizontal drying between coatings. The emulsion is often a mixture of poly(vinyl acetate) and poly(vinyl alcohol), and a dye is added to make the stencil readily visible on the screen.

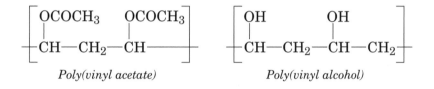

Poly(vinyl acetate)   *Poly(vinyl alcohol)*

During the mixing process, the emulsion is sensitized with a diazo compound (such as 4-diazodiphenylamine sulfate/formaldehyde condensation resin). The photosensitized emulsion remains stable for up to three months at room temperature or up to six months in the refrigerator. Diazo sensitizers are biodegradable and relatively nontoxic. Ammonium dichromate ("bichromate"), $(NH_4)_2Cr_2O_7$, formerly used, is highly toxic and nonbiodegradable. It is now replaced with diazo chemistry.

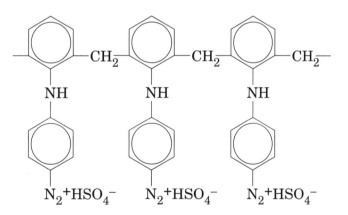

Resin from 4-diazodiphenylamine sulfate and formaldehyde

(Note the CH_2 or methylene bridges joining the units. These come from the formaldehyde.)

There are also "diazo-acrylic" or "dual-cure" emulsions; these use a light-sensitive acrylic monomer in the base emulsion to which separate diazo sensitizer is added. The term "dual-cure" is used because both components are acted upon by light energy.

In the ***direct-indirect method,*** a sheet of polyester film is coated on one side with an emulsion, often with a mixture of poly(vinyl alcohol) and poly(vinyl acetate), which is not light sensitive. The screen is placed on top of the film, coated-side up. Then a sensitized emulsion (the same as is used in the direct method) is squeegeed over the top of the screen. This emulsion sensitizes the coated film underneath and also binds it to the screen.

When the emulsions on both sides are thoroughly dry, the polyester film backing sheet is removed. An exposure is made through a film positive, the same as with the direct method, and the unexposed film is dissolved in water and washed away.

The advantage of the direct-indirect method over the direct method is that the stencil on the print side is uniform in thickness and only one coat is required on the top side. The screens print longer than if made with the indirect method, but not as long as with the direct method.

In the ***indirect method,*** most of the steps outlined above are performed on a special film before it is applied to the screen. These films are of various types; one of them consists of a thin backing of cellulose acetate or vinyl film coated with a ferric salt sensitizer in a gelatin base.

Exposure hardens the gelatin in the nonprinting areas as defined by the transparent areas of the film positive. The film is placed in a bath of 2% hydrogen peroxide (H_2O_2) for approximately one minute and developed with warm water to remove the unexposed gelatin from the image areas. A cold-water treatment hardens the remaining gelatin stencil.

The wet gelatin film is placed, emulsion side up, in contact with the print side of the screen. Blotting excess moisture from the screen using clean (unprinted) newsprint and a soft roller causes the gelatin to adhere to the screen. After thorough drying, the thin backing is removed, leaving the gelatin stencil attached to the screen. The screen is ready for printing.

The original knifecut films (solvent-adhered) are still manufactured for use in the indirect method. These are a lacquer-nitrocellulose blend. A newer, water-adhered variety is a blend of poly(vinyl alcohol) and poly(vinyl acetate).

Virtually all indirect photographic films are gelatin-based. At one time, these were a simple adaptation of carbon tissue (used in gravure) and required cumbersome sensitizing with dangerous bichromates. Now, the gelatin-based films are "presensitized" and use the safe iron-salt sensitizing system. There are a few non-gelatin "no-developer" indirect system photographic films that have poly(vinyl acetate)–poly(vinyl alcohol) bases and use diazo, pure photopolymer, or leuco-indigo dye as the light-sensitive systems.

The fourth stencil system, and the most recent, ***capillary direct film*** (CDF), combines long run lengths of direct emulsion and the sharp-edge resolving power of indirect films with one added feature. The ability to generate several thicknesses of emulsions for various mesh counts from very coarse to extremely fine in one application provides a substantial reduction in coating time.

A capillary direct film is a uniform coating of direct emulsion applied to a polyester film backing, very similar to indirect films. Capillary films are placed in contact with a clean, degreased screen (print side), and the dry emulsion is attracted into the wet fabric by capillary action. After the film has thoroughly dried, the polyester film backing is removed, a film positive is placed in contact with the stencil, and a normal exposure is made. For exceptionally long run lengths (300,000–400,000 impressions), the capillary film can be strengthened by encapsulating the fabric with a coating of direct emulsion on the squeegee side. This is similar to the direct/indirect method with the following exceptions:

- Capillary films are presensitized; direct/indirect films are not.
- Capillary films adhere to the fabric by water through capillary attraction.
- Capillary film emulsions embed deeply into the fabric for superior adhesion; direct/indirect remains fragile with limited run lengths, a maximum of 50,000 to 150,000 impressions.

Like emulsions, capillary direct films (CDF) use poly(vinyl acetate)–poly(vinyl alcohol) bases and use the same three light-sensitive systems. With films, the expertise of the film coater minimizes the need for expertise on the part of the stencil maker. Solvent- and water-based are the two broad categories. The type of ink determines the selection of the stencil chemistry. Water-based inks must be used with solvent-adhered films. Solvent-based inks are used with water-adhered stencil films, but many types of stencil are compatible with solvent-adhered films also. Gelatin-based and no-developer indirect system photographic films are water-soluble and cannot be used with water-based or water-containing inks.

Some diazo-sensitized emulsions are compatible with both solvent- and water-based inks. Pure photopolymer and diazo-sensitized capillary films are compatible with solvent-based inks only. "Dual-cure" capillary films can be used with most solvent-based and water-based inks. They are also ideal for inks containing both solvent and water components such as water-based UV-cured inks.

Diazo photopolymer emulsions or "dual-cure" diazo/acrylic emulsions feature faster exposure speeds, exceptional edge definition, and solvent- and water-resistance for water-based ink systems. The dual-cure system is more expensive than other systems, but the ability to use other ink chemistries and to take advantage of the other qualities listed above frequently justify the cost.

The actual composition of these systems depends on the intended use; different formulas are used for water-based and for solvent-based inks.

Substrates and Ink for Screen Printing

Paper and ink for screen printing are discussed in chapters 7 and 8. The nature of the screen printing process makes it possible to print on almost any material, and the thick ink film covers most surfaces very well, giving the printer a wide choice of substrates suitable for printing.

The ink must be appropriate for its intended use: it must adhere to the substrate and meet end use requirements. The screen fabric and the imaging material must be appropriate for the ink.

Inks used for screen printing are as varied as the surfaces they are printed on. The principal categories are water-based, evaporative, oxidizing, catalytic, and UV. Aqueous inks dried by radio-frequency radiation also provide opportunities for high-speed production of screen-decorated prints. UV technology, combined with continuous web printing, has permitted a rapid growth in the volume of screen printing in recent years.

Screen inks must be very short so that threads of ink do not form to ruin the image as the screen is removed from the print.

Letterpress Plates

Letterpress has largely been replaced with lithography in the job shop, lithography and flexography in newspapers, gravure and lithography for long-run publications, and lithography and flexography in package printing. Letterpress is a popular method for printing labels in rolls, and plastic letterpress plates are used with automatic book manufacturing systems.

Letterpress and flexography are very different systems. Although both print from relief (raised image) plates, they differ in the inks and printing presses used and the hardness of the plates. The hardness of flexo plates is measured in terms of Shore A hardness, while letterpress plates are measured on the Shore D scale. The demise of letterpress can be attributed to the difficulties of preparing halftone plates and the long makeready times required on press.

Photopolymer plates have replaced the old metal plates, overcoming the lengthy procedure of etching a halftone into metal, overcoming the handling of heavy metal plates, and avoiding working with the hot molten type metal.

Photopolymer Letterpress Plates

Photosensitive plastic plates are used to produce letterpress relief plates. These are called ***wraparound plates*** and are made in one piece to be wrapped around the plate cylinder of the press. They range from 17 to 30 mils (0.017 to 0.030 in. or 0.43 to 0.76 mm). Photopolymer letterpress plates have been used on offset presses in an offset letterpress system sometimes called ***dry offset*** or ***letterset,*** but even this process has gradually disappeared.

Photopolymer plates are prepared from a photosensitive resin or sometimes a combination of two or more such resins. These are compounded with a photoinitiator and other additives. Exposing the plates to UV light through a negative, polymerizes the resin molecules and hardens image area of the plate. The resin molecules combine with each other, both by joining end-to-end to form chains and by cross-linking between the chains. Exposure to light not only hardens the resin but makes it insoluble in the washout solution.

Various light-sensitive resins are used, depending on the type of plate. One plate system involves a mixture of urethane-based polyene and a tetrathiol.

$$C(CH_2OH)_4 \ + \ HS-CH_2-COOH \ \rightarrow \ C(CH_2O\overset{\overset{\displaystyle O}{\displaystyle \|}}{C}CH_2SH)_4$$

Pentaerythritol   *Thioglycolic acid*   *Pentaerythritol tetrakis-thioglycolate*

The urethane polyenes are complicated substances with this feature in common: they all have carbon-to-carbon double bonds so situated in the molecules that they can easily link together when these double bonds are opened by irradiation. Each thiol group is converted to a free radical by a photo-initiator, and it both cross-links the polyene and forms additional free radicals. The extensive cross-linking produces the hard resin required of the letterpress plate.

Other letterpress plates are produced from polyamides, poly(vinyl alcohol), and other low-molecular-weight polymers that contain carbon-to-carbon double bonds. Plates often contain a small amount of a plasticizer to ensure plate flexibility.

Such systems may be oxygen-sensitive; that is, oxygen can terminate the polymerization process. Sensitive plates are stored in an atmosphere of carbon dioxide (CO_2) to replace the oxygen on the surface of the plates.

Another method of reducing oxygen is to pre-expose these plates through a green-colored filter sheet. This sheet removes most of the ultraviolet light that is responsible for polymerization, but allows some of the monomer to combine with the oxygen present. Then the filter is removed, and the plate is exposed as usual through a negative. With thinner versions of these plates, no carbon dioxide storage or filter pre-exposure is needed.

Following exposure, the nonimage areas of the plates are washed out with an alkaline aqueous solution. One washout solution consists of alcohol and water. These solutions are formulated to dissolve the unhardened parts of the plate.

Liquid photopolymer letterpress plates can be developed by exposing the heavy liquid to UV light, then blowing away the unexposed nonimage area, just like liquid photopolymer flexo plates.

Printing Impression

Unlike the liquid flexo inks, letterpress inks are stiff and pasty (like lithographic inks), and (like lithography) they require a complicated inking system to work and distribute them on the press. Flexography prints with a "kiss" impression (minimal impression) while letterpress requires significant pressure to transfer the stiff, pasty inks. The pressure is commonly measured as "thousandths of squeeze" or the actual distance (in inches) that the plate bites into the substrate and the impression cylinder. With letterpress this amounts to 0.002–0.003 in. (0.050–0.075 mm), while with flexo it is kept as close to 0.000 as the precision of the system will allow.

Type Metal

Much interesting chemistry of the lead (Pb), antimony (Sb), tin (Sn) alloy has gone by the board with the death of type metal for letterpress printing. The alloy has the curious property of expanding on solidifying so that the stereotype is fully filled with type metal, faithfully reproducing serifs and other small details.

Inkjet Printing

Chemical interest in inkjet printing has been focused mostly on the inks (see chapter 8), but advanced metallurgy is involved in forming the print heads. Techniques for making an image with ink droplets are conveniently divided into continuous jet and impulse jet (drop-on-demand or DOD). Two techniques are used for printing with continuous inkjet printers—the *Sweet*-type device and the *Hertz*-type device.

The Sweet-type printer creates a continuous jet of uniform droplets and guides the droplets to the desired position on the substrate by controlling the charge on the droplets (figure 6-16) or by altering the electrical field through which they pass. It is also possible to reverse this technique and print with uncharged droplets, deflecting and collecting the charged droplets.

The Hertz-type printer uses an electrode to deflect charged drops onto an aperture plate from which ink can be recycled. Uncharged drops pass through the aperture plate to create an image on the substrate. This technique does not require that all of the droplets be of uniform size.

Figure 6-16.
Principle of a continuous-inkjet (CIJ) printer. *(Courtesy Marconi Data Systems, Inc.)*

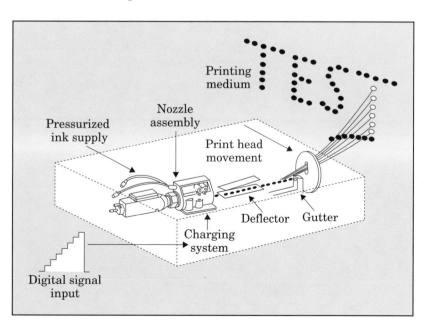

Figure 6-17.
Drop-on-demand ink-jet printers:
A. Principle of piezo-electric DOD printer
B. Principle of bubble-jet DOD printer
C. Principle of hot-melt DOD printer
(Courtesy Pira International)

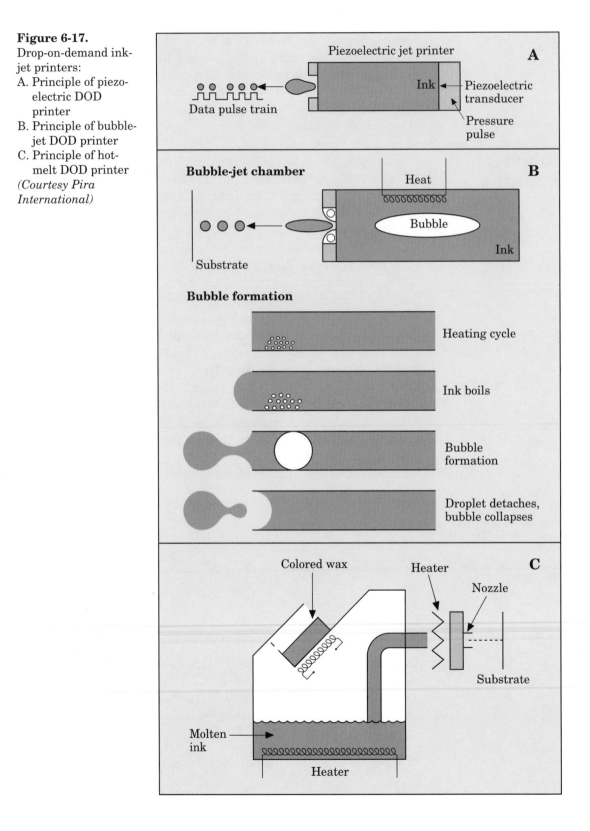

Three major techniques for impulse or DOD printers are illustrated in figure 6-17: the piezoelectric, thermal bubble-jet, and hot-melt printers.

In the **piezoelectric printer,** materials such as barium titanate or lead zirconate titanate change their dimensions when exposed to an electric field. The charge expels an ink droplet from the ink chamber.

In the **thermal bubble-jet printer,** materials such as HfB_2, ZrB_2, Ta_2Al, and TaN use an electrical impulse to generate heat and form a bubble, expelling an ink droplet. The number and timing of the droplets are controlled by a computer.

The **hot-melt printer** works in a similar manner, but the ink is supplied as a solid wax and melted in the printer to supply the ink for the droplets. This ink is not absorbed by a paper substrate, and it is capable of producing brilliant color when it hardens on the surface of the sheet.

An additional DOD technology, **spark jet,** ejects a droplet of ink as a result of the high temperature and pressure generated by a spark or electrical surge.

The drastic conditions that generate thermal bubbles require not only materials that resist heat and erosion, but materials that can be shaped with great precision. Processes for shaping inkjet printing heads include electroforming and photolithographic etching of silicon.

Heat Transfer Printing

Heat transfer printing is not a basic printing process. The printing is done by gravure, flexography, screen printing, or lithography or letterpress. It is what happens to the printed paper after printing that makes heat transfer a special process.

There are two general methods used in heat transfer printing—melt transfer and dry transfer. In the **melt transfer** method, dyes or pigments are incorporated into a thermoplastic binder, and this colored binder (ink) is printed on paper. When the paper is brought into contact with a textile fabric, and heat and pressure are applied, the entire binder film transfers to the fabric. This method is used for the printing of cotton and other fabrics. If the ink contains dyes, the dyeing of the fabric is completed by conventional dye fixation methods. If pigments are used (decalcomania process), no fiber dyeing occurs. The pigments are held on the fiber surface by means of the binder in the ink.

The heat transfer inks look flat and dull when printed on paper. When they are transferred to the fabric, the colors become bright and clear. There must be close cooperation between the printer and the textile firm to achieve the desired results for a particular fabric.

The *dye transfer* method is also known as the sublimation transfer process. The following discussion is concerned with this method, and any further reference to heat transfer printing implies dry transfer.

In the dye transfer process, paper is printed with inks that contain sublimable dyes. Any material, including a dye, is said to sublime when, upon heating, it changes from solid to vapor without going through the liquid stage. The printing can be in a single color, in multiple colors, or in process color. The printed paper is placed in contact with the fabric. As heat and pressure are applied to the back of the paper, the dyes in the ink sublime, condense on the fiber surface, and then diffuse into the interior of the fiber.

As heat is applied, the dyes begin to sublime into the narrow space between the printed sheet and the fabric. The air becomes supersaturated with the dye vapor, and the vapor condenses on the surface of the fabric. Then the dyes begin to diffuse into the interior of the fabric. The efficiency of dye transfer from the ink on the paper to the fabric depends on the pressure applied, the temperature, the dwell time, and the shade or color.

The efficiency increases as the pressure increases, up to a certain value; then further increase in pressure has little effect. The efficiency also increases as temperature and dwell time increase. For certain dyes, the optimum dwell time ranges from 30 sec. at 410°F (210°C) to 20 sec. at 445°F (230°C). The color transfer efficiency varies from about 90–95% for light shades to about 40–60% for heavy shades.

Suitable Fabrics   Heat transfer printing is not suitable for cotton or wool. It can be used on certain synthetic fibers but not on others. The dyes must be heated to at least 390°F (200°C) to obtain adequate sublimation, and this is above the softening point of some synthetic fibers. The softening and melting points of several common fibers are shown in table 6-II.

Table 6-II makes it apparent that cellulose acetate and nylon 6 cannot withstand the required transfer temperatures. Triacetate, nylon 66, and acrylic could be used, but they exhibit poor washfastness when printed by heat transfer. Furthermore, nylon 66 and acrylic do not give colors as

Table 6-II.
The softening and melting points of common synthetic fibers.

	Softening Point		Melting Point	
	°F	°C	°F	°C
Cellulose acetate	375	190	450	230
Cellulose triacetate	435	225	575	300
Nylon 6	340	170	420	215
Nylon 66	455	235	480	250
Acrylic	420–490	215–255	—	—
"Qiana" nylon	about 445	about 230	535	280
Polyester	445–465	230–240	495	255

bright as can be obtained by the direct printing of acid dyes on nylon 66 and of cationic dyes on acrylic.

"Qiana" nylon is suitable for the heat transfer printing of light to medium shades of color. Polyester has none of the disadvantages mentioned; it is the most suitable fiber for printing by this method.

Suitable Dyes

The sublimable disperse dyes are mostly primary, secondary, or tertiary amines of three main types: amino-azobenzene, amino-anthraquinone, and nitro-diaryl-amine. They sublime in a relatively narrow range: 390–420°F (200–215°C). Dyes are available that sublime at lower temperatures, but they generally have poor washability and lightfastness. Dyes that sublime at higher temperatures cannot be used because the temperatures exceed the softening point of the synthetic fibers.

Printing Processes for Heat Transfer Printing

Heat transfer printing on paper of seamless patterns for transfer to wide textile rolls is mainly done by rotogravure. It is also done sometimes on rotary screen printing presses, and in increasing amounts by flexography.

Printing on sheets of paper is carried out largely on sheetfed lithographic presses. These sheets are used to make polyester fashion shirts and pictorial T-shirts. A high percentage of such printing uses process color.

Gravure inks contain fast-drying solvents that make it possible to print heat-transfer gravure inks on relatively nonporous paper. As a result, the dye crystals stay mostly on the surface of the paper, producing maximum dye transfer to the fabric.

Heat transfer inks for flexography and rotary screen printing are often water-based. Such inks require a more porous paper to prevent puckering and to avoid smearing in multicolor printing. In this case, more dye is absorbed into the paper, resulting in somewhat poorer color transfer to the fabric.

Other Products

Besides synthetic textiles, heat transfer printing is used for decoration of polyester-surfaced table and counter tops, polyester lacquer-coated metals, wall hangings, and other specialties. Heat transfer accounts for only a small fraction of the total textile printing market.

Melt transfer is used to prepare novelties—printed papers that are sold or distributed to people who can apply the pattern to a textile with a flat iron.

Collotype

Collotype is a continuous-tone process. It has the advantage of high resolution, but it lacks reproducibility. Indeed, every print varies a little, and if the operator does not keep close watch on the press, variation in color can be considerable. No dampening is applied during printing, and the presses are direct printing.

Collotype printing has been largely replaced by fine screen or screenless lithography. The process is labor-intensive, it suffers lack of science or technology, and it lacks reproducibility. Nevertheless, the process is technically interesting.

In collotype, gelatin sensitized with dichromate is exposed through a continuous-tone negative. The gelatin is hardened in relation to the amount of ultraviolet light received. When the plate is printed under the proper conditions, it accepts ink in inverse proportion to the amount of hardening of the gelatin. Thus a continuous-tone print is obtained when the ink is transferred to paper. Actually, the gelatin is not strictly continuous-tone; under a microscope, fine cracks or reticulations in the gelatin are visible.

Originally, the sensitized gelatin coating was applied to a lithographic stone. In modern collotype, the gelatin coating is applied to a grained aluminum plate. The gelatin solution can be sensitized either with ammonium dichromate, $(NH_4)_2Cr_2O_7$, for a fast emulsion, or potassium dichromate, $K_2Cr_2O_7$, for a slow one. A mixture of the two sensitizers makes it possible to control the required length of the exposure. Besides gelatin and sensitizers, the coating solution may contain chrome alum or glyoxal, $O=CH-CH=O$, as hardeners.

Exposure time through a continuous-tone negative is critical. It depends on the light source used and the speed of the coating. The gelatin must be exposed correctly in each tonal area. The negatives used have a density range of about 1.1–1.3 and a gamma value close to 1.0.

After exposure, the plate is immersed in water or an alcohol-water mixture at about 50°F (10°C). This immersion removes the dichromates, making the coating no longer light-sensitive. Then the plate is dried and stored for a day, which toughens the gelatin to increase the plate life on press.

Before printing, the plate is soaked for about 25 minutes in a tank containing a mixture of three parts water to one part glycerol or ethylene glycol. On the press, it is treated with a mixture of glycerol and formaldehyde. The formaldehyde tans the gelatin, giving the plate the oleophilic nature that it needs to hold ink. (Be cautious in handling formaldehyde; it has been shown to cause cancer.)

Water swells the gelatin. The more it swells, the more hydrophilic and the less oleophilic it becomes. During printing, a balance must be maintained to keep the gelatin swollen just the right amount. If too little water is present, the plate takes too much ink. If too much water is present, the plate takes too little ink and the prints look washed out.

During printing, the paper gradually removes water from the plate. Therefore, the water must be replenished at intervals. The glycerin helps to hold water on the plate. If the relative humidity of the pressroom air is 40–50%, the glycerol absorbs moisture from the air, maintaining the water level on the plate. A multiple-spray system can also add water containing a little glycerol to the plate.

Inks for collotype are usually made with drying-oil varnishes, often of the alkyd type, with a tack much higher than sheetfed lithographic inks. It is essential that their pigments be very finely ground. There is a wide choice of pigments, the only restriction being that they must not bleed into the water-glycerol solution. A manganese drier is commonly used; a cobalt drier is occasionally employed.

Papers for collotype must have the proper moisture content and have a pH close to neutral. High pick resistance is needed to withstand the pull of the smooth gelatin, particularly in nonprinting areas. Pick resistance is not quite so important when the printing is done on an offset press. Uncoated paper must be used for direct printing; a supercalendered paper is best. Coated paper can be printed if an offset press is used.

Collotype is used for short runs of from 100 to 5,000 full-color copies with an image size up to about 40×60 in. (1.0×1.5 m). It is used to reproduce fine-art paintings and to produce posters, murals, banners, point-of-purchase displays, maps, ad blowups, etc. Translites for light-box displays are produced by printing both sides simultaneously on opaline paper or vinyl sheets.

Screenless lithography (chapter 5) produces continuous-tone images, and it has virtually replaced collotype. Dot gain is of maximum significance with fine-screen printing, and a small gain can quickly plug the shadow areas of a print.

Thermography

Thermography is a method of producing ink with a raised image using an offset press. The raised image gives the appearance of engraved printing at a fraction of the cost. It is useful, for example, in business cards or stationery.

The principle is simple. A heavy paper or light board is printed with a lithographic ink, and the wet ink is heavily dusted with a thermoplastic powder. The product is placed in an oven at the appropriate temperature, which melts the powder into the ink, producing a raised image of the color of the ink.

Further Reading

Adams, J. M., D. D. Faux, and L. I. Rieber. *Printing Technology.* 4th ed. (656 pp) 1991 (Delmar Publishers, Albany, N.Y.).

Crouch, J. Page. *Flexography Primer.* 2nd ed. (192 pp) 1998 (Graphic Arts Technical Foundation, Sewickley, Pa.).

Cusdin, George. *Flexography, Principles and Practices.* 5th ed. 2000 (Foundation of the Flexographic Technical Association, Ronkonkoma, N.Y.).

Ingram, Samuel T. *Screen Printing Primer.* 2nd ed. (156 pp) 1997 (Graphic Arts Technical Foundation, Sewickley, Pa.).

"Inkjet Printing." *Ullmann's Encyclopedia of Industrial Technology.* 5th ed. 1989. Vol. A-13, pp. 588–637.

Kasunich, Cheryl L. *Gravure Primer.* (112 pp) 1997 (Graphic Arts Technical Foundation, Sewickley, Pa.).

Odiotti, M.E., V.I. Colaprico, and E.K. Gillett III. *Gravure Process and Technology.* (490 pp) 1991 (Gravure Association of America, Rochester, N.Y.).

Thompson, Bob. *Printing Materials: Science and Technology.* (567 pp) 1998 (Pira International, Leatherhead, Surrey, U.K.).

7   Chemistry of Paper

Introduction

Papermaking is an ancient art. It was invented by the Chinese about A.D. 100. At first, it was used only for handwriting, but after the invention of printing, paper was also used for printing. Paper was first made by hand. The practice was costly and time-consuming, making paper a luxury item for the rich. The paper machine was invented early in the nineteenth century; since then, hand methods have almost disappeared. Now, many different papers are produced, designed specifically for lithography, gravure, flexography, and for other printing processes, duplicators, copiers, and for packaging. Some paper is made from cotton linters, rags, or other cellulosic or synthetic fibers, but about 98% is produced from various kinds of wood or recycled wood fibers. Straw, kenaf, and bagasse (sugar cane fiber) are sometimes used in making paper or paperboard.

Wood Chemistry

Wood pulp is produced from two types of trees, deciduous or hardwood trees that yield short fibers (1.0–1.5 mm long) and from coniferous or softwood trees that yield fibers about three times as long (figure 7-1). Short fibers give smoothness

Figure 7-1.
Two types of trees.

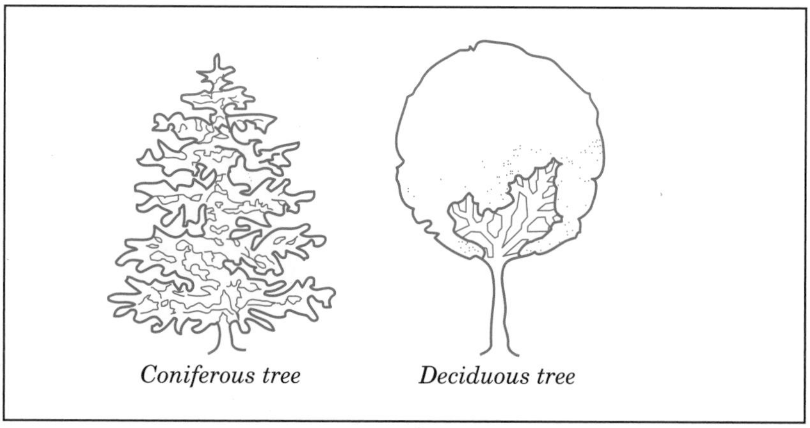

Coniferous tree   *Deciduous tree*

and opacity to a sheet while the long fibers contribute
strength. Accordingly, printing papers are made from a furnish containing both hardwood and softwood pulps.

The trunk (or stem or bole) of the tree, from which timber
and pulpwood chips are obtained, contains about 50% cellulose, around 25% lignin (a polymeric, phenolic compound
that binds the cellulose fibers together in the wood), and
around 25% hemicellulose, carbohydrate polymers that contribute strength and stiffness to paper. Hemicelluloses,
however, are more or less soluble in alkali, and they are
largely lost during alkaline pulping.

Wood also contains a small amount of mineral salts, which
yield ash when wood is burned, and a little resin, or pitch;
the percentage varies considerably, depending on the species
of wood.

The bark and the leaves of the tree contain cellulose and a
wide variety of other chemicals. To make paper pulp, the
bark is removed, and the wood is converted to paper pulp. (In
whole-wood chipping, the entire tree is chipped, and the bark
is dissolved during pulping. This increases the chemical
requirements of pulping but also yields more cellulose.)

Bonding the cellulose molecules in the paper fiber is
another polymer called lignin. It is a complex polymer of
coniferyl alcohol, p-coumaryl alcohol, and sinapyl alcohol.

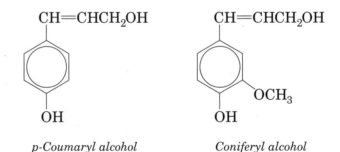

p-Coumaryl alcohol   *Coniferyl alcohol*

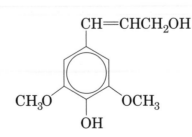

Sinapyl alcohol

The polymer is broken down to soluble fragments during chemical pulping. In mechanical pulping, the lignin molecule is softened with the heat generated by the grinding, making it possible to defiber (pull apart) the wood fibers. Lignin contributes opacity to paper, but it darkens with age reducing the brightness of the paper.

The wood fiber has a complicated structure in which cellulose molecules are wound together in three layers (figure 7-2).

Figure 7-2.
Structure of a wood fiber.

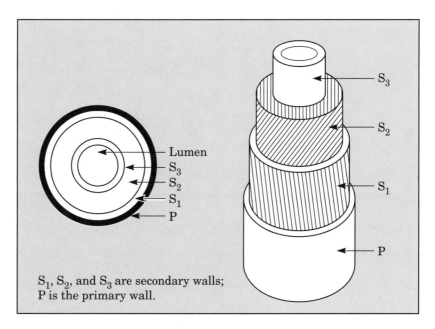

S_1, S_2, and S_3 are secondary walls;
P is the primary wall.

Chemistry of Cellulose and Starch

Two of the most important chemicals in paper and papermaking are cellulose and starch, both polymers of glucose. Cellulose is a pure, regular, crystalline polymer. Starch is a mixture of two polymers: amylose, and amylopectin. Amylose, like cellulose, is a regular crystalline polymer but in natural starches, it is mixed with amylopectin, a branched, irregular and noncrystalline product (figure 7-13).

Cellulose and starch have the same empirical formulas $(C_6H_{10}O_5)_n$, but cellulose has a much higher molecular weight than amylose. The remarkable differences in the properties of cellulose, amylose, and amylopectin result from the molecular weights and the way the glucose units are connected. In cellulose, the glucose units are linked with a configuration called an α-glucosidic structure. Glucose units in starch are linked in a β-glucosidic configuration. Cellulose, like amylose, is highly insoluble in water. Low-molecular-weight cellulose polymers called "hemicellulose" are soluble in caustic.

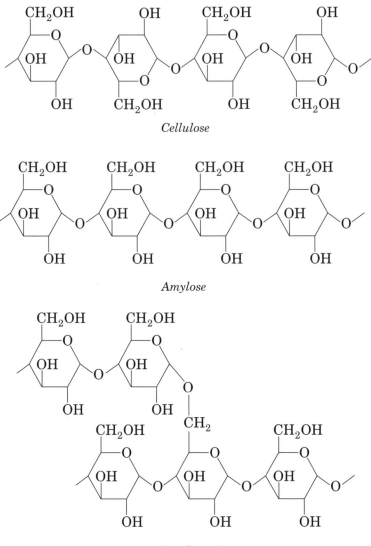

Cellulose

Amylose

Amylopectin

Groundwood or Mechanical Pulp

Wood can be converted to paper pulp by grinding. Groundwood is made by removing the bark and forcing the logs against a revolving grinding stone in the presence of water to yield a slurry containing minute particles of both fibrous and non-fibrous portions of the wood. The lignin in the groundwood contributes high opacity to the paper, so that lightweight publication papers often contain groundwood or mechanical pulp. Groundwood papers also have excellent smoothness and ink receptivity. On the other hand, lignin turns yellow in a matter of hours when exposed to sunlight or within months when stored in the dark so that groundwood and mechanical papers lack permanence. They are suitable for newspapers, paper towels, and most magazines but not for Bibles, dictionaries, encyclopedias, or other "archival" materials (documents requiring permanence).

Although the heat generated by the process softens the lignin, the grinding process damages the fibers, creating a great deal of fines that slow drainage of the sheet on the forming wire of the paper machine. Addition of 20–25% chemical pulp facilitates papermaking and yields a stronger sheet.

"Alphabet" Pulps   Groundwood is supplemented by a variety of pulps in which the mechanical separation of the fibers is assisted by heat or chemicals. One of the first was refiner groundwood (RMP). The wood is chipped, and the chips are separated into fibers in a disk refiner.

More frequently, the chips are softened by heating with high-pressure steam before they are defibered in a disk-type refiner. This product is called thermomechanical pulp (TMP). The fibers produced in this way are more completely separated from each other and suffer less damage than those produced by grinding. TMP is cleaner and stronger than groundwood, and the process gives a higher yield of fiber than chemical pulping; it uses no chemicals and creates no water or air pollution. TMP is used for newsprint for lightweight coated publication papers and for uncoated supercalendered papers—SC'A', SCA, SCB, and SCC (where SC'A' has the highest quality). These papers are produced from TMP without addition of any chemical pulp.

Another pulp that includes all of the wood is chemi-thermomechanical pulp (CTMP) in which caustic is used to soften the wood before it is defibered at a high temperature. If the pulp is bleached, it is referred to as BCTMP. The process yields a pulp that is even stronger than TMP. Neutral sulfite semichemical pulp (NSSC) comes from wood treated with sodium sulfite to soften the lignin before it is defibered. NSSC is commonly used as the corrugating medium in corrugated board. Other variations of mechanical pulping carry similar alphabetical designations, and this is the origin of the name "alphabet pulps" for pulps made from whole wood. The alphabet pulps are usually strong enough to make satisfactory paper without the addition of chemical pulps.

Chemical Pulps

Because lignin turns dark with age, it is removed to produce paper that retains its brightness over the years. In fact, high-brightness papers are always made from lignin-free pulps.

In chemical pulping, the wood is reduced to chips, which are placed in a pressure reactor and cooked with a material (a "pulping liquor") that converts the lignin to soluble products. The pulping liquor also dissolves most of the hemicellulose, so that a yield of 50% of the wood is the normal result of chemical pulping.

During digestion, the chips are converted into a pulp that consists mostly of cellulose fibers. After the process is completed, pressure is removed from the digester, and the pulp is washed. The pulp is still dark and contains undigested knots. The knots are screened out, and the pulp is bleached. After treating in a mechanical refiner, the pulp is ready to go to the paper machine.

The Sulfate or Kraft Process

Most chemical pulp is now produced by the sulfate or kraft process. The word "kraft," which means "strength" in German and in Swedish, reflects a principal characteristic of this pulp.

The principal reactant, sodium sulfide (Na_2S), is produced by adding sodium sulfate (Na_2SO_4) to the spent liquor from the previous batch, evaporating, and burning the mixture. Heat and the carbon contained in the digested lignin convert the sulfate to sulfide. This product, called "green liquor" (although it is very black), is used to digest the next batch of wood chips.

$$Na_2SO_4 + 2C + 4NaOH \rightarrow Na_2S + 2Na_2CO_3 + 2H_2O$$

Sodium sulfide converts the lignin to soluble materials, but it generates methyl mercaptan (CH_3SH), dimethyl sulfide (CH_3SCH_3), dimethyl disulfide (CH_3SSCH3), and other organic sulfides, among the most foul-smelling materials known. An immense amount of work has been devoted to removing organic sulfides from stack gases, but even the most modern of kraft mills still emit an unmistakable odor.

The process is commonly carried out in a continuous digester (figure 7-3).

High alkalinity attacks pulp, and sodium sulfide and sodium carbonate are buffers that keep the process from becoming too strongly alkaline, one reason that kraft pulp is strong.

The carbonate reacts with the resins in the wood, forming water-soluble soaps. In one way or another, the nonfibrous materials are dissolved and separated from the cellulose fibers.

Kraft pulp was difficult to bleach, but research produced bleaching processes that yield a strong, bright pulp. The first kraft paper was produced in 1909. Initially it was used only

Figure 7-3
Continuous chemical
pulping system.

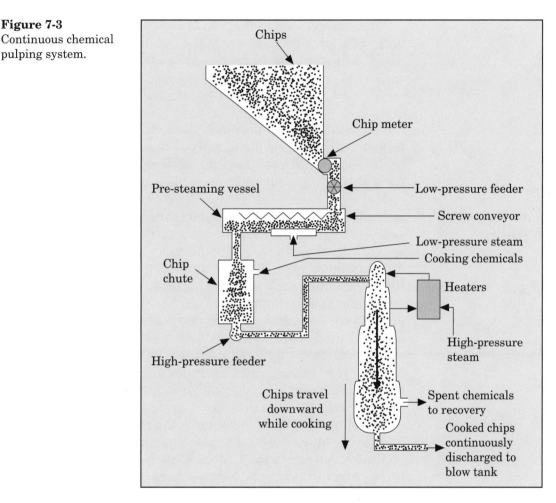

for such things as wrapping paper, paper bags, and corrugated
cartons, where whiteness was not demanded but where
strength was important. Bleached kraft pulp appeared in the
1930s, and in 1950, the first chlorine dioxide multistage
bleaching plant was built for producing pulp with a brightness in the high 80s.

**The Sulfite
Pulping Process**

Sulfite pulping was developed in the 1800s by a chemist
named Mitscherlich who was studying lignin. Learning
about the shortage of rags for papermaking, he turned his
attention from the lignin to the cellulose that remained after
dissolving the lignin. For many decades, his process was the
chief chemical pulping method. Since most of the resins in
the wood are not dissolved during pulping, this process is
limited to woods that are low in resin content.

Mitscherlich cooked the wood fibers with a mixture of lime and sulfur dioxide, converting the lignin into soluble materials. The reagent can be represented as calcium bisulfite:

$$CaCO_3 + 2SO_2 + H_2O \rightarrow Ca(HSO_3)_2 + CO_2$$

The calcium liquor presents severe waste disposal and stream pollution problems, and calcium based pulping has been replaced with magnesium, ammonium, or neutral sulfites. Kraft pulping is replacing sulfite pulping.

Recycled Fiber

It is often less costly to collect paper or paperboard and to recover the fiber than it is to grow and harvest trees and to convert the wood to pulp. Furthermore, paper represents a major component of landfills, and space for additional landfills is becoming scarce. Paper buried in landfills often remains intact for many years. Some attention has been given to converting paper into mulch.

The United States collects about 50% of its wastepaper and board, exporting large amounts of it. Paper and board made in the United States averages around 30% recycled fiber. The percent of recycled fiber is higher in Japan and many European countries.

Recycling fiber usually requires the removal of ink from printed paper, a process discussed in chapter 8.

Most pulpwood in the United States is grown on tree farms, and it has become a major agricultural crop. Although it is not often considered to be recycling, burning the paper to recover its heat value and releasing the carbon dioxide to the atmosphere where trees can once more convert it to wood is a form of recycling. The effect of the carbon dioxide on the atmosphere is still a matter of intense debate.

Bleaching the Pulp

Whiteness is the equal reflectance of all wavelengths plus the lack of grayness. If all wavelengths are reflected equally, paper may be white, gray, or black, so that high reflectance is one of the two requirements of whiteness. Whiteness is desirable to develop high print contrast between paper and print and also to avoid altering the color of the print.

Brightness, as defined by standard tests, is the reflectance at 457 nm, a wavelength in the blue part of the spectrum. This wavelength is particularly useful for measuring yellowness, the characteristic color in pulp or paper that is not fully bleached or in paper that has aged.

Brightness may be measured by comparing the reflectance of the paper with the reflectance of a block of freshly scraped magnesium oxide, which is assigned a brightness of 100. (In practice, a calibrated white panel is used as a reference.)

Brightness of newsprint ranges from the high 50s to the low 60s, Many bleached printing papers range from 78 to 85, and some very bright papers will reach the low 90s. High-brightness papers are costly to manufacture.

To make white paper, pulp must be bleached. Different bleaching procedures and chemicals are used. In one, chlorine gas is passed into the pulp-water mixture. The chlorine reacts with the small amount of lignin still remaining. The chlorinated lignin is then removed by treatment with caustic soda, NaOH. The final bleaching is carried out with sodium or calcium hypochlorite [NaOCl or $Ca(OCl)_2$], produced by passing chlorine gas into NaOH or $Ca(OH)_2$. Either of these chemicals oxidizes the small amount of colored materials in the fibers to colorless materials. This process, however, produces traces of dioxin, a highly toxic and persistent material, and hypochlorite bleaching has been replaced with other methods.

Chlorine dioxide, ClO_2, is now used to bleach kraft. It produces no dioxin. Hydrogen peroxide (H_2O_2) and ozone (O_3) are also used.

Groundwood pulp cannot be bleached with chlorine or chlorine dioxide because the high lignin content overwhelms the chlorine used. The brightness of mechanical pulps is increased by bleaching them with hydrogen peroxide (H_2O_2) or sodium peroxide (Na_2O_2), sodium dithionate ($Na_2S_2O_4$), or zinc dithionate (ZnS_2O_4) (sometimes called zinc hydrosulfite). These bleaching agents remove color without removing the lignin.

Stock Preparation

Following bleaching, the pulp-water slurry is adjusted so that it contains about 5% fiber and 95% water. The stock is treated by running it through a disk refiner, followed by a conical refiner called a *jordan* shown in figure 7-4.

Other materials are added, including fillers and coloring materials, to improve the optical properties of the paper. Internal size is added to control its resistance to water.

Treatment in the refiner increases the strength of the paper about tenfold by unraveling and fraying the individual cellulose fibers, increasing their surface area and increasing the area available for bonding when the sheet is formed. Furthermore, the beating breaks down the water-resistant outer wall

Figure 7-4.
Conical-type (jordan) continuous refiner. Fibers are refined as they pass between the refining bars of the rotating plug and the stationary bars of its surrounding shell. *(Courtesy Black Clawson Co.)*

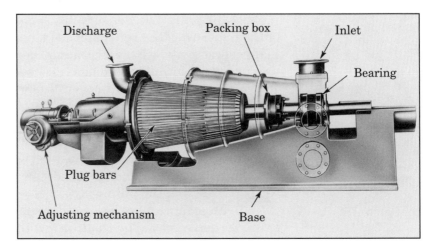

Discharge   Packing box   Inlet

Bearing

Plug bars

Adjusting mechanism   Base

of the fibers, exposing the inner fibrils and causing the fibers to swell and take on water. The effect is called ***fibrillation*** or ***hydration.*** As the refining process continues, the fibers become more hydrated and the resulting paper becomes stronger.

Different types of papers are refined to different extents, depending on how the paper is to be used. This varies from little or no refining (for making blotting paper and sanitary tissue that are weak but absorbent) to prolonged refinement for making a transparent paper such as glassine.

Every paper is a compromise of interdependent properties; it is important to understand the problems facing the paper-maker. Increasing refining increases bursting strength, tensile strength, and folding endurance, owing to the greater internal bonding of the fibers. There is also an increase in sheet density, hardness, rattle, and resistance to pick during printing. These qualities are generally considered to be advantages unless a soft, high-bulk paper is desired.

There are also some disadvantages to increased refining. In addition to the high cost of power to run the refiner, the tear strength and opacity of the paper are reduced. Highly refined pulp has poor dimensional stability when the relative humidity changes. Paper expands as it picks up moisture from the air. The amount of expansion increases with paper made from pulp that has been highly refined. Expansion around exposed edges causes ***wavy-edged paper*** in rolls or skids. If the edges lose moisture, the resulting contraction causes ***tight-edged paper.***

Papermaking is a compromise. It is impossible to make low-basis-weight paper that has both high bursting strength and high tear strength, or to make paper that has high tensile strength and good dimensional stability. Because dimensional stability is important for printing papers, the pulp is refined only enough to produce a paper strong enough to resist forces of the printing press.

Fillers

Materials called fillers are added to the pulp during stock preparation. Printing papers may contain 15–25% of fillers by weight. Other papers, designed for strength and rugged use, such as bond and ledger papers, may contain only 2–6% of fillers.

The three most common fillers are clay, a naturally occurring aluminosilicate, limestone or calcium carbonate, $CaCO_3$, and titanium dioxide, TiO_2. Precipitated calcium carbonate is becoming popular owing to a combination of low cost, excellent whiteness and opacity, and its control of ink absorption. It often eliminates the need for expensive TiO_2.

Fillers are added to increase opacity, brightness, whiteness, and smoothness, to reduce ink strike-through, and to reduce cost. Fillers also improve texture and feel, and they make the paper more dimensionally stable.

Fillers also contribute some disadvantages. As the percentage of filler increases, the burst, tensile, and tear strengths decrease. Folding endurance and stiffness of paper also suffer, another example of compromise required of the papermaker.

Some fillers have special properties. Titanium dioxide—more expensive than clay—is unexcelled for increasing the brightness and opacity of paper. Papers containing calcium carbonate have an excellent affinity for ink. Calcium carbonate also contributes to paper permanence, by neutralizing any acidity in the paper.

Coloring Materials

Coloring materials are also added. Large amounts of dye may be added to create colored papers, but white papers usually contain small amounts of coloring materials to make them whiter or brighter. Blue coloring materials help to overcome the slight yellow color of the pulp. Fluorescent dyes, called optical brighteners or optical bleaches, are often used. The action of these brighteners depends on ultraviolet light. Paper containing them glows when viewed under a "black light" (ultraviolet radiation without visible light) in a dark room. These papers have maximum brightness when viewed under fluorescent lamps or on clear, bright days. Their brightness appears lower when viewed under incandescent lamps.

Paper Sizing

Sizing controls the penetration of water or ink into the paper. Two methods of sizing are used: internal sizing employs an agent introduced into the furnish (the suspension of fibers and other materials furnished to the paper machine). Surface sizing applies the sizing agent to the surface of the sheet on the paper machine. Papers with internal sizing are said to be "hardsized." Sized papers are not waterproof, but they possess significant resistance to wetting by water or by water-based inks.

Rosin, a common internal sizing agent, is added to the furnish before it is fed to the head box of the paper machine. Rosin is one of the materials obtained, along with turpentine, from pine trees. It is a mixture of complex acidic compounds. Abietic acid is a major component.

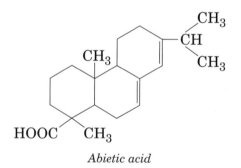

Abietic acid

Rosin size is prepared by boiling rosin with aqueous soda ash (sodium carbonate) or caustic soda (sodium hydroxide) to form soaps that emulsify the rest of the rosin and keep it in a finely divided state.

The rosin size is then "set" by adding papermakers' alum [aluminum sulfate, $Al_2(SO_4)_3$], which is acidic enough to lower the pH of the fiber slurry to 4.0–4.5. The alum converts the rosin soap back into free rosin in a finely divided colloidal form. The rosin becomes fixed on the cellulose fibers in tiny globules, and when the paper is later dried on the dryers of the paper machine, these globules melt or fuse together. Paper made this way usually has a pH between 4.5 and 6.0.

Rosins modified with maleic anhydride or fumaric acid are more efficient and are easier to handle than the natural rosin.

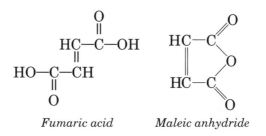

Fumaric acid   *Maleic anhydride*

For many purposes (such as the daily newspaper), paper has served its purpose after a day or so, but printed materials, such as books and office records must remain in good condition for years or even centuries. Acid, and especially acid plus aluminum ions, rapidly depolymerize cellulose causing the "silent fire" that destroys books and documents after ten or twenty years. Neutral and alkaline sizes are now used to size printing papers. No alum is required, and the paper so produced resists aging.

One popular product that requires no alum is alkyl ketene dimer; another is an alkenyl succinic acid. These products become chemically bonded to the cellulose fibers.

Virtually all printing paper is now manufactured on the alkaline side. Not only does the higher pH reduce deterioration, but paper made at pH above 7 has some 15% greater tensile strength and bonding than paper made at pH 4.0–5.5 (required for the formation of the Al^{3+} ion). This makes it possible to reduce the basis weight of the paper or to add more filler.

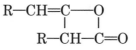

where $R = C_{14}H_{29}$ or $C_{16}H_{31}$

Alkyl ketene dimer

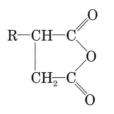

where $R = C_{14}H_{27}$ or $C_{16}H_{31}$

Alkenyl succinic anhydride

Alkyl ketene dimer and alkenyl succinic anhydrides are used for sizing of graphic arts papers. (The alkyl groups typically have 16 carbon atoms.) The emulsified sizing agents are added to the furnish before it is sent to the headbox. (The furnish is the dilute suspension of fiber and additives furnished to the paper machine.)

Wax emulsion and fluorochemicals are also used as internal sizing additives.

The Papermaking Process

After the pulp has been refined, and the fillers, coloring material, and size have been added, the mixture is ready to be made into paper. It is diluted with water to make a slurry that contains only about 0.5–1% of cellulose fibers.

The dilute slurry is pumped to a headbox on the wet end of a fourdrinier papermaking machine (figure 7-5). It is spread out on a fine plastic screen (called a "wire") with about 65 meshes per inch (2.6 meshes per millimeter). Fine papers are made at speeds of 3,000 ft/min (15 m/sec), and the machine runs even faster for other grades. Newsprint machines have operated at speeds greater than 5,000 ft/min (26 m/sec), and speeds are continually increasing. The machine may be as wide as ten meters (400 in or 33 ft) Much of the water drains through the screen, leaving the interlocked fibers on top. After the wire has advanced a short distance from the headbox, suction boxes remove more water, and the web of paper reaches about 20% solids at which level it is strong enough to be lifted from the wire and fed into the press section. The

Figure 7-5.
Twin-wire paper machine.
(Courtesy Beloit Corporation)

press section presses the web between felts, raising the solids to 33%. Most of the remaining water is removed by passing the web over steam-heated drums so that two tons of water must be evaporated to produce one ton of paper. Printing paper contains around 5% water. The web next passes between heavy steel calender rolls to smooth it, and it is then rolled up.

Surface Sizing

Two thirds of the way down the dryer section, a unit called a size press (figure 7-6) is installed to add surface size to the sheet. The surface size is a solution of dextrinized starch or other special starches. The remaining heated cylinders dry this surface sizing. Instead of the size press, a coating station may be installed at this point on the machine to coat the paper as discussed below.

Figure 7-6.
Size press.
(Courtesy Beloit Corporation)

Surface sizing seals the paper surface fibers increasing the surface strength and reducing picking on the printing press. Surface sizing also increases ink holdout. With writing papers, surface sizing improves the writing surface and the erasability.

Felt and Wire Sides

Paper produced on a fourdrinier machine has different properties on the top and bottom sides, called the felt side and wire side, respectively. The wire side is in contact with the moving wire on the machine, and the felt side or top side is lifted from the machine by the felt.

Modern papermaking machines, especially twin-wire machines, produce paper that is much less two-sided (that is, the properties of the two sides are more similar) than older machines. Nevertheless, for some printing applications it is important to know which is the top or felt side and which is the wire side.

The wire side has more fibers oriented in the machine direction than does the felt side. As a result, the wire side has a more pronounced grain direction. The wire side sometimes has the impression of the wire mesh left in its surface. The top side has finer fibers than the wire side, and when a sample of paper is placed in an oven, it will curl towards the top side, the side with the most fibers—the side that shrinks the most.

Shrinkage is greater across the grain than in the grain direction so that warming a sample of the paper in an oven usually shows both grain direction and wire and felt sides (as shown in figure 7-7). In the twin-wire forming of paper, a jet of fibers dispersed in water enters the converging zone of two wires. Water drains simultaneously from both sides of the web being formed. Paper made this way has a more even distribution of fibers in the structure and has essentially two wire sides. Twin-wire newsprint has less tendency to lint on the web offset press.

Figure 7-5 shows a modern twin-wire machine. Figure 7-8 diagrams the "former" (where the sheet is formed), and figure 7-9 diagrams all sections of that machine. The functions of

Figure 7-7.
Curl, felt side, and grain direction.

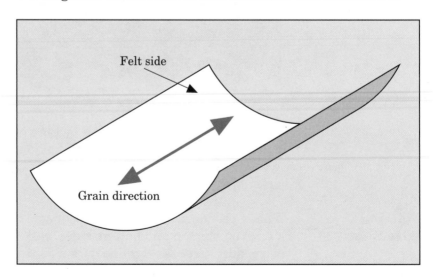

Figure 7-8.
Twin-wire former
(Bel Baie).

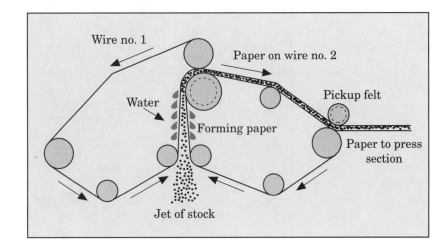

the various parts of a fourdrinier paper machine are presented in figure 7-10.

Coated Papers

Papers are coated with a mineral pigment to improve optical properties and print quality. Packaging papers and board may also be coated with plastics and resins, which are referred to as functional coatings.

Pigments used in coatings include clay, ground limestone or precipitated calcium carbonate, TiO_2, and, for photographic papers, barium sulfate ($BaSO_4$). Microscopic plastic spheres (called "plastic pigment") are sometimes used on lightweight coated papers to increase bulk, gloss, and brightness without increasing basis weight.

Coated printing papers are produced on a coating machine using an air-knife coater, a trailing blade coater, or frequently both (figures 7-11 and 7-12).

Pigment-coated papers include enamels, matte-coated, cast-coated, and sizepress-coated (also called film-coated). The amount of coating and its treatment after application differs with each process.

Enamel Papers

Enamel (glossy) papers are coated with a slurry of water, pigment, and binder, then dried and supercalendered. This treatment creates a smooth surface and increases opacity. The dried coating weight varies from about 10 to 20 lb./ream on a 25×38-in. basic size (15–30 g/m^2). The reported basis weight of a coated paper includes the weight of the base stock and the coating. The base stock provides most of the strength, so a 60-lb. coated paper has lower tensile and tearing strength than a 60-lb. uncoated paper.

Figure 7-9. Diagram of a paper machine. *(Courtesy Beloit Corporation)*

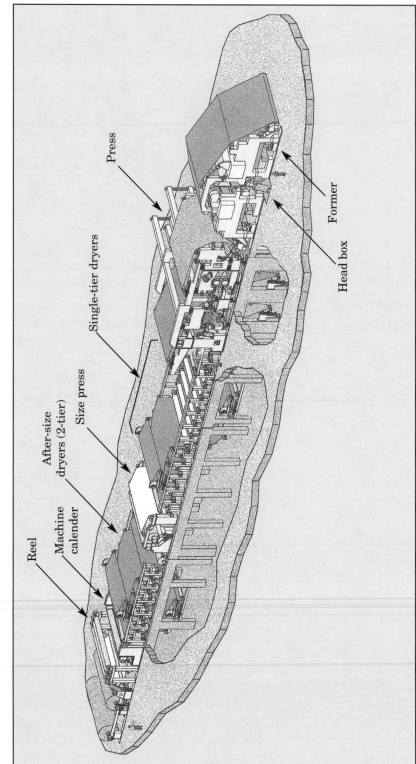

Figure 7-10. A fourdrinier papermaking machine.

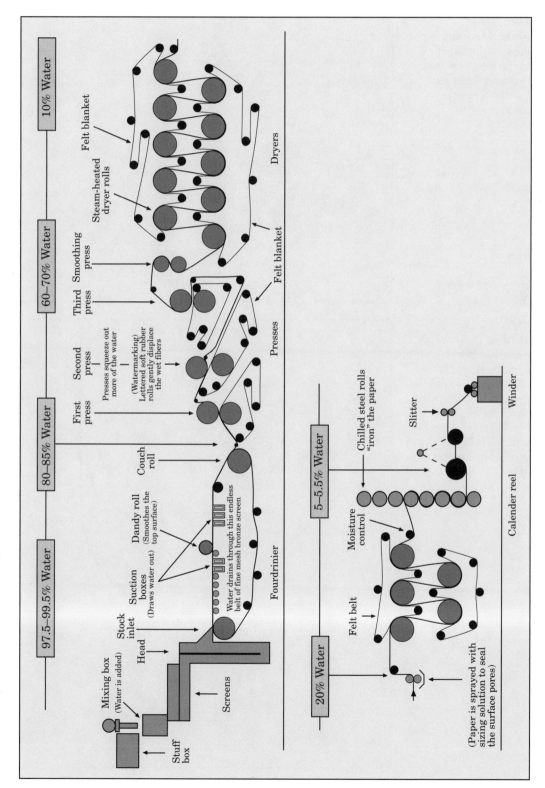

Figure 7-11.
Air-knife coater. Surplus coating, applied to the web, is removed by an air-knife jet, leaving a smooth coating.

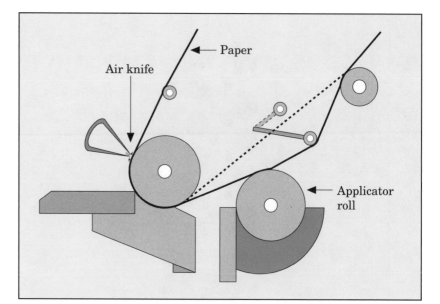

Figure 7-12.
Trailing blade coater. An applicator roll applies a surplus of coating to the paper web. A flexible steel blade evens off ("doctors") the excess coating, producing a very level coated surface.

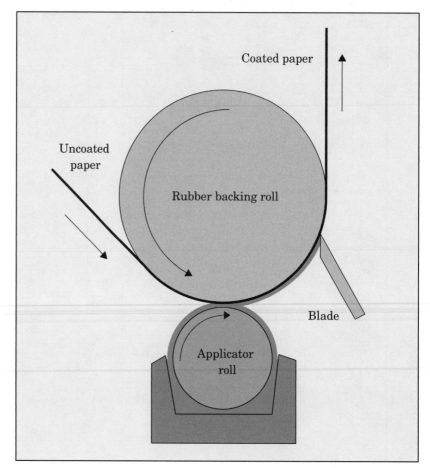

Matte-Coated Papers

Matte-coated (low-gloss) papers are sometimes used in offset printing. They are not supercalendered, and therefore retain their brightness and opacity—some of which are lost in supercalendering. The color strength of inks is higher when printed on matte-coated paper than on uncoated paper.

Cast-Coated Papers

Special coatings are used for cast-coated papers. While the coating on the paper is still wet, it is pressed and dried against a polished chrome roll yielding a coating with very high gloss. The process is somewhat like making glossy photo prints by pressing a coated photo paper onto a ferrotype plate.

Sizepress-Coated Papers

Sizepress coatings often contain some pigment, usually clay. The binder is either starch or a mixture of starch and synthetic binder that provides water resistance for offset papers. The amount of coating is considerably less than is used on fully coated papers, but even this small amount of coating improves paper smoothness and the printing of halftones.

Starch

Many materials have been used as binders for coatings, but most formulas include starch. Corn, potato, or tapioca starches are the most common starches. Starch is a mixture of two polymers, **amylose,** a straight chain starch, and **amylopectin,** a branched or "ramified" starch (figure 7-13). After starch is cooked, the amylose has a tendency to crystallize, forming a tough, insoluble film. If this occurs, the starch paste is useless, and the starch is said to "retrograde." To prevent crystallization, the starch maker alters the structure of the amylose by enzymatic digestion of the amylose or by chemical modification. These modified starches are called **converted starches.**

Figure 7-13.
Structure of amylose and amylopectin.

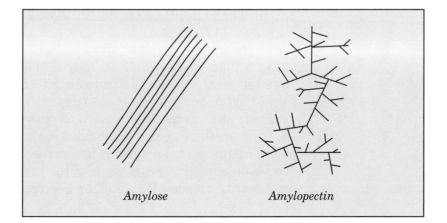

Amylose   *Amylopectin*

Latex Adhesives

A latex is a milky white fluid consisting of tiny droplets of rubber or elastomer dispersed in water. The synthetic latex adhesives used in paper coating include styrene-butadiene, styrene-acrylic, and vinyl-acrylic elastomers.

Starch is frequently blended with a synthetic latex to make the coating more pliable, to reduce cracking of the coating at folds in the paper, and to produce cleaner edges during trimming and slitting. Without starch or protein, the rubbery polymers decrease the stiffness of printing papers, making them limp. Paperboard has sufficient stiffness of itself so that synthetic binders may be the only adhesive.

Paper Coating Machines

Paper can be coated either "on-machine" or "off-machine." On-machine coating reduces the amount of handling required, but if the coater goes down, the production of the entire machine is lost. Off-machine coating is faster but requires additional investment in drying equipment. Both processes are in use, and the finest-quality papers are double-coated—first on-machine and then off-machine. Off-machine coaters are operating in excess of 5,000 fpm (25 mps). On-machine coaters have an air-knife coater (figure 7-11) installed about one-half to two-thirds of the way down the dryer section of the paper machine.

Effects of Coating on Printing

The coating applied to the web contains 50–70% solids (by weight). Blade coating fills the valleys in the paper surface producing a very smooth, coated surface. The air-knife coater applies a uniform coating over the hills or valleys of the surface, but if the first coat is applied by blade, the double-coated surface is very smooth.

Sodium silicate, used to disperse the pigment in the starch, raises the pH of coatings (to 7.5–8.5, sometimes as high as 10.0). The high pH promotes the drying of inks containing drying-oil varnishes.

In general, colored inks appear grayer ("dirtier") on uncoated papers than on coated paper. They also may not have the same hue (shade). A magenta ("process red") appears warmer on uncoated paper than on coated paper. A spectrophotometric curve of such a magenta, proofed on coated and uncoated paper, shows almost equal reflectance in the green and red portions of the spectrum, but the ink on the uncoated paper shows lower blue reflectance. Because of the lower reflectance of blue, a magenta appears less blue, or warmer, on uncoated paper.

In addition, the smoother surface yields a halftone dot that is less ragged. Accordingly, coated papers give both better color and sharper halftone dots than do uncoated papers. With the growth of four-color process printing, the demand for coated papers has continued to grow.

Calendering

A *supercalender* increases the smoothness of either coated or uncoated paper. It contains a series of rollers that are alternately steel and cotton. The cotton rollers are made by threading disks of the material onto heavy shafts, putting the disks under hydraulic pressure, and fastening them with steel heads.

The paper passes—under great pressure—between several pairs of these rollers while a small amount of steam is applied. The rolls impart a high finish to the paper. Such paper is called supercalendered paper.

Calendering techniques, including "hot-soft calendering" and "extended nip calendering," improve the uniformity of the paper surface and reduce a print problem called "mottle."

Paper mills are replacing the supercalender with an on-line *soft-nip calender* (figure 7-14). The paper is calendered in nips formed between polymeric or resin-filled cotton rolls. In addition to offering improved paper properties, the on-line machine avoids an off-line conversion operation in the paper-making process.

Functional Coatings

Besides pigment coating to improve printing, papers are coated to provide special functions. For example, *carbonless papers* (figure 7-15) are coated on one side with a dye that is microencapsulated in gelatin. The microencapsulated dye is in its *leuco* or colorless form. The adjacent side of the next sheet is coated with an attapulgite clay. This clay is abrasive and acid, and when a pencil or ball-point pen presses the microcapsules, they are broken, and the acid clay turns the colorless, leuco dye into its colored form, developing an image of the writing. Dyes are selected to give blue or black copies of the writing.

Functional coatings are also used to make packaging papers. The most familiar coating is polyethylene that is extruded onto the paper by a "converter," a company that processes the paper after it is manufactured. Polyethylene-coated papers have the moisture and air resistance of polyethylene film and the stiffness and opacity of paper. They are commonly printed by flexography.

Figure 7-14.
On-line calender,
JANUS MK2.
*(Courtesy Voith Sulzer
Finishing GmbH)*

Wet-Strength Papers

Most papers lose much of their strength when wet completely with water. Individual cellulose fibers do not weaken when they are wet, but they lose fiber-to-fiber bonding. Wet-strength resins overcome much of this loss by forming chemical bonds that are not destroyed by water.

A wet-strength paper is one that maintains at least 15% of its dry tensile strength when wet with water. Some retain 50% or more of their wet strength. Wet strength is imparted by the addition of about 3% of a melamine-formaldehyde or urea-formaldehyde resin to the papermaking furnish.

Figure 7-15.
Carbonless paper.

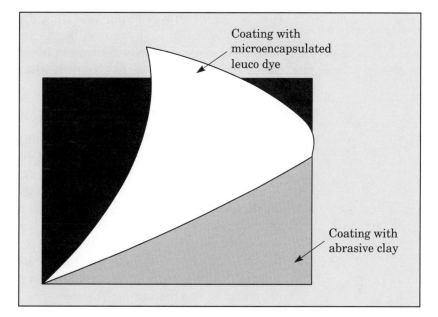

Figure 7-15.
Carbonless paper.

Coating with microencapsulated leuco dye

Coating with abrasive clay

Wet-strength papers are used where they must resist the application of water such as hand towels, industrial wipes, table covers, grocery bags, wet-type sandpaper, wallpaper, beer and wine labels, and photographic paper. Wet-strength papers are not necessarily water-repellent. Some, such as those used for paper towels, are in fact very water-absorbent.

The Effect of Moisture on Paper

Only a few of the properties of paper are discussed here: the effect of the relative humidity and temperature of the air, the printing consequences of grain in paper, and the effect of pH of paper or coating on the drying of inks. Other properties of paper are treated in chapter 9, "Chemistry in the Pressroom."

Bonding in Cellulose and Paper

Cellulose molecules are bonded, one to the other, with hydrogen bonds. A hydrogen bond is an example of a "secondary valence bond." It is much weaker than a "primary valence bond" such as those discussed in chapter 1, but secondary forces play a major role in paper chemistry. They result from the polar or electrostatic attraction of molecules for each other.

A hydrogen bond may be depicted as a hydrogen atom sharing its electron with two different oxygen atoms:

$$\text{Cellulose—O}\cdots\text{H}\cdots\text{O—Cellulose}$$

The strength of a hydrogen bond is 5 kcal. per mole as compared with the strength of a carbon-to-carbon, which is around 85 kcal. per mole. It is these weaker hydrogen bonds that break when wet paper is torn or dispersed in water. The cellulose-to-cellulose hydrogen bond can easily be replaced with a cellulose-to-water hydrogen bond, which explains why paper loses so much of its strength when it is wet with water.

$$\text{Cellulose---O} \cdots \text{H} \cdots \text{O---H}_2$$

Cellulose fibers, in paper, are bonded to each other with hydrogen bonds, and the greater the degree of refining, the greater the area over which hydrogen bonds can form and the greater the number of hydrogen bonds.

Other liquids that form weak hydrogen bonds, such as alcohol, somewhat reduce the strength of paper, but those that do not form hydrogen bonds, such as toluene or acetone, do not weaken the paper. Furthermore, a slurry of cellulose fibers in acetone will form fluff, not a sheet, when the acetone is drained away on a screen. Water is an essential chemical in the formation of paper.

Water in Paper

When cellulose fibers in a sheet of paper are exposed to water vapor, the water molecules are absorbed and the bonded areas expand. Although paper appears to be dry, it contains about 4–6% (by weight) of water under normal room conditions. This is called hygroscopic water, meaning that it is not combined with cellulose by primary valence bonds. It can be driven off by exposing the sheet to dry air.

If a sheet of paper is exposed to air of a given temperature and relative humidity (RH), it loses water or gains water until it reaches an equilibrium. At equilibrium, the water content and size of the sheet will remain constant indefinitely if the temperature and relative humidity of the air remain constant; but if either the temperature or the RH of the air is changed, the sheet will again gain or lose water and change size. The importance of RH is often emphasized while the effect of change in temperature is neglected.

A skid or roll of paper brought into a pressroom from a cold warehouse should be left wrapped until it comes to room temperature. This conditioning may require up to several days, depending on the volume of the paper and the initial difference in temperature between the paper and the pressroom (figures 7-16 and 7-17). If the cold paper is unwrapped before conditioning, moisture condenses on the sides of the sheets or roll, often causing wavy-edged paper.

Studies with single cellulose fibers show that as they expand or contract with change in moisture, the percent change in

Figure 7-16.
Temperature conditioning chart for paper (premetric units).

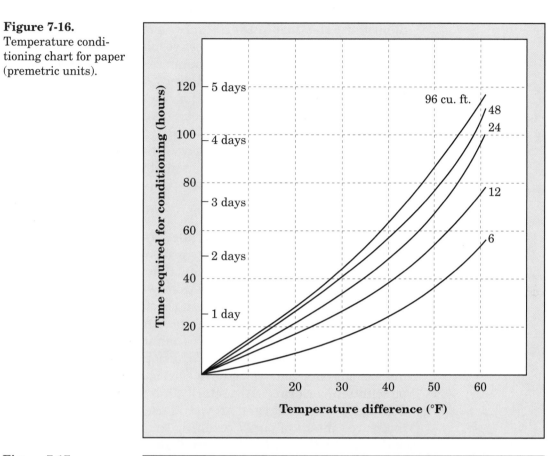

Figure 7-17.
Temperature conditioning chart for paper (metric units).

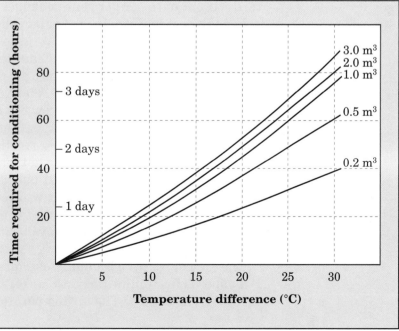

diameter of the fiber is 15–20 times as much as the change in its length. Grain direction in paper results from the alignment of paper fibers, and the paper therefore expands more across the grain than with the grain (figure 7-7).

This leads to misregister of color work printed on single- or two-color litho presses. If sheets are on a pile, or if a roll of paper is involved, absorption of water from the air leads to wavy edges because only the paper at the edge of the skid or roll grows in size. On the other hand, if water is lost from the sides of sheets or the ends of rolls, the result is "tight-edged" paper.

Hysteresis

Paper exhibits a characteristic called ***hysteresis;*** that is, its dimensions depend on its history or its previous condition. A strip of paper at room conditions that is taken to 90% RH and then returned to room conditions will be slightly longer than it was originally. Similarly, if the same strip of paper is equilibrated at 10% RH and then brought back to room conditions, it will be slightly shorter than it was. Because of hysteresis it is impossible to restore paper to its original dimensions once conditions have been changed. Reconditioned paper often performs satisfactorily, but sometimes not, depending on how severely it was abused. The changes in moisture content of paper resulting from changes in the RH of the air are illustrated in figure 7-18, and changes in moisture content of the paper cause changes in its dimensions.

For the best performance of paper, sheetfed pressrooms should be air-conditioned. (Web pressrooms depend on printing the paper faster than it can change its moisture content and accordingly its size.) Adequate air conditioning includes control of both temperature and relative humidity. In fact, if air coolers are used alone, they dry out the air and cause a great deal of trouble with the paper. They are an important source of paper trouble in the summertime.

The printer must understand that while it is possible to make paper tight or wavy after it has been manufactured, it is impossible to make paper that will remain dimensionally stable when it is exposed to air that is not in equilibrium with the paper. Paper must always be protected from wet or dry air.

The idea of relative humidity of paper needs explanation. If a paper is in equilibrium with air of, say, 45% RH, it is said to have a RH of 45%. The air temperature should be included

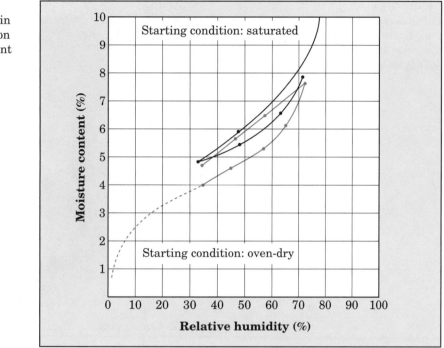

Figure 7-18.
Effects of changes in relative humidity on the moisture content of paper.

in such a reference. If the specified RH exists at 70°F (21°C), the complete description is "45% RH, 70°F" paper.

It is desirable to purchase paper already in equilibrium with the pressroom RH and temperature, but the merchant does not always have the desired grade available at those conditions. However, if the paper is properly wrapped, it should perform satisfactorily if it is allowed to achieve pressroom temperature before it is unwrapped. A properly air-conditioned pressroom improves paper performance. Air conditioning of the pressroom and temperature control of paper storage areas are usually profitable investments.

Instruments are available to determine the difference between the RH of the air and that of a skid of paper (figure 7-19).

Grain in Paper

Individual cellulose fibers (like pipes) are cylindrical. When paper is formed, considerable crisscrossing and interlocking of the fibers occur, but they tend to align themselves in the direction that the paper web travels through the paper machine (machine direction).

This fiber alignment creates grain in paper. The grain direction is parallel to the edge of the paper web. Because moisture causes the fibers to swell more across the diameter

Figure 7-19.
Portable moisture and
temperature test set.
*(Courtesy Beckman
Industrial Corporation)*

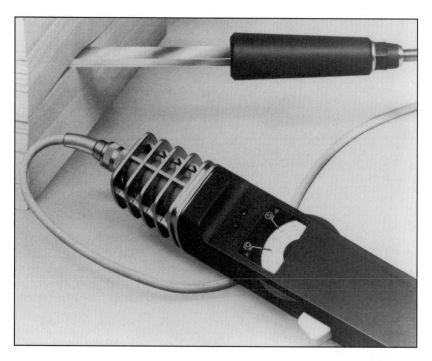

than in length, a sheet of paper that absorbs water expands
from two to eight times as much across the grain as in the
grain direction (depending on the degree of fiber orientation
in the paper).

For this reason, paper to be printed on a sheetfed press is
usually cut so the grain direction parallels the long side of
the sheets. On an offset press, the short side of the sheet
usually goes around the press cylinder, while the long side
runs the length of the cylinder (figure 7-20). When grain-long
sheets are printed, expansion due to water pickup is greater
around the cylinder (the short way of the sheet) and can be
compensated by changing the packing under the plate and
blanket.

Some sheetfed presses are equipped with roll sheeters. If
the cut sheets feed directly into the press, they are printed
grain short. Certain roll sheeters are designed to cut the
sheets and then turn them 90° before they are fed into the
press, creating grain-long printing.

Webfed presses, obviously, must be fed with paper that is
grain long—that is, paper with the grain running in the
direction of the paper machine. However, if the web press
has a cutoff that is longer than the web width, the resulting
printing is called "grain-long."

Figure 7-20.
Grain direction of sheetfed paper on the cylinder of the offset press.

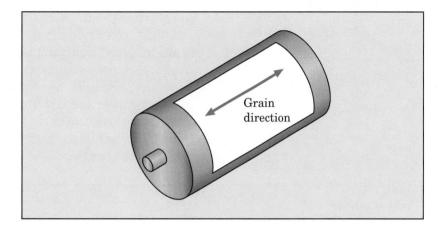

Fluting

One paper problem that develops on a heatset web offset press is *fluting*. The printed paper develops corrugations that run in the web direction and remain there after the product is cut, bound, and distributed. The cause of fluting is poorly understood, but it is known that fluting decreases with:

- Increasing basis weight
- Decreasing coat weight
- Decreasing air permeability of the paper
- Improved uniformity of basis weight, caliper, ash, and moisture.
- Decreasing temperatures in the dryer

The Effect of pH on Ink Drying

Whether the pH of paper or coating affects ink drying depends on the type of ink. The pH of the paper is not involved in the drying of nonheatset, heatset, or ultraviolet-curing inks, and it has nothing to do with the setting of a "quickset" type of ink. The pH does have a significant effect on the drying time of inks containing varnishes that dry by oxidation and polymerization. This effect on drying includes the final hard drying of quickset inks.

Decreasing the pH of the paper decreases the rate of drying of inks that dry by oxidation-polymerization. In one test, an ink printed on uncoated paper dried in 4 hours on paper with a pH of 7.1, 5 hours when the paper pH was 5.3, and 18.5 hours when the paper pH dropped to 4.4. Thus, as the paper becomes more acid, the drying time of the ink increases.

On coated papers, the drying time of inks is determined by the pH of the coating and not that of the base stock. In one test, an ink printed on a coating with a pH of 7.5 dried in 7.3 hours, on a coating of pH 8.0 in 4.8 hours, and a coating of pH 8.7 in 4.6 hours. Thus, as the surface of the coating becomes more alkaline, the drying time of the ink decreases.

Other factors are involved in drying time. As the relative humidity of the air in the pressroom increases, ink drying time increases considerably. In some cases, it may take twice as long for an ink to dry after the RH increases. Also, a highly acid fountain solution retards the drying of litho inks containing varnishes that dry by oxidation and polymerization. The effect of pH and RH on drying time is shown in figure 7-21.

Figure 7-21.
Effects of constant relative humidity (RH) and changing pH on ink drying time.

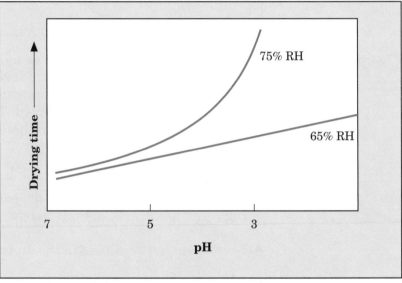

Other Properties of Paper

Many other properties of paper are important to the papermaker and the printer. These include tensile strength, bursting strength, folding endurance, tearing resistance, stiffness, internal bond strength, porosity, density, opacity, brightness, curling tendency, pick resistance, and ink absorbency. Laboratory instruments are available to measure these properties.

Basis Weight or Substance

The premetric measuring system used in the United States, once called the "English system," is no longer used in Great Britain. In this system, basis weight is the weight in pounds of one ream (usually 500 sheets) of a particular size. The trouble is that this basic size varies, depending on the type of paper or board. It is 25×38 in. for book and offset papers, 25½×30½ in. for index, and 17×22 in. for bond and ledger. Therefore, the basis weights of different papers do not represent the true relationship of weight per unit area.

There is only one metric basis weight for any kind of paper or board. It is the weight, in grams, of one sheet that has an area of one square meter. Thus the unit of measure is grams per square meter (g/m^2).

A factor *(F)* can be used to convert premetric basis weights to metric basis weights, or the reverse. The equations for doing this are:

$$\text{Premetric basis weight} \times F = \text{Basis weight (in } g/m^2)$$

$$\frac{\text{Basis weight (in } g/m^2)}{F} = \text{Premetric basis weight}$$

The value to use for *F* depends on the basic size for the paper or board. Values of *F* for various types of paper are given in the table 7-I.

Table 7-1.
Factors to convert premetric basis weights to metric basis weights, and vice versa.

Class of paper or board	Basis size (inches)	F
Bond, ledger	17×22	3.76
Book, offset	25×38	1.48
Blotting	19×24	3.08
Index	20½×24¾	2.77
Index	25½×30½	1.81
Cover (antique, coated)	20×26	2.70
Bristol, plate vellum	22½×28½	2.20
Bristol, smooth vellum	22½×35	1.78
Newsprint, kraft, tag	24×36	1.63
Postcard	28½×45	1.10

Using these factors, the metric basis weight of a 50-lb. offset paper is:

$$50 \times 1.48 = 74 \text{ } g/m^2$$

For a 20-lb. bond paper, it is:

$$20 \times 3.76 = 75.2 \text{ } g/m^2$$

A 20-lb. bond paper is a little heavier than a 50-lb. offset paper. Boxboard is usually sold on a 1000-sq-ft basis, "x" pounds per 1000 sq ft. The metric system makes comparison much easier.

Further Reading

Bennett, Kathleen M., et al. *How Paper Is Made* (multimedia training tool). 1997 (Technical Association of the Pulp and Paper Industry, Atlanta, Ga.).

Eklund, Dan, and Tom Lindström. *Paper Chemistry, An Introduction.* 1991 (Technical Association of the Pulp and Paper Industry, Atlanta, Ga.).

Wilson, Lawrence A. *What the Printer Should Know about Paper.* 3rd ed. (336 pp) 1997 (Graphic Arts Technical Foundation, Sewickley, Pa.).

8 Printing Inks, Toners, and Coatings

Introduction

Inks are complicated mixtures of chemical compounds. Their composition varies greatly depending on the following: whether an ink is to be used for lithography, flexography, gravure, or screen printing; whether it is for sheetfed or web printing; the substrate; whether it is for publication or package printing; and the type of drying. Inks are often formulated for special purposes, such as resistance to scuffing or fading, and for use in contact with various kinds of foods. Because of all these variations, the manufacture of inks is a complicated business, requiring a high degree of technology and skill. A great deal of printing ink is made by a batch process, although inks for long-run publications are usually made by a continuous process.

Inks are often divided into liquid and paste inks. Gravure and flexo inks are called liquid inks; litho and screen inks are called paste inks because their viscosity is much higher. News ink is usually thought of as a paste ink because, in spite of its relatively low viscosity, it is used mostly in lithography.

This chapter specifically excludes writing inks, the inks used in fountain or ballpoint pens. Inks for inkjet printing rather resemble writing inks, and they are included in this chapter.

A printing ink consists of a pigment and a vehicle. Although the words "varnish" and "vehicle" are sometimes used synonymously, the vehicle consists of a varnish, solvent, and additives required to give proper performance. The vehicle is also defined as the mixture of all printing ink ingredients except the pigment. Typical chemicals are listed in table 8-I, although there are far too many materials in common use to list them all. The greater number of materials used for flexo, gravure, and screen printing reflects the wide variety of substrates on which they are used.

	Heatset web offset	Sheetfed offset	Flexography	Rotogravure	Screen
Binder	Rosin ester Hydrocarbon resin Waxes Rosin-modified phenolic	Long-oil alkyd Hydrocarbon resin Waxes Drier	Polyamide Nitrocellulose Shellac Acrylic resin	Rosin ester Metallated rosin Cellulose ester Hydrocarbon resin	Long-oil alkyd Epoxy resin Nitrocellulose Cellulose ester
Solvent/ diluent	Hydrocarbon oil	Hydrocarbon oil	Alcohol Water	Ester/ketone Toluene Aliphatic hydrocarbon	Hydrocarbon Alcohol Ester/ketone
Typical product	Publications Catalogs	General commercial	Packaging	Packaging Publications	Textiles Posters

Table 8-I.
Typical components of printing ink vehicles.

The vehicle transports the pigment from the ink fountain to the paper or other substrate and binds the pigment to the substrate. Choice of vehicle depends on the printing method (gravure, offset, flexo, or screen, sheetfed or web), on the substrate (coated paper, newsprint, film, or metal), on the method of drying (heatset, nonheatset, radiation, or air drying), and on end-use requirements.

Varnishes must wet the pigment. Wetting means that the varnish molecules form an adsorbed film around the pigment particles. If a pigment is not wet by the varnish, the ink will be very short (unable to form a string). Such an ink is said to be "buttery," and it has poor flow properties so that it fails to transfer properly from one ink roller to the next (see figure 8-4). Generally the wetting of pigments by varnishes increases somewhat as an ink ages. Wetting of a pigment is strongly affected by the coating or finishing given to the pigment during its manufacture.

Process Color Printing

Before discussing ink pigments, it is necessary to discuss the nature of process color printing and the requirements of process color pigments.

In process color printing, the various colors are produced by printing halftone dots with yellow, magenta, cyan, and black ink. In most areas, one ink overprints another to create the secondary colors of red, green, and blue in various shades.

Transparent pigments are used in process inks. All of the colors that the eye will see are reflected from the surface of the paper. Each ink absorbs part of the incident light and transmits the remainder. If one ink overprints another, then each absorbs part of the incident light, and the reflected color is different than the color of either ink. This explains why the color of the paper, the illumination, and the viewing conditions are so important in observing the color of a printed sheet.

The wavelengths of visible light can be considered to consist of three segments: blue, green, and red. Table 8-II shows what happens when white light strikes a white paper on which different process inks are printed.

A person cannot tell by looking at a block of color whether it is a single color or a mixture. In this way, sight is different from sound. The ear can tell the difference between a single note and a chord; the eye cannot.

What the eye interprets as yellow is a combination of the green and red part of the spectrum. Magenta, a bluish red

Table 8-II.
Process ink
overprinting.

Color of Ink	Absorbs	Transmits or Overprints
Yellow	Blue	Green and red (equals yellow)
Magenta	Green	Blue and red (equals magenta)
Cyan	Red	Blue and green (equals cyan)
Yellow and magenta	Blue and green	Red
Yellow and cyan	Blue and cyan	Green
Magenta and cyan	Green and red	Blue

color, results from the reflection of a combination of blue and red light, and cyan, a greenish blue color, is a reflection of blue and green light.

One often hears magenta referred to as "red" and cyan referred to as "blue." Magenta (sometimes called "process red") is a different color than red, and cyan (sometimes called "process blue") is a different color than blue.

White paper transmits all colors. If magenta is printed over yellow on white paper, the yellow absorbs blue light and the magenta absorbs green light. This leaves only red light to be reflected to the viewer's eyes. A yellow-cyan combination absorbs blue and red and transmits green. A magenta-cyan combination absorbs green and red and transmits blue.

Process color printing produces other combinations, which may be called in-between colors. While yellow plus magenta gives red, increasing the amount of yellow or decreasing the amount of magenta produces an orange. The shade of an in-between color depends on the size of the halftone dots as well as the hue of the pigment and the ink film thickness. The shade also depends on trapping of the overprint color discussed later in this chapter.

The discussion should make it apparent why buyers like to buy paper with a high reflectivity (high whiteness). If the paper does not reflect the color, the hue of the print is altered, and brilliant colors become gray or dull.

While process inks are highly transparent, they are not completely so. A transparent yellow has the highest transparency. Magenta is a little less transparent, and cyan still less. Since black absorbs almost all color, it is opaque.

Different magenta and cyan pigments used in process inks have different hues. When one process ink prints over another, the cleanliness of the in-between colors depends on the cleanliness and hue of the process colors. It is important to choose process inks carefully and to maintain a single, constant source of supply. The practice of buying the lowest price ink makes it impossible to maintain constant color from one job to the next.

Pigments

A wide variety of pigments is available for use in printing inks. Many are usable in inks for every process, although a few are suitable only for lithographic inks and some others are suitable only for nonlithographic inks.

Lithographic printing usually deposits a thinner ink film than other printing processes. To obtain sufficient color strength on the printed material, lithographic inks must be more highly pigmented than other inks.

Typical ink film thicknesses are listed in table 8-III.

Table 8-III.
Ink-film thicknesses of major printing processes (wet film, full solid, on coated paper).

Process	mils	mm	microns
Sheetfed offset	0.2	0.005	5.0
Web offset	0.3	0.008	7.5
Flexography	0.6	0.015	15
Gravure	1.2	0.030	30
Screen	0.5–5.0	0.012–0.125	12–125

It is convenient to divide pigments into four classes: carbon black, organic pigments, white inorganic pigments, and colored inorganic pigments.

Carbon Black

Black printing inks and toners are made with carbon black. Since more black ink is used than any other color, it is the most important pigment.

Carbon black consists largely of a crystalline form of carbon called ***graphite.*** It has a very high absorption of light and not only produces an excellent black color, but it also is an effective opacifier. Another crystalline form of carbon—diamond—is neither an effective pigment or opacifier.

There are several ways of making carbon black, and the properties of the pigment vary somewhat depending on the process. Chemically, carbon black consists mostly of elementary carbon, a small percentage of ash (mineral matter), and a somewhat higher percentage of volatile matter—mostly compounds of carbon, hydrogen, and oxygen.

The principal carbon black is furnace black, made from residues of petroleum refining that are sprayed into a furnace with an insufficient amount of air to burn them completely. The furnace chamber is lined with firebrick. The burning oil raises the temperature to 2,300–2,600°F (1,250–1,425°C). The burning oil forms carbon monoxide, hydrogen, nitrogen, water vapor, and suspended particles of carbon black, which are cooled with a water spray. The furnace black is separated with electric precipitators and cyclone collectors.

Organic Pigments The growing demand for process color printing makes these pigments increasingly important. Organic pigments provide a wide selection of colors. Many have high color strength, and inks made with them print clean tones. Most of them have a low specific gravity and small particle size that help produce inks with good working properties. Most organic pigments are easily wet by varnishes—more easily than with water, which is essential for lithographic inks. Most organic pigments are easy to grind, and they produce inks with low abrasion characteristics.

Some organic pigments cause problems. A few produce inks that are short; it is necessary to remedy this with the proper varnish or extender. Pigments that dissolve or bleed into water and pigments that are easily wet with water are unsatisfactory for litho inks. Some pigments sublime (change from a solid to a vapor) because of the heat generated in the delivery pile. The vapor may transfer to the next sheet or even penetrate several nearby sheets before it cools and resolidifies, producing a tint of the pigment. Sublimation is usually undesirable, but heat-transfer printing inks for textiles require pigments that sublime.

If a pigment and the varnish into which it is ground have nearly the same index of refraction, a thin film of the resulting ink is transparent. If the index of refraction of the pigment and varnish are widely different, a thin film of the resulting ink is opaque. (The refractive index is a measure of the amount that a beam of light is bent as it enters or leaves a substance.) Inorganic pigments usually produce opaque inks because their refractive index is much higher than that of the organic varnishes that carry them. Organic pigments usually produce transparent inks, used for process colors. If opacity is required, an opaque pigment such as titanium dioxide (TiO_2) must be added.

**Chemical
Structure**

The chemistry of organic pigments is very complicated. Most of them are derivatives of benzene, naphthalene, or anthracene, all of which are carbon ring compounds. Some pigments contain four or five rings in one molecule. In addition, pigment molecules contain certain groupings of atoms, called ***chromophoric groups,*** which are responsible for the pigment color. Examples of such groups are =C=NH, −CH=N−, and −N=N−. These chromophoric groups must be in such positions in the molecules that the electrons can change energy levels when they are illuminated. The aromatic rings are major contributors to the chromophore. Electrons in the chromophoric group may exist in several energy levels, and when white light strikes a film of ink containing an organic pigment, electrons change energy levels, absorbing specific wavelengths of light. The pigment color, then, is due to the wavelengths that are *not* absorbed by the chromophoric group but instead are reflected to the viewer's eyes.

A dye is soluble in common solvents, while a pigment is insoluble. Pigments are usually preferred in printing inks to prevent bleeding of the printed product. Dyes are converted to pigments through several different chemical reactions.

Basic dye pigments are made by reacting basic dyes with phosphomolybdic acid ($H_3PO_4 \cdot 12MoO_3$) or phosphotungstic acid ($H_3PO_4 \cdot 12WO_3$). When reacted with both, they are known as PMTA pigments. The resulting pigments have good fastness to light, yet still have strong, brilliant colors. Pigments in this group include the rhodamines, Victoria blue, and the red, violet, blue, and green PMTA pigments.

Azo pigments are a large group of organic pigments. They are made by a chemical reaction between an aromatic amine and nitrous acid, followed by a coupling of the diazonium salt. Azo pigments include diarylide yellow, lithol rubine, red lake C, toluidine red, Hansa yellow, DNA orange, and naphthol red. All contain the azo group −N=N−.

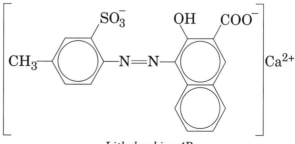

Lithol rubine 4B

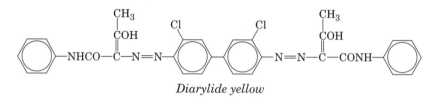

Diarylide yellow

The most important cool-shade ink pigments are the phthalocyanine pigments, which include cyan blue and cyan green. Copper phthalocyanine, called phthalocyanine blue, phthalo blue, or cyan blue, is produced in two shades, one called red-shade and the other green-shade cyan blue. The green-shade cyan blue makes an excellent cyan for process color work. If chlorine is introduced into the molecules of cyan blue, the result is a brilliant green pigment, called cyan green. This, too, is produced in two shades. All the phthalocyanine pigments are fast to light and alkalies. They are highly insoluble, resisting bleeding. Inks made with them dry easily.

Phthalocyanine pigments are made by heating phthalic anhydride with urea and cuprous chloride, Cu_2Cl_2. The structure includes four six-membered rings and eight nitrogen atoms.

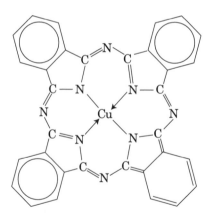

Phthalocyanine blue

Flushed Pigments   Organic pigments for printing inks are usually prepared as flushed pigments. The color is commonly prepared as a dye, then converted to a pigment by precipitating from a water solution. The mixture is filtered, washed, and pressed to remove most of the water. The water in the wet pigment filter cake (also called press cake) can be removed by a process known as flushing. The wet pigment is mixed with a viscous varnish, such as a linseed alkyd, in a large dough mixer. The varnish gradually replaces the water adsorbed on the pigment surface, and the pigment passes from the aqueous phase to the oil phase. The displaced water collects during the mixing operation and can be poured off by tilting the dough mixer. The rest of the water is removed by heating under vacuum. The mixture remains very short until practically all of the water is removed, when it suddenly changes to the consistency of a heavy-body paste ink. The flushed pigment is sold in this form to the ink manufacturer.

Flushed pigments have advantages and disadvantages. Flushed pigments do not contain grit or unground, dry particles, and they are easily mixed with other ink ingredients. They have high color strength and low moisture content.

Some pigments, such as alkali blue, dry to a hard horny texture with reduced color strength when prepared as a dry pigment. Flushed pigments suffer a lack of flexibility in vehicle formulation, the requirement of a relatively large inventory to ensure uniformity of color, an opacity insufficient for many uses, and a number of color and body changes that occur during manufacture and storage.

Most of the inks made in North America are made using flushed pigments, and flushed pigments have become popular in Europe. Organic pigments are commonly flushed with an alkyd or rosin ester varnishes. The use of such flushed pigments improves the fast-setting properties of quickset inks.

Dry Color

If instead of flushing, the filter cake is dried in an oven, and crushed and ground, the powdered product is referred to as dry color. These must be ground with a varnish to prepare a printing ink.

Pigment Chips

In addition to dry color and flushed pigment, pigments are sold in the form of chips, pigment dispersed in a solid resin. Chips are mixed into solvent or varnish to dissolve the resin binder, dispersing the pigment and making the ink. Chips are usually used for preparing liquid inks—flexo and gravure inks—for packaging. They produce inks with exceptional color strength, transparency, and gloss.

Pigment chips containing an organic pigment (60% by weight) in acrylic resin (40%) produce water-based flexo inks with greater gloss, transparency, and color than inks from dry color. Use of these chips, however, is more costly than the more familiar dispersions that are let down (diluted) to form a press-ready ink.

Coated or Rosinated Pigments

Pigment suspensions are usually treated with rosin or a rosin derivative before the pigment is made into a press cake. Flushing or drying causes the rosin to coat the individual particle, improving transparency of the pigment and giving a darker masstone to the ink. The pigments are also called "rosinated" pigments. (A *resin* is an amorphous, translucent, fusible, or thermoplastic substance; *rosin* is a natural resin obtained from pine trees.)

White Pigments

The major white pigments are titanium dioxide, calcium carbonate, and clay. There are no white organic pigments.

The most important white pigment is titanium dioxide, TiO_2. It is insoluble and chemically inert with excellent resistance to heat and light. It comes in two crystalline forms, *anatase* and *rutile.* Rutile is bluer, more opaque, and harder than anatase. Rutile is used to make opaque tints from transparent inks. Anatase is preferred in gravure inks because it is less abrasive. Many grades are available.

Calcium carbonate ($CaCO_3$) is available as ground limestone or as precipitated calcium carbonate. Ground limestone, available in many grades, ranges from very hard to soft. Printing inks usually use the precipitated form, a product with controlled, uniform properties, which is made by reacting a solution of calcium hydroxide (lime) with carbon dioxide.

$$CaCO_3 \rightarrow CaO + CO_2$$

$$CO_2 + Ca(OH)_2 \rightarrow CaCO_3 + H_2O$$

Clay is a complex aluminosilicate, found extensively in nature. The hexagonal platelike form, *kaolin,* is the principle form used in printing inks. It has low opacity and is used largely as an extender for letterpress and screen inks.

Like calcium carbonate, clay, treated clays, fumed silica, and blanc fixe (barium sulfate) are sometimes used as extenders. Other white pigments, such as alumina hydrate, gloss white, magnesia, and zinc oxide are rarely used in printing inks.

Inorganic Colored Pigments

Before discovery of the synthetic organic dyes in the mid-19th century, printers and painters had only inorganic pigments and natural dyes extracted from plants and some animals. The inorganic pigments were usually dull and dirty in hue, while the organic pigments were costly and fugitive to light.

Most of the colors used now are synthetic organic colors, which have overcome the disadvantages. Some old inorganic colors are still used, and many new synthetic inorganic colors have been developed.

The principal colored inorganic pigments used in printing inks are the iron blues. While the iron blues have different shades, they are essentially the same chemically, consisting of ferric ferrocyanide. They are produced by adding a solution of potassium or sodium ferrocyanide to a solution of ferrous sulfate. A light-blue precipitate of ferrous ferrocyanide is formed:

$$K_4Fe(CN)_6 + 2FeSO_4 \rightarrow Fe_2[Fe(CN)_6] + 2K_2SO_4$$

The light-blue precipitate is converted into a deep-blue precipitate of ferric ferrocyanide by the addition of hydrochloric acid (HCl) and an oxidizing agent such as sodium dichromate ($Na_2Cr_2O_7$) or sodium chlorate ($NaClO_3$). The net effect of this reaction is to oxidize the ferrous ion, Fe^{2+}, to ferric ion, Fe^{3+}. The oxidation changes the precipitate to ferric ferrocyanide $Fe_4[Fe(CN)_6]_3$. The different shades of iron blue are produced by changing the acidity of the solution, temperature, and heating time during the oxidation.

The iron blues are lightfast and have high tinctorial strength, but less than that of some organic blue pigments. Dilute acids do not affect iron blues, but alkalies destroy their color. They may be mixed with most other pigments, except those that are reducing agents or are alkaline in nature. The iron blues include Prussian blue, bronze blue, Milori blue, and Chinese blue.

In the past, chrome and lead pigments have been used (chrome yellow, molybdated chrome orange, and others), but the potential health hazards and the problems of disposing of materials containing lead and chromium have virtually eliminated their use in printing inks. They have been replaced in most applications by organic pigments.

Pigment Flocculation and Bleeding

A good pigment consists of very tiny particles. These are dispersed throughout the varnish when the ink is ground on an ink mill. If something causes these particles to clump together to form fewer but larger particles, the pigment is said to *flocculate.* Flocculation may occur when a lithographic ink contacts water on the press. If a pigment flocculates, the ink becomes short or pasty, causing stripping of the ink from the ink rollers and piling on rollers and plate. Luckily, most pigments do not flocculate.

Pigment flocculation may be explained as follows: pigment particles are surrounded by an adsorbed film of varnish molecules, even after the dampening solution becomes emulsified in the ink. Some pigments, however, are more easily wet with water than varnish, and in the presence of water, they take on an adsorbed film of water. The pigment particles then become more polar, since water molecules are highly polar. Polar particles have an electrical attraction for each other, and they join together or flocculate (figure 8-1).

Figure 8-1.
Pigment flocculation.

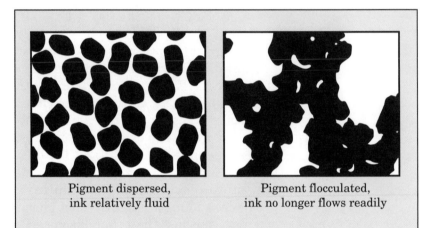

| Pigment dispersed, ink relatively fluid | Pigment flocculated, ink no longer flows readily |

Pigments for lithographic inks must be chosen so that they do not bleed into the dampening solution—they must not be soluble in water or in dilute acid solutions. This requirement is no problem with inks for other printing processes, where the inks do not come in contact with water.

Varnishes

The varnish is the part of an ink in which the other materials—pigments, driers, waxes, and modifiers—are suspended or dissolved. The varnish must dissolve or disperse these ingredients and enable inks to dry rapidly and produce high scuff resistance.

The many types of varnishes include mineral oil, soya oil, resin-oil, and resin-solvent varnishes and drying oils modified with alkyd, urethane, or phenolic resins. Besides these, there are special varnishes for inks that dry with ultraviolet light or electron beams, and for inks that dry on various types of plastic films.

Drying-oil Varnishes

The chemistry of drying oils is explained in chapter 3, "Chemistry of the Compounds of Carbon." The main drying oils used to produce varnishes for lithographic, letterpress, and screen process inks are linseed oil, tall oil (from kraft pulping of wood), soybean oil, safflower oil, and china wood oil (tung oil). These oils can be made into varnishes and modified by bodying them (increasing their viscosity) by heating for various lengths of time. The longer they are heated, the greater the viscosity of the varnish.

Except for soya oil, unmodified drying-oil varnishes are little used in modern inks. But the drying oils or fatty acids derived from them are used to make the commonly used alkyd, urethane, quickset, and gloss varnishes.

Alkyd Varnishes

Alkyd varnishes are used in inks for lithography, screen, and letterpress printing. They are the main varnishes used to make inks for metal decorating. Inks made with alkyd varnishes dry rapidly, and the dried films are tough, with good scuff resistance.

One way to make an alkyd varnish is to react phthalic anhydride (phthalic acid minus a molecule of water) with glycerol (glycerin) to produce a glyptal resin. The product is polymeric because each phthalic anhydride molecule can react with two glycerol molecules, and each glycerol molecule can react with three phthalic anhydride molecules.

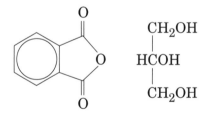

Phthalic anhydride   Glycerol

The phthalic anhydride and glycerol are reacted with a drying oil, such as a linseed oil, to produce a linseed alkyd. If a high proportion of drying oil is used, the result is called a ***long-oil alkyd*** varnish. This is the type normally used for ink making.

Urethane Varnishes

Urethanes are produced by reacting an isocyanate with an alcohol.

$$R—N{=}C{=}O + HOCH_2R' \rightarrow R—NH—CO—OCH_2R'$$

Materials such as toluene diisocyanate (TDI) and tri-methylol propane (TMP) are typical reactants.

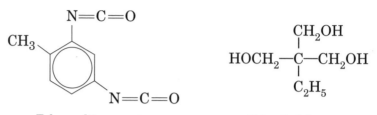

Toluene diisocyanate   *Trimethylol propane*

These components react with drying oils to form a urethane-modified drying-oil varnish. Such varnishes have good pigment wetting characteristics and are used in lithographic sheetfed inks. They are little used in metal decorating inks, as they tend to yellow when baked.

Phenolic Varnishes

Under proper conditions, phenol and formaldehyde react to produce an insoluble resin, such as Bakelite. Resins used for printing inks are usually made from phenol derivatives.

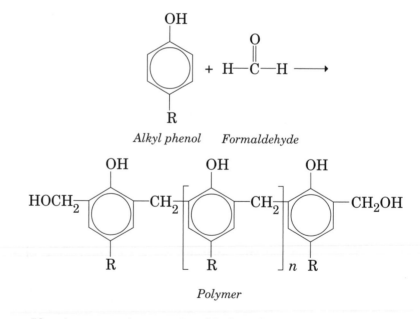

Alkyl phenol   *Formaldehyde*

Polymer

If rosin or a rosin ester is added to the phenol and formaldehyde, and the mixture heated, a solid resin is formed. The rosin introduces oil solubility and, when it is heated with a drying oil, makes the resin suitable for use in an ink varnish. These resins are used to make high-gloss inks and varnishes.

Rosin Esters

A common method of making a rosin ester is to react rosin with pentaerythritol, a solid with a melting point of 504°F (262°C). Its formula is $C(CH_2OH)_4$: one carbon atom surrounded by four $-CH_2OH$ groups. To make an ink varnish, the rosin ester is dissolved in a high-boiling hydrocarbon solvent. One that is commonly used has a boiling-point range of 460–520°F (238–271°C).

Maleic-Modified Rosin Esters

Upon heating, maleic acid loses a molecule of water to produce maleic anhydride (anhydride means "minus water").

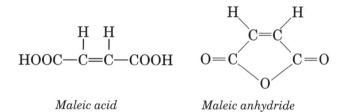

Maleic acid   *Maleic anhydride*

Rosin and maleic anhydride combine upon heating to produce a resin. The product has a high acid content, and it is esterified with a polyol to produce a maleic-modified rosin ester. To make varnishes for gloss inks and quickset inks, the modified rosin ester is heated with a drying oil to produce a long-oil varnish. To make a resin-solvent varnish for heatset inks, the resin is dissolved in a high-boiling hydrocarbon solvent.

Hydrocarbon Resins

Hydrocarbon resins are produced as byproducts from the manufacture of materials such as ethylene, butadiene, and indene that are used to make polymers. These resins can be dissolved in a hydrocarbon solvent to produce an inexpensive varnish for use in publication and commercial inks.

Shellac

Owing to shortages and high prices, shellac, a natural resin obtained from the lac beetle, has been largely displaced with acrylic emulsions and dispersions or by alcohol-soluble maleics and phenolics. Shellac is still used in some flexographic inks.

Soybean Oil

Refined ("food grade") soya oil is used in sheetfed offset inks to replace up to 20% of the more expensive vehicle, such as linseed alkyd. This improves ink/water balance and ink transfer properties. In heatset web offset inks, some of the hydrocarbon oil can be replaced with less-volatile soya oil.

The reduction in volatile organic compounds (VOCs) is especially important where printers face stringent environmental limitations. Soya oils in some ways are better solvents than the heatset oils.

In newspapers, soya oils improve the appearance of colored prints. The cost of soya oil has been a deterrent to its use in black news inks. Soybean oil has lower volatility than many ink oils, and its lower volatility helps printers cope with tight VOC limitations.

Gloss Inks

It is the varnish that makes an ink glossy. Rosin-modified phenolic and rosin-maleic anhydride varnishes are commonly used. The drying oil in such varnishes enables them to polymerize and become solid (or dry) in the presence of air. The rosin-maleic anhydride varnishes do not give quite as much gloss as the phenolics, but they have less tendency to yellow as printed matter ages.

Quickset Inks

The varnishes used in quickset inks vary from one ink manufacturer to another. The varnishes often include rosin-modified phenolics, alkyds, urethane alkyds, maleic-rosin esters, and hydrocarbon resins. In addition to these resins, the varnishes contain a high-boiling hydrocarbon solvent (a heatset solvent). The quickset ink is printed on a coated paper, and the solvent soaks into the coating, leaving the high-viscosity varnish on the surface. The viscosity rises rapidly, and the ink has "set." The ink film then dries by oxidation.

Heatset Inks

The fastest and surest way to get an ink to dry on coated paper is by evaporation of the solvent from the ink film. Heatset inks consist of a pigment plus resins dissolved in a specially purified hydrocarbon solvent. After printing, the solvent is evaporated in a high-velocity, hot-air dryer, and the melted resin is set by cooling the web on a chill roll. (Actually, as much as 25% of the solvent remains in the printed paper so that the "drying" process increases the viscosity of the ink film, and the remaining solvent keeps the ink film flexible.) For gravure and flexo inks, a lower temperature is required, and the binder is sufficiently set without using a chill roll.

Heatset varnishes are made by cooking a solid synthetic resin with a heatset oil. The oil dissolves the resin, forming a varnish suitable for making heatset inks. Varnishes of different viscosities are made by using different resins and different proportions of resin and oil. In addition to varnish, heatset inks contain a considerable percentage of "free" heat-set oil to give them the correct body for good printing.

For heatset web offset inks, typical varnishes are those made with rosin esters, maleic rosins, and/or hydrocarbon resins. These varnishes are made with high-boiling specially purified hydrocarbon solvents, and no drier is needed. Sometimes a small amount of drying oil is added to a maleic-rosin varnish to improve the grease resistance of the dried ink.

Heatset inks once contained solvent equal to about 40% of the weight of the ink, including the solvent in the varnish. Such inks are still being used on web offset presses. Use of hydrocarbon solvents treated by hydrogenation produces low-odor, low-smoke heatset inks. Modern varnishes make it possible to reduce the solvent content of the inks to 25–30%. Using such inks helps to reduce air pollution.

Varnishes for Ultraviolet Curing Inks

An ink formulated for drying with ultraviolet (UV) light contains *oligomers* and a photoinitiator. (Monomers are sometimes used.) Oligomers are low-molecular-weight polymers (say three to five or ten monomer units) produced by controlled polymerization. On irradiation with UV they polymerize to form high-molecular-weight polymers. Because these oligomers do not readily polymerize simply with irradiation, UV inks must contain a photoinitiator that, when illuminated with UV light, produces free radicals to initiate the polymerization reaction. All of these materials are involved in the drying process, and they are described in the "Mechanism of Ink Drying" section of this chapter.

It is important that the oligomer be free of monomeric acrylates that are highly irritating to the skin and eyes.

Varnishes for Electron-Beam Curing Inks

A beam of electrons is a powerful means for the curing or drying of coatings or inks that are formulated with the same types of oligomers used for UV inks, but no photoinitiator is required. Electron beam (EB) is used for the drying of inks and coatings on packaging. When mounted on printing presses, EB units must be shielded to absorb X-rays that are produced.

Although the original unit is more expensive than a UV installation, EB offers some advantages. EB is "color blind," that is, unlike UV, it does not dry one color faster than another. A nitrogen blanket is required to prevent the inhibiting effect of oxygen upon drying.

Printing on Plastics

Most plastic films are printed by flexography or gravure. Two types of inks are used: "aggressive" inks that contain a solvent that softens the plastic surface, helping the ink to adhere, and "nonaggressive" inks that do not attack the surface but adhere solely because of mechanical and chemical bonding.

Ink must be formulated for the particular type of plastic. Thus polystyrene and sometimes vinyl plastics require an aggressive ink. (However, some vinyl fabrics become too fragile when printed with an aggressive ink. In these cases, a nonaggressive ink must be used, even though it does not bond as well to the vinyl.) Nonaggressive inks are formulated for polyethylene, polypropylene, and polyester. In order to obtain ink adhesion on polyethylene and polypropylene, the surface must first be treated with a corona discharge. The extruder of the films usually applies this treatment, but it is sometimes done in the printing plant on the press.

Coating of Wet Ink Films

An overprint varnish is one type of coating that can be applied over printed inks, but this section refers to the use of a water- or alcohol-based film-forming coating solution to cover the wet ink. Such a solution must not dissolve the inks, and the solvent in the overcoating must be volatile.

The use of an overcoating allows inks without solvents to be used on web presses to avoid air pollution. Overcoating can also be used on sheetfed presses, where it has found application particularly for the printing of board. The overcoating is dried rapidly with an air blower or mild heat. After overcoating, printed board can be stacked high on a sheetfed press delivery pile without setoff or blocking; no antisetoff spray is needed. Sheets can be diecut only a few hours after printing.

A typical overcoating contains about 30% resin, such as an alcohol-soluble form of the propionate ester of cellulose. In this case, the solvent consists of about 70% alcohol and 30% water. Water-soluble acrylic resins are also used.

Work to be overcoated can be printed with conventional inks if they are formulated without silicones or hard waxes, which prevent the overcoating from forming a complete film over the wet inks. When printed sheets are protected with an overcoating, they can be handled within a few minutes, even though the inks may require several hours to dry. There is no problem with the recycling of sheets printed in this way.

As an alternative to water- or alcohol-based overcoatings, a clear, ultraviolet-curable coating may be used over dry or wet UV inks. This is being done on web offset presses. Such a coating is also used to a limited extent over wet conventional sheetfed inks, but this use reduces the gloss of the coating. A better method is to allow the sheetfed inks to dry and then apply the UV varnish.

While overcoatings of one type or another have several advantages, they have the disadvantage of greater cost. This is not important in sheetfed operations, since the cost is less than that of an overprint varnish. But when a fast-moving web must be completely covered on both sides, the cost is high. The cost of overcoating compared with the cost of using UV inks varies depending on the percentage coverage of the UV inks.

Resins for Gravure Inks

There are many types of gravure inks, each requiring a different type of formulation and using different resins. Gilsonite, a dark hydrocarbon resin that is mined from the earth, is commonly used for inexpensive black inks. Colored publication inks use ***metallated rosins.***

Rosin, a mixture of compounds, is obtained from pine trees. It contains such materials as abietic acid, which can be reacted with alcohols such as pentaerythritol to form high-melting esters. More frequently, it is reacted with a metal derivative to form a rosin soap (a salt of the rosin acid). Zinc, magnesium, or lithium reduce its acidity and raise its melting point. The product is called metallated rosin. The name ***limed rosin*** reflects the historical use of lime to neutralize the rosin.

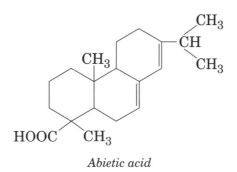

Abietic acid

A variety of ***cellulose esters*** are used to prepare gravure and flexo inks. When cellulose is reacted with nitric acid, an ester properly called cellulose nitrate is produced, although it is commonly called "nitrocellulose" or even "cotton." With small amounts of nitric acid, the product is soluble in hydrocarbons (spirit soluble); with larger amounts, the product is soluble in alcohol. These materials are highly flammable.

Nitrocellulose explosive or guncotton has been reacted with even more nitric acid. Ethylcellulose and ethylhydroxyethyl cellulose (EHEC) may be used in publication inks, but cellulose nitrate is used mostly in packaging gravure.

Gravure inks for packaging may also be made from chlorinated rubber, vinyl polymers, and polyamides. Vinyl polymers are used for ink to print on vinyl film.

Resins for Flexographic Inks

Since flexo and gravure are both used on high-speed webs for packaging and publication, flexo inks use many of the same resins used in gravure inks. Rosin esters, shellac, polyamides, and alcohol-soluble nitrocellulose are commonly used for flexo packaging inks. Styrene-maleic resins provide an inexpensive caustic-soluble binder used in flexo for paper bags and corrugated boxes.

Polyamides, widely used in flexo inks, are produced from dimer acid or trimer acid. A fatty acid, such as linoleic acid, is dimerized or trimerized and then reacted with a polyamine such as diethylene triamine.

$$CH_3(CH_2)_4CH{=}CHCH_2CH{-}CHR(CH_2)_7COOH$$
$$CH_3(CH_2)_4CH{=}CHCH_2CH{-}CHR(CH_2)_7COOH$$

Dimer acid

$$CH_3(CH_2)_4CH{=}CH{-}CH_2{-}CH{=}CH(CH_2)_7COOH$$

Linoleic acid

$$CH_3(CH_2)_4CH{=}CHCH_2CH{-}CHR(CH_2)_7COOH$$
$$CH_3(CH_2)_4CH{-}CHRCH_2CH{-}CHR(CH_2)_7COOH$$
$$CH_3(CH_2)_4CH{-}CHRCH_2CH{=}CH(CH_2)_7COOH$$

Trimer acid

$$H_2NCH_2CH_2NHCH_2CH_2NH_2$$

Diethylene triamine

Polyamides are dissolved in a hydrocarbon/alcohol cosolvent, but some are soluble in alcohol alone.

Ethylene-acrylic acid copolymers are soluble in ammonia and volatile bases, providing excellent binders for flexo news inks and other water-based flexo inks. The polymers precipitate on contact with the acid newsprint, and the dried inks have good rub resistance and holdout.

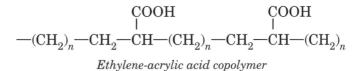

$$-(CH_2)_n-CH_2-\overset{\overset{\displaystyle COOH}{\displaystyle |}}{CH}-(CH_2)_n-CH_2-\overset{\overset{\displaystyle COOH}{\displaystyle |}}{CH}-(CH_2)_n$$

Ethylene-acrylic acid copolymer

Solvents

The most familiar solvent is water, and its low toxicity, low cost, and low impact on the environment make it increasingly popular in flexo and gravure inks. Because of the ink-water relations on the offset plate, it is not used as a solvent for offset inks.

Water is the only inorganic solvent used in printing. Organic materials used as solvents were discussed in chapter 3: hydrocarbons, alcohols, glycols, glycol ethers, esters, and ketones.

Heatset Solvents

High-boiling-point hydrocarbon solvents (frequently called **Magie Oils,** a familiar trade name) are used in the manufacture of heatset inks for web offset and quickset inks for sheetfed offset. Heatset inks are dried very rapidly by passing the printed paper web through a dryer. Much of the heatset solvent in the inks is vaporized and passes out of the dryer into the air unless it is condensed or recycled. As a result, the ink on the web becomes dry as soon as the paper web is cooled.

Cosolvent Inks

Sometimes a mixture of two liquids is a more powerful solvent for a resin than either liquid alone. An alcohol or glycol added to a hydrocarbon is called a cosolvent, and the mixture is used in inks or varnishes, especially for screen printing. When inks containing such a varnish are heated in the press dryer, the alcohol evaporates (along with much of the hydrocarbon solvent) while the resin precipitates. The exact balance of screen stability, fast drying, and cost may require two or even three solvents such as propylene glycol and an aromatic or aliphatic hydrocarbon.

Environmental regulations usually prohibit the discharge of evaporated solvent to the air. Sometimes, the solvent can be recovered by using condensers to chill it, but often it is more economical to burn the effluent and to use the heat to dry the ink. Gravure ink solvents are usually recovered, while flexographic and lithographic heatset solvents are usually burned.

Driers

An ink containing drying-oil varnishes dries by a slow, complicated chemical reaction called ***oxidative-polymerization.*** A drier is added to such inks to accelerate their drying. The drier is not changed chemically during the reaction. It acts as a catalyst to speed the oxidation and polymerization of the varnishes.

Chemical Composition of Driers

Driers consist of liquid driers and paste driers. Paste driers are salts of fatty acids, while liquid driers are salts of ***naphthenic acids,*** aliphatic acids obtained by distilling petroleum.

The best driers are compounds of cobalt, manganese, cerium, or zirconium. These metals combine with organic fatty acids (such as linoleic acid) to form soaps (salts) that are soluble in the varnishes. Other driers are cobalt, manganese, cerium, or zirconium resinates or salts of acids found in soya oil. Octoate soaps are now the most used because of their low odor. Octanoic acid has the formula $C_7H_{15}COOH$; a cobalt octoate drier has the formula $(C_7H_{15}COO)_2Co$.

Metal salts of naphthenic acid are soluble in aliphatic hydrocarbons and form low-viscosity "naphthenates." Naphthenic acids are the acidic constituents obtained from distillation and hydrogenation of petroleum. They contain such materials as cyclohexane carboxylic acid or caproic acid (hexanoic acid).

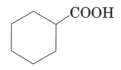

$$CH_3(CH_2)_4COOH$$

Cyclohexane carboxylic acid   *Hexanoic acid*

Amount of Drier Needed

Natural varnishes contain some oxidation inhibitors, and they dry very slowly. Addition of small amounts of drier overcomes this inhibitor but does not accelerate drying. As more drier is added, the ink dries faster. The rate of drying increases rapidly with the first small additions of drier. As more drier is added, it has less effect on the drying time. A point is finally reached where addition of more drier has little or no effect on the drying time (figure 8-2).

Figure 8-2.
Effect of drier content
on ink drying.

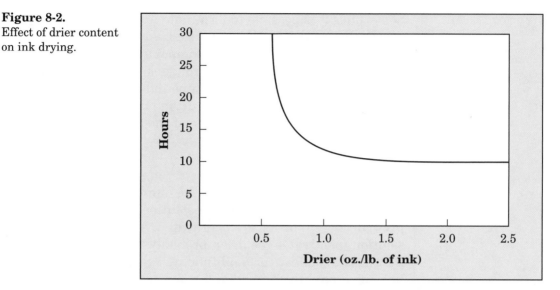

The amount of drier needed to dry an ink in, say, four hours depends on the nature of the drier (cobalt, zirconium, etc.), on the drying properties of the varnishes in the ink, and on the pigments that are present. An ink with colored pigments may dry in four hours when it contains 0.5–1% of a 12% cobalt drier, while a black ink may require 1–2% of the same drier for the same drying speed. An ink containing only an iron blue pigment will sometimes dry satisfactorily with no drier. The pigment itself acts as a drier.

Carbon black, a pigment with very fine particles, adsorbs more drier than do the colored pigments that are much larger in size. The adsorption inactivates the dryer, making it necessary to use more drier when such pigments are present. The additional drier ensures that sufficient active drier remains to dry the ink in a reasonable length of time.

Drier Dissipation

As inks containing drying oils age in the can, the drying time often increases. This effect is called ***drier dissipation.*** The drier, being nonvolatile, cannot leave the ink, so it is likely that the pigment slowly adsorbs more and more drier as the ink ages in the can, making the drier inactive. Inks more than one year old should not be used before checking for rate of drying.

If an ink has suffered drier dissipation, the remedy is to add more drier just before the ink is used on a job.

Drier Accelerators

Materials such as cobalt acetate and magnesium perborate are sometimes added to the dampening solution to accelerate the drying of offset inks.

Feeder Driers

Sometimes it is possible to formulate an ink so that part of the dissipated drier will be replaced with new drier. This is the purpose of *feeder driers.* Compounds such as cobalt borate, manganese borate, and magnesium perborate can serve as feeder driers. These compounds are only slightly soluble in varnishes containing drying oils. As part of the original drier is dissipated, some of the feeder drier dissolves in the varnish to replace it.

Drying Problems

The greatest single cause of slow drying in sheetfed offset inks is high acidity (low pH) in the dampening solution. When the pH of the dampening solution is less than around pH 4, drying problems are aggravated. Acid in the dampening solution inactivates the drier, probably through hydrolysis; that is, the linoleate or naphthenate salt is converted to linoleic or naphthenic acid and a phosphate by the phosphoric acid in the dampening solution.

Modifiers

Some inks are made with nothing but pigment, varnish, and drier. Web inks have pigment and varnish but no drier. Usually, however, small amounts of other materials, called modifiers or compounds, are also included. These are ingredients such as antiskinning agents, wax compounds, reducers, and solvents.

Antiskinning agents are added to some inks to help prevent skinning of the ink in the ink fountain and on the press rollers. Such agents will not prevent the ultimate skinning of the ink, but they can in many cases increase considerably the time required for an ink to skin in the can or ink fountain. The addition of an antiskinning agent to an ink always increases its drying time somewhat. A good antiskinning agent is one that increases the skinning time greatly without much increase in drying time. Materials such as methyl ethyl ketoxime, butylated hydroxy toluene (BHT), and hydroquinone are used.

Waxes are added to printing inks to improve slip, mar-resistance, and water repellency. They reduce tack without changing the viscosity of offset inks. Waxes usually decrease gloss and increase drying time so that they must be added with care. They can also interfere with dry trapping of inks.

Many kinds of waxes are used: polytetrafluoroethylene (PTFE), polyethylene wax, fatty acid amides such as stearamide ($C_{17}H_{35}CONH_2$), and a variety of hydrocarbon waxes including paraffin and microcrystalline waxes.

Table 8-IV.
Ink modifiers and
additives.

Type	Example	Specific Function
Defoamers and antifoam	Silicones	Reduce foam
Waxes	PTFE	Reduce scuff and marking
Dispersants	Nonionic detergents	Improve pigment dispersion
Photoinitiators	Benzoin methyl ether	Promote curing of UV inks
Adhesion promoters	Organic titanate	Promote adhesion to film
Wetting agents	Polyglycols	Improve spreading of inks and dispersion of pigments
Driers	Cobalt octoate	Speed sheetfed ink drying
Antiskinning agents	Hydroquinone	Retard formation of skin on ink on the press

Petroleum solvents are also used to provide quick setting properties. Petroleum solvents are sometimes added to reduce the tack of an ink. A light varnish, such as a 00 litho varnish, also reduces ink tack. It is much easier to reduce the tack of an ink than it is to increase it. Tack can be increased somewhat by the addition of a No. 7 or No. 8 litho varnish, or by the addition of a heavy gloss varnish.

Modifiers must be handled intelligently or they can cause more harm than good. Modern practice is to add necessary modifiers at the ink manufacturing plant. Thus inks supplied to the printer are ready to be run on the press.

Mechanisms of Ink Drying

Inks dry in several ways depending on the type of varnish and whether the drying is accelerated by the use of a gas-fired dryer, infrared radiation, ultraviolet light, or electron beams. Each of these methods is better suited for some printed products than for others (table 8-V).

Drying by Absorption

Ink for nonheatset web offset printing of newspapers consists mainly of carbon black dispersed in a petroleum solvent chemically similar to fuel oil. With such inks, drying occurs by the absorption of the solvent into the newsprint. These inks never dry hard—the reason that newspaper inks rub off easily onto the hands or clothes. In addition to newspapers, a great deal of commercial nonheatset web offset printing is done using news inks on newsprint and similar papers. Dry-

Table 8-V.
Ink drying methods
for some common
printed products.

Product	Method of Printing	Method of Drying
Newspapers	Nonheatset web offset, Flexo	Absorption
Newspaper inserts/ supplements	Heatset web offset Flexo, Gravure	Evaporation Absorption
Books	Sheetfed Web offset Flexo Web letterpress	Oxidation/polymerization Evaporation Absorption Evaporation
Magazines	Heatset web offset Gravure Sheetfed offset	Evaporation Evaporation Oxidation/polymerization
General commercial (brochures, ads, etc.)	Sheetfed offset Sheetfed offset Nonheatset web offset Heatset web offset	Oxidation/polymerization Radiation curing Absorption Evaporation
Packaging Film Foil Paperboard	Flexo, Gravure Flexo, Gravure Flexo, Gravure Sheetfed offset Heatset web offset	Evaporation Evaporation Evaporation Oxidation/polymerization Evaporation
Corrugated metal	Flexo Offset Offset Flexo	Absorption High-temperature baking Radiation curing Evaporation
Specialties Textiles Gift wrap Plastic laminates	Screen Flexo, Gravure Gravure	Oxidation/polymerization Evaporation, absorption Evaporation

ing oils or other varnishes are sometimes added to the ink to reduce the amount of ruboff.

Some absorption occurs when any paper is printed. On coated paper, quickset inks set largely by absorption of the low-viscosity component of the varnish by the paper coating.

Drying of Heatset Inks

Resin-solvent varnishes, used in heatset inks, contain solvent for viscosity adjustment, and some heatset web offset inks contain a certain percentage of water.

When the printed web passes through the dryer, the heat evaporates the solvent and water (figure 8-3).

Figure 8-3.
Nozzles of a hot-air,
high-velocity dryer.

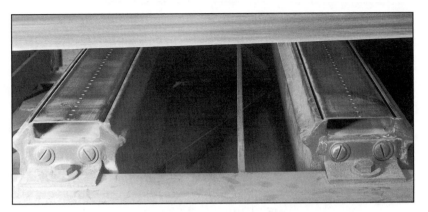

However, the heat also softens the resins so the ink film is still soft. As the web emerges from the dryer, it passes over chill rollers that cool it to a temperature of about 90°F (32°C). This hardens the resins to complete the drying. The entire drying process occurs in a second or less.

Not all of the solvent is removed from the inks; 5–15% of the weight of the dried ink film consists of solvent. But from a practical point of view the inks are dry and jobs are ready to be trimmed and collated.

Drying of Inks Containing Drying-Oil Varnishes

Drying oils dry by a polymerization of the varnish molecules, aided by oxygen from the air and a catalyst called a drier. This is true for the straight drying-oil varnishes, the alkyds, the urethanes, phenolics, and the gloss varnishes.

Oxygen in the air adds to the molecules of the drying oil to form hydroperoxides. The amount of peroxides formed increases to a maximum and then decreases. In other words, there is first ***peroxide formation*** and later ***peroxide decomposition.*** Water is one of the by-products of peroxide decomposition.

- Addition of oxygen to the carbon adjacent to the double bond.

$$R-CH_2-CH{=}CH-R' + O_2 \rightarrow R-\overset{\displaystyle OOH}{\underset{\displaystyle |}{CH}}-CH{=}CH-R'$$

- Cobalt-catalyzed decomposition of hydroperoxide.

$$R-\overset{\displaystyle OOH}{\underset{\displaystyle |}{CH}}-CH{=}CH-R + Co^{2+} \rightarrow RO^\circ + OH^- + Co^{3+}$$

The hydroperoxides decompose to form free radicals (RO°) that initiate the ***polymerization,*** causing a rapid increase in the viscosity of the ink film. The free radicals are very reactive. A free radical attacks another molecule, forming a carbon-to-carbon bond and a new free radical that attacks another molecule, adding to the growing molecule and, in turn, generating another free radical. Because many of the molecules in the growing chain have more than one reactive site, one chain can react (or cross-link) with another chain. Oxygen continues to be absorbed, and the viscosity increases until a hard, rub-resistant film is obtained. This stage may continue, at a continually decreasing rate, for a week or two as the ink film becomes harder and more rub-resistant.

The final drying stage is called ***film deterioration.*** Late in the drying process, some of the varnish molecules split into small volatile molecules including water, carbon dioxide, acetic acid, and formic acid. Some of these volatile materials create the odor of the drying ink.

Drying of Quickset Inks

Typical varnishes for quickset inks contain a resin, drying oil, and a high-boiling hydrocarbon solvent. The inks usually contain some free solvent in addition to that present in the varnish. The total solvent content is about 15%.

When a quickset ink is printed on paper, some of the solvent is absorbed into the paper, causing a rapid increase in the viscosity of the ink film, which results in setting of the ink. The oil is absorbed into the coating of coated paper faster than into an uncoated paper. For this reason, inks are more quicksetting on coated papers than on uncoated.

Modern inks for sheetfed lithography may contain pigments that have been flushed with a quickset-type varnish containing resins of limited solubility. Consequently, when the ink is printed and a small amount of the solvent is absorbed by the paper, the resin precipitates and sets the ink.

While quickset inks set rapidly, the drying process may take a few hours. The mechanism of oxidation and polymerization is the same as that for all inks containing varnishes with drying oils.

Infrared Setting and Drying

Most chemical and physical processes are speeded by increasing the temperature. Both setting and drying of inks accelerate at increasing temperature, and infrared (IR) radiation speeds the setting of inks. Furthermore, sheets exposed to IR are warmed, and ink dries faster in the pile. Temperatures in the printed pile should not exceed 110–120°F (45–50°C), or the ink will soften enough to cause blocking of the prints. The drying time required for sheetfed inks can be reduced by 75% by careful use of IR on the press.

Inkmakers use different approaches to the formulation of inks for infrared setting. One is to use quickset inks with a low drying oil content. IR radiation drives the ink solvent into the paper, precipitating the resins in the varnish and setting the inks rapidly. They may become smear-free in 30–60 seconds. Drying continues by oxidation and polymerization over a period of a few hours.

The use of IR dryers allows jobs to be backed up rapidly, while reducing the need to use antisetoff spray powder. Two types of IR dryers are manufactured. The type most popular in the United States is nearly "color-blind" (meaning that yellow inks set almost as rapidly as black inks), while the type most popular in Europe is not color-blind. One color-blind infrared dryer operates at 1,100–1,200°F (600–650°C). At this temperature, it emits radiation in wavelengths of 2.2–4.0 microns, a range that corresponds closely to the natural frequency of vibration of the molecular bonds of resin-based varnishes. Decreasing the wavelength (bringing the radiation nearer to the visible) increases the energy and dries inks more quickly, but it is much more effective for blacks than for yellows (it is not color-blind.)

Drying of Ultraviolet Inks

Drying of UV inks is largely different from the drying of other ink systems. The UV light rapidly polymerizes the liquid vehicle in the ink, and the polymerized material is solid; the ink is dry.

UV inks dry by an acrylic polymerization, a reaction that is widely used in the preparation of industrial polymers. The reaction is usually initiated by a catalyst, and it is complete in a few seconds. However, the ink must remain stable for a long time on press, and UV radiation is used to initiate the reaction.

The polymerization reaction involves four steps. The first is generation of a *free radical,* in which UV light acts on an ingredient called a *photoinitiator,* a material requiring far less energy for reaction than that required to activate the molecules of the vehicle. For example, UV radiation splits benzoin methyl ether into two free radicals that react rapidly with the molecules in the ink vehicle.

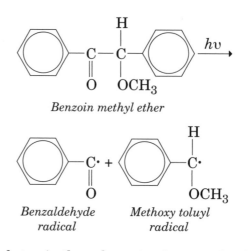

Benzoin methyl ether

Benzaldehyde   Methoxy toluyl
radical   radical

The second step in the polymerization reaction is ***initiation*** of the polymerization, a step that proceeds rapidly.

$$R-CH=CH-R' + \cdot R'' \rightarrow R-\overset{\overset{\displaystyle R''}{|}}{CH}-\overset{\displaystyle \cdot}{C}H-R'$$

The third step is ***propagation*** in which the highly reactive free radicals seek out a vinyl (–CH=CH–) or acrylic (–CH=CR–) group, adding it to the growing molecule and forming a new, reactive free radical. This is the essence of the ***chain reaction.***

$$R-\overset{\overset{\displaystyle R''}{|}}{CH}-\overset{\displaystyle \cdot}{C}H-R' + R-CH=CH-R' \rightarrow R-\overset{\overset{\displaystyle R''}{|}}{CH}-\overset{\overset{\displaystyle CH-\overset{\overset{\displaystyle R}{|}}{\cdot C}H-R}{|}}{CH}-R' \text{ (etc.)}$$

The final step is ***termination.*** Certain materials can inhibit (interfere with) the propagation, terminating the polymerization. If termination occurs before the desired degree of polymerization is achieved, additional UV energy may be required to accomplish it. If the polymerization proceeds at a much higher rate than termination, the product is a high-molecular-weight polymer. If termination proceeds rapidly, the product is of lower molecular weight.

One inhibitor is the oxygen of the air that is absorbed on the surface of the wet inks. Much of the irradiation is used to overcome this inhibition. Oxygen inhibition is reduced by passing sheets wet with UV inks through a chamber filled with nitrogen gas. Satisfactory drying is obtained with considerably less UV energy. This process is in use for the drying of inks on sheets of metal and also on offset printing paper webs that are traveling at a fairly low speed. Oxygen inhibition of EB curing is even more serious than inhibition of UV drying. A nitrogen blanket is required for EB curing.

One of the ingredients of a typical UV ink vehicle is called an **oligomer.** Oligomers are low-molecular-weight polymers that have acrylic groups at the ends. (Acrylic monomers can be used in UV inks if their volatility is low enough to prevent evaporation from the press.)

Polyfunctional acrylates have lower viscosity than oligomers, and they cure faster. The term **polyfunctional** refers to two or more acrylic double bonds in the molecule. Examples are pentaerythritol triacrylate, trimethylolpropane triacrylate, and hexanediol diacrylate. These materials are responsible for the cross-linking of the linear polymer chains.

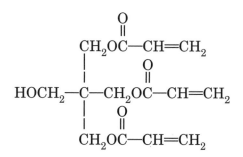

Pentaerythritol triacrylate

These polyfunctional ingredients greatly increase the molecular weight of the polymer by cross-linking polymers of lower molecular weight. One of the acrylic groups becomes part of one polymer chain and another acrylic group becomes part of another growing chain.

Coalescence

Rubber-based inks used for office duplicating and some sheet-fed commercial work, as well as **plastisol inks** used for screen printing dry by the flowing together (coalescence) of particles suspended in the varnish. Rubber-based varnishes contain particles of a chemically modified rubber dispersed in a high boiling hydrocarbon. As long as the liquid remains in the ink, the rubber particles remain dispersed, but when the ink is printed, the solvent drains into the sheet and the rubber particles flow together to form a film.

Plastisols are suspensions of vinyl resins in plasticizer. They remain stable when cold, but when they are heated, the resin dissolves, forming a film of plasticized vinyl resin. When used in a pigmented ink formulation, the inks remain stable until they are heated. The resulting tough vinyl film provides good properties for the dried ink.

Properties of Lithographic Inks

Ink properties differ according to the materials used to formulate them. Inks usually have some odor. They may be "long" or "short." They exhibit what is called "tack," which can vary greatly from one ink to another. Some have greater stability on press ink rollers than others. Sometimes the ink pigments flocculate, or they bleed into the lithographic dampening solution.

Length

A *long ink* is one that can be stretched into a long string when it is pulled away from the ink slab with an ink knife or a finger. A *short ink* breaks sooner under the same conditions. Usually a long ink has good flow characteristics (see figure 8-4). Lithographic inks need to have good flow to transfer well from one roller to another and to tolerate the emulsification of water. Screen inks need to be short so that they will not form strings when the screen is pulled away from the print.

Working an ink on the slab makes it longer. This is related to the breaking down of the internal forces in the ink. Different inks vary in the amount of increase in length with working. Since they vary with the amount of working, inks should be

Figure 8-4.
Length of ink.

compared for length only after they have been worked equally on the slab.

An offset ink must not be too long, however, because this promotes ink flying or misting.

Figure 8-5.
Ink flying and misting.

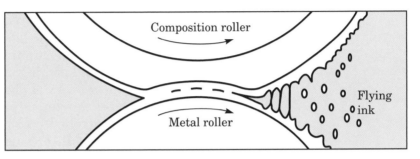

Composition roller

Metal roller

Flying ink

Yield Value

The yield value of an ink is the force required to start it to flow. Many liquids, like water and alcohol have no yield value—that is, an infinitely small force will begin flow. (This can be observed in a simple experiment: take two bottles, one with a little catsup and one with a little honey and invert them. The honey, with a very low yield value, will flow back to the bottom of the jar; the catsup, with a higher yield value, will not.) Figure 8-6 compares inks with different yield values: the yield value of one is great enough to resist the force of gravity; the yield value of the other does not. If the yield value of an ink is too high, it will not flow out of the ink fountain. The printer refers to the ink "hanging back" or "backing away from the roller."

Litho inks are thixotropic, that is, their viscosity decreases when they are worked or stirred. Working them with a conical ink agitator in the ink fountain improves their flow.

Figure 8-6.
Two inks with different yield values. The one on the left has the higher yield value.

Ink Tack

Tack is the resistance to splitting of a thin film. It can be described as "stickiness." It is judged by the pull on one's finger, wet with a thin film of ink on the slab, when it is pulled quickly away from the slab. Tack manifests itself during printing in the power required to drive the rollers of the ink-distributing system and the pull required to separate the paper from the printing areas of the offset blanket. Excessive ink tack results in excessive consumption of power, generation of heat in the rollers, and picking or tearing of the paper.

On the other hand, a low-tack ink does not give sharp halftone dots on the plate, blanket, or paper. Tack, therefore, must be measured and controlled.

In measuring tack, both surfaces must first be wet by the ink. When the two surfaces are pulled apart rapidly, the film between them splits. Usually about half of the ink adheres to each surface. However, when ink is transferred from an offset blanket to paper, more than 50% is transferred to the paper, due to the absorption of some of the ink vehicle by the paper, which increases the tack on the paper side.

It was found in the GATF research laboratory that the force required to split an ink film when it is printed on paper is almost twice as great as that required when splitting from smooth metal.

Measuring Ink Tack

Several instruments are available to measure ink tack. Tack is commonly measured using an instrument called an ***Inkometer*** (Figures 8-7 and 8-8).

Figure 8-7.
The MBC Inkometer.
(Courtesy Thwing-Albert Instrument Co.)

Figure 8-8.
Model 106 Electronic
Inkometer.
*(Courtesy Thwing-
Albert Instrument Co.)*

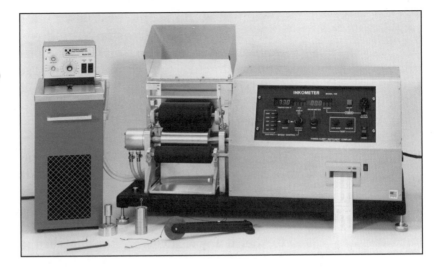

The Inkometer has three rollers, a driven chrome roller, a rubber rider roller, and a rubber vibrating roller to distribute the ink evenly. The splitting ink film produces a torque on the rider roller. A device, called a transducer, converts force into an electrical signal that is displayed as a dial reading or digital number on the instrument. This number is the ***tack number,*** and its dimensions are gram-meters (of torque).

Two instruments are available, a mechanical instrument on which the force or torque is measured by a sliding weight on a beam and an electronic model that has a digital readout.

Ink manufacturers use tack meters to assure that the ink will meet the printer's requirements. Tack readings vary, depending on the instrument used and the speed of the rollers—the higher the speed, the higher the tack reading. The tack of sheetfed ink is often measured at 800 revolutions per minute (rpm) and that of web offset ink at 1,200 rpm. Therefore, stating that an ink has a tack of, say, 16 is meaningless if the testing speed and temperature are not indicated. The use of a tack-measuring instrument is especially valuable for comparing the tacks of inks measured at the same speed and temperature on a particular instrument.

Importance of Tack

If the tack of an ink is too high, it may cause picking of paper or the pulling of coating from coated paper. Low tack contributes to dot gain, poor dot sharpness, drying problems, and low gloss.

The most important reason for controlling tack is to control trapping on a multicolor press. The best trapping is achieved when inks for four-color, wet-trapping process work are adjusted so that the first-down ink has the highest tack, and the succeeding colors have progressively lower tack. This adjustment is made so that the inks printed later will trap well on the inks printed first. A high-tack ink on the paper will trap a lower-tack ink from the blanket. If the first-down ink has a lower tack, poor trapping occurs, which causes variations in the color of the print.

A set of inks for four-color process work has a certain tack as received from the inkmaker. But, once the inks are on the press, several things begin to happen that either increase or decrease the tack of each ink. Unless such changes are the same with each ink, poor trapping will cause color variability.

Evaporation of the solvent increases the tack. Tack increases further if any of the solvent is absorbed into rollers or blankets. As the press operates, the ink rollers heat, heating the ink and decreasing the tack. As one ink is printed on paper, a little of the solvent is absorbed into the surface of the paper, increasing the tack of the printed ink film.

On lithographic presses, the printing plates are kept moist with water. Some of the water becomes emulsified in the ink, decreasing its tack.

Because of all of these variables, good quality demands that the entire process be kept under close control. To overcome trapping problems, the tack sequence can be altered, but this solution generates additional problems.

The optimum tack of inks has changed over the years. When ink varnishes were heat-bodied drying oils and lithographic plates were not as good as they are now, it was common to use inks with tack numbers varying from 25–30 (800 rpm). Later, when quickset inks were developed, it was possible to produce inks with tacks varying from about 16–25 for sheet-fed printing. If the tack was reduced much below 16, the ink became "soupy" and did not print sharp.

Modern "low-tack" inks for sheetfed printing have tack readings varying from 16 down to 10. The heavy body and low tack are brought about by using gel varnishes. It is not necessary that a heavy-bodied ink must have a high tack. In fact, high-tack inks cannot be used at modern press speeds without picking the paper.

Generally, the tack range of inks must be decreased as speeds increase, and web offset press inks have very low tack. Higher tacks are usually required for sheetfed offset inks. Low-tack inks may perform satisfactorily on sheetfed presses that operate at higher speeds.

Emulsification of Water in Ink

The statement "ink and water don't mix" is not strictly true. It is true that water does not dissolve in ink, but tiny droplets of water become dispersed in ink running on a lithographic press. This produces a water-in-ink emulsion (see figure 8-9). To perform well, a litho ink must emulsify a significant amount of water. An ink that does not will not transfer properly on the ink rollers, and they become stripped of ink. The tendency of the ink to emulsify water is determined by all of its ingredients, but predominately by those in the varnish.

Figure 8-9.
Water-in-ink emulsion *(left)* and ink-in-water emulsion *(right)*.

If an ink emulsifies too much water, or if it emulsifies large droplets, it becomes short and "buttery" and fails to transfer well. The ink is said to be ***waterlogged.*** This condition can lead to a washed-out print and piling of the ink on the rollers and plate. It can also cause gum streaks when a plate containing such a "waterlogged" ink is gummed up.

During the operation of a lithographic press, water is transferred to the plate along with the ink in two forms: a water-in-ink emulsion (emulsified water) and droplets of water on the surface of the inked image areas of the plate (surface water). It is important that some surface water exist if scumming is to be avoided. Excessive surface water, however, causes a weak or "snowflaky" appearance to the print due to failure of the ink to transfer well because of the large amount of droplets on the ink surface. Thus the balance between too much and too little surface water is critical. This can only be achieved when the ink and dampening solution are properly matched to work as a team.

Lithographic inks should be printed using as little water as possible—just enough to keep the nonprinting areas from beginning to accept ink. This keeps the amount of water emulsified in the ink at an acceptable level, with little surface moisture on the ink.

Surface water greatly decreases tack (as compared to when no water is present). In GATF tests, tack readings (made on the press) dropped to about half the original value as soon as the dampening rollers were put onto the plate. The effect largely disappeared by the time the ink reached the paper, since the ink had lost most of its surface moisture.

Measurement of Emulsification . Much more important than the amount of water that the ink will adsorb is the shape of the curve when ink emulsification is measured over a 10-minute period. A device called a Duke Tester, which looks like a cake mixer, is used. Ink and a small amount of water are placed in the tester, power is turned on, and samples are withdrawn every minute. The amount of emulsified water is determined and plotted against time. Typical curves are shown in figure 8-10.

Figure 8-10. Surland water-emulsification curves for five inks.

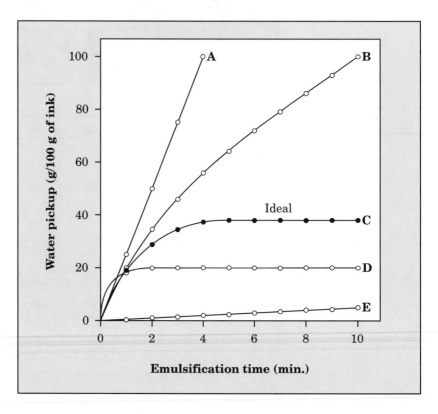

Inks that take up too much water and continue to take up water for ten minutes, as shown by curves A and B, are poor inks, as are those that fail to take up enough water (curves D and E).

Flexographic Inks

Flexographic inks are liquid inks—that is, inks with low viscosity. They are formulated with pigments (or sometimes dyes), resin-solvent varnishes, plasticizers, and an organic solvent or with aqueous varnishes and water. Many of the pigments used with lithographic inks are also used in flexographic inks. Resins used in flexo inks include shellac, vinyl, rosin esters, maleic-modified rosin esters, styrene-maleic resins, polyamide, and nitrocellulose. Many of the early flexo inks were made with shellac, but it is now very expensive and is used mainly for inks to print on polystyrene.

Flexographic inks are suspended in water or organic solvents. Solvents used for flexo inks depend on the resin used in the ink varnish and the substrate to be printed. The principal ones are shown in table 8-VI. The ink manufacturer recommends a "normal" solvent, or mixture of solvents, to be used for normal running speeds on the press. If a press is running fast, then a "fast" solvent is used, so that it will evaporate faster when the ink is printed on the substrate. If a press is running slowly, a "slow" solvent is used.

Table 8-VI.
Solvents used in flexographic inks.

Name	Formula	Boiling Point °F	°C
Anhydrous ethanol	C_2H_5OH	173.1	78.4
Hexane	C_6H_{14}	155.7	68.7
Normal-propyl acetate	$CH_3–COOC_3H_7$	214.9	101.6
Isopropyl acetate	$CH_3–COOC_3H_7$	191.1	88.4
Ethyl acetate	$CH_3–COOC_2H_5$	171.0	77.2
Normal-propanol	$C_3H_7–OH$	207.0	97.2
Isopropanol	$C_3H_7–OH$	180.3	82.4
Heptane	C_7H_{16}	187.0	86.1
Toluene	$C_6H_5–CH_3$	231.1	110.6
Cellosolve	$HOCH_2CH_2OC_2H_5$	275.2	135.1
Naphtha	Hydrocarbon mixture	243–293	117–145

Table 8-VII shows typical resins or resin mixtures used; solvents or solvent mixtures used for "slow," "normal," and "fast" operation; and the principal substrates to which the ink adheres well. To produce a "slow" solvent, a solvent with a high boiling point is added to the "normal" solvent. Addition of a solvent with a low boiling point produces a "fast" solvent.

Table 8-VII.
Resins and solvent mixtures for flexographic inks.

| Resins | Solvent Mixture | | | Principal Substrates |
	Slow	Normal	Fast	
Polyamide and nitro-cellulose	Anhydrous ethanol; normal-propanol (high percentage)	Anhydrous ethanol, 80%; normal-propanol, 20%	Anhydrous ethanol, 90%; normal-propanol, 10%	Polyethylene Polypropylene Polyester, coated Paper Board Nylon Glassine Cellophane, coated with polyvinylidene chloride
Nitro-cellulose and maleic-modified rosin	Anhydrous ethanol; normal-propyl acetate; normal propanol	Anhydrous ethanol, 85%; normal-propyl acetate; 15%	Anhydrous ethanol, normal-propyl acetate; ethyl acetate	Cellophane, moisture-proof Cellophane, PVDC-coated Polyethylene Polyester, coated Paper Board Glassine
Vinyl	Small amount of Cellosolve added to normal mixture	Anhydrous ethanol, 50%; normal-propyl acetate, 50%	Small amount of ethyl acetate added to normal mixture	Polyester, coated Polyester, uncoated Nylon Cellophane, PVDC-coated
Shellac	Small amount of Cellosolve added to normal mixture	Anhydrous ethanol, 65%; normal-propyl acetate 35%	Part of normal-propyl acetate replaced with ethyl acetate	Styrene
Acrylic		Water	Water Ethanol (5–10%)	Newsprint Corrugated board

Organic solvents used in flexo inks (e.g., alcohols and esters) have fairly low boiling points, and prints can be dried with a dryer temperature of 120°F (50°C) or lower. This is below their boiling points, but they evaporate rapidly at this temperature. Drying can be achieved with a steam-heated roll rather than the high-velocity, hot-air dryer required for web offset.

Water-based flexo inks are used for printing on paper (especially newsprint) and paperboard. They dry by precipitation of the acid resin by the acidic substrate and by absorption into the paper, not by evaporation.

On newsprint, these inks have the important property of rub resistance; the inks do not rub off on hands and clothing. Water-based flexo inks contain acrylic resins with free carboxyl groups. The resin is thought to combine with alum in the paper to form a paper-ink bond.

Water-based flexo inks are sometimes used to print on foil and plastic films, but the inks dry at a higher temperature than solvent-based inks. To enable the inks to wet the substrate, the solvent consists of about 80% water and 20% solvents such as glycols and alcohols. The formulation must be suited to the substrate, and it varies depending on whether foil, polyethylene, polyester, or poly(vinylidene chloride) is to be printed.

The mechanism of flexographic ink drying is the same as that of heatset web offset ink drying except that flexo inks dry at a lower temperature and chill rolls are not required. When the solvents are vaporized, the solid resin remains to bind the pigments to the substrate. The inks contain no driers, and drying by oxidation and polymerization is not involved.

Gravure Inks

Gravure inks, like flexographic inks, are liquid inks; that is, they are inks with low viscosity. They are formulated with pigments, resin varnishes, plasticizers, and solvents. Many of the pigments used in lithographic inks are also used in gravure inks. The different types of gravure inks use different types of resins.

The mechanism of gravure ink drying is the same as that of flexographic inks. When the solvents are evaporated in the dryers, the solid resin remains to hold the pigments and to bind the ink to the substrate. The inks contain no driers. (Rotogravure news inks require no heat, as they dry by absorption.)

Publication gravure inks are used to print long editions on coated or uncoated paper. Here, ink cost is a factor, so relatively inexpensive resins are used for the resin-solvent varnishes.

These include ester gums, zinc, magnesium or lithium rosin salts, and some hydrocarbon resins. Ester gums are chemically modified rosin, such as the pentaerythritol ester of rosin. The zinc resinates are rosins treated with zinc compounds, and the hydrocarbon resins are the same as those described for heatset web offset inks.

With gravure packaging inks, other resins and solvents are used, depending on the material to be printed. There are many types of such inks, sometimes designated with letters: A, B, C, D, etc. Some of the resins employed, depending on the ink type, are spirit- or alcohol-soluble nitrocellulose, chlorinated rubber, and vinyl resins.

Gravure inks are shipped in a concentrated form that decreases the tendency of the pigment to settle out. The inks are thinned at the press with a mixture of solvents suitable for use with the particular type of ink. The solvent mixture can also be varied in accordance with the desired rate of evaporation. As an example, consider a gravure packaging ink containing spirit-soluble nitrocellulose. One ink manufacturer suggests the following solvent mixtures for various evaporation rates:

Slow	Regular	Fast
1 part normal-propyl acetate	2 parts ethyl acetate	1 part ethanol
1 part ethanol	2 parts ethanol	4 parts ethyl acetate
	1 part toluene	

Thinners used for other types of gravure packaging inks include some of the solvents mentioned above as well as lactol spirits, naphtha, heptane, isopropyl acetate, methyl ethyl ketone (MEK), and others.

As a gravure ink is printed, the solvents evaporate from the ink remaining in the ink pan, the viscosity of the ink increases, and the ink becomes poorer in the lower-boiling solvent.

A makeup solvent mixture must be added. The idea is to use a makeup solvent mixture that will bring the percentage of each solvent in the ink back to its original concentration. Gravure inkmakers supply specifications and recommended makeup mixtures.

To keep the ink printing density constant, it is necessary to control the viscosity of the ink in the pan. Various viscometers are available. Many gravure presses are equipped with devices that control ink viscosity automatically.

Screen Printing Inks

Screen printing inks must adhere well to a wide variety of substrates. Therefore, many different types of inks are needed. Some dry by solvent evaporation; others dry by the slower process of oxidation and polymerization of the varnish. One type polymerizes with heat, and ultraviolet drying screen inks are useful.

In general, the inks contain pigments, varnishes, and solvents. Some contain driers; others do not. Screen inks are sometimes supplied ready to be printed, but more often they require the addition of solvent to bring them to the desired consistency. The screen printer must know what solvents are compatible with the resins in a particular type of ink.

One type of screen ink is referred to as *lacquer.* Such inks use nitrocellulose as a base, with various ketones for solvents. Lacquers can be thinned with lacquer solvent. They dry by solvent evaporation, requiring 20–40 minutes in the air, or a few seconds in a dryer. They are used for printing on paper, paperboard, foil, book covers, wood, and certain plastics.

Modified ethyl cellulose can also be used as a base. Such inks are usually not called lacquers. They contain mineral spirits or aliphatic naphtha with a high flash point and dry by solvent evaporation. Some formulations will air-dry in 5 minutes or less, but most require 20–40 minutes. They are used mostly for printing on paper and paperboard.

Gloss enamel inks are made with a long-oil alkyd varnish. They contain a cobalt drier, and dry by oxidation and polymerization. They are used principally for printing on paperboard, treated polyethylene, and metal.

When good adhesion and solvent resistance are required on difficult surfaces, an epoxy ink is used. This employs an epoxy resin, a catalyst to cure the resin, and solvents such as glycol ethers. With a two-part ink, the catalyst is added to the ink and well mixed. The mixture is allowed to stand for 30–35 minutes as an induction period. Then the ink can be used for about 5–6 hours before it begins to gel. One-part inks are also available.

Two-phase epoxy inks will dry at room temperature, but single-phase inks must be baked for curing. Typical curing procedures are to bake 3 minutes at 400°F (200°C), or 7 minutes at 350°F (175°). Epoxy inks are used for printing on glass, metals, ceramics, and certain plastics such as phenolics, polyesters, and melamines. Epoxy inks are not recommended for outdoor displays or for printing on paper.

Most screen printing inks for printing on plastics dry by solvent evaporation. A few dry by oxidation and polymerization. The correct type of ink must be selected for each substrate. In some cases, the solvent in the ink softens the surface of the plastic to achieve bonding. With inks that dry by oxidation and polymerization, the bond is mechanical in nature.

As an example, for printing on a vinyl substrate, the ink contains poly(vinyl chloride) and solvents such as cyclohexanone, isobutyl ketone, and diacetone alcohol. For printing on acrylics, cellulose butyrate, and polystyrene, a modified acrylic lacquer is used. Inks of this type also dry by solvent evaporation. With other plastics, epoxy inks are used.

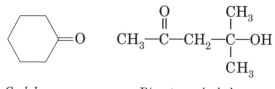

Cyclohexanone   *Diacetone alcohol*

Most screen printing inks contain pigments, not dyes. It is estimated that 80–90% of the inks use opaque pigments. Transparent pigments are used for process color work.

Fluorescent inks are made with ***fluorescent pigments.*** A fluorescent material absorbs some of the shorter wavelengths of radiation (violet and ultraviolet) and emits light of longer wavelengths. This light is emitted only while the material is receiving these shorter wavelengths. (Materials that emit light after the source is turned off are called phosphorescent.) Fluorescent pigments have limited lightfastness, although modern pigments are better than earlier products. The fluorescent pigments should not be mixed with other pigments, as mixtures greatly reduce the fluorescence.

Dyes are used in inks that are to be printed on textile material. Some of these inks use water instead of organic solvents, along with binders such as gum arabic, dextrin, glue, and casein.

Cotton garments, such as T-shirts, can be printed with a 100%-solids ink containing ***plastisol resins,*** which are resins suspended in a plasticizer in which they are insoluble while cold. When heated, the plasticizer dissolves the resin, converting it from a suspended powder to a plasticized film. Two methods are used. In one method, the garments are printed directly with the plastisol ink, then heat-cured for 3 minutes at 300°F (150°C). In another method, the printing is done on transfer release paper, which is then heated to form a film. Additional heat is applied to transfer the printed film to the garment.

Inks that dry with ultraviolet light are used in screen printing of glass, paper, printed circuits, and metal decorating.

Glass bottles are often decorated with screen printing. A *glass frit,* a suspension of colored powdered glass in an adhesive, is printed onto the glass. When the bottle is heated toward the melting point, the frit color melts into the bottle, leaving the printed design.

The solvents used in screen printing inks vary in solvent power, boiling range, flash point, and relative rate of evaporation. Screen printers often classify solvents as fast, medium, or slow in evaporating. A relative evaporation rate system uses 100 for butyl acetate. Figures above 100 indicate a faster rate of evaporation than that for butyl acetate, while figures below 100 indicate a slower evaporation rate. For example, the figure for methyl ethyl ketone is 165 (b.p. 79.6°C), for cyclohexanone, 23 (b.p. 155°C), and for diacetone alcohol, 14 (b.p. 168°C).

News Inks

Tremendous amounts of inks are produced for printing newspapers and other products printed on newsprint.

News inks are simple, inexpensive inks. Litho and letterpress news inks "dry" entirely by penetration of the ink oils into the newsprint. Low-rub news inks may include some drying oil in the formulation. Flexo news inks are water based.

The main ingredients in offset and letterpress news inks are carbon black and mineral oil. The carbon black is a type called furnace black, produced by burning atomized mineral oil in brick-lined furnaces. It is common to use about 1–2% of a wetting agent, such as gilsonite (a soluble hydrocarbon that is mined from the earth) or pitch (residue from petroleum refining), to produce a news ink that has good flow.

Because news ink is very "soupy" (low viscosity) and is used on high-speed presses, it tends to "fly" or "mist." A small amount of anti-misting additive, such as bentonite clay or silica, reduces misting.

In the United States, soybean oil is added to colored news inks where it improves lithographic performance and gives brighter colors than mineral oil.

Low-rub News Inks

Because of the objectionable tendency of news inks to rub off onto the hands and clothing, inkmakers have tried to find formulations that hold the carbon black onto the paper. Exact formulations are confidential, but addition of hydrocarbon resin or even oxidation-polymerization-type varnish decreases the tendency of the ink to rub off (but they increase the cost of the ink). Addition of soybean oil is reported to reduce rub-off of news inks.

Flexo inks, which are much more complicated in formulation, have excellent rub resistance. They are formulated with water-based varnishes that contain acrylic resins with free carboxyl groups. These resins are dissolved in water by the addition of a base. Upon contact with newsprint, which always contains some alum (aluminum sulfate), the resin is precipitated. Precipitation not only makes the pigment adhere to the sheet, but it increases ink hold out which improves print opacity and the brilliance of the ink.

Inks for Inkjet Printing

Inkjet printers can be divided into two groups: continuous inkjet (CIJ) (figure 8-11) and drop-on-demand (DOD) (figure 6-17).

CIJ produces a stream of droplets from a jet of liquid ink, and digital electronic processes direct the droplets to produce dots on a substrate or to the recycle trough. DOD techniques generate a droplet when it is needed to form a dot on the substrate. CIJ produces images at a faster rate than DOD,

Figure 8-11.
Principle of a Videojet continuous inkjet (CIJ) printer.
(Courtesy Marconi Data Systems, Inc.)

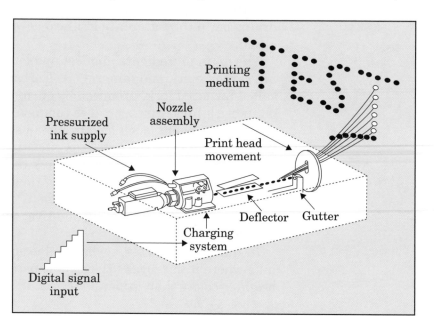

and it is used to personalize form letters on presses moving at 500 m/min (1500 ft/min). In general, CIJ printers are used in packaging applications, while DOD printers are used in office printing.

Inkjet printing, whether CIJ or DOD, is especially suited to producing images from digital data. Inkjet printers can print on anything—delicate, uneven, or recessed surfaces—and one early manufacturer is said to have demonstrated this capability by printing an image on the yolk of a raw egg.

In CIJ printing, a charge applied to the ink droplet guides it through an electrostatic field to the desired position on the substrate. The position on the substrate is determined by the amount of charge applied to each droplet. The ink droplets must break cleanly from the jet stream to form precisely uniform droplets if they are to follow the prescribed path. Formation of uniform drops depends on ink chemistry, design of the system, and the interactions of the ink chemistry and printer structure and chemistry. Accordingly, ink formulations vary broadly depending on the application and system.

In commercial printers, the jet travels at speeds of 60 mi/hr (1600 m/min), and within 1 mm of the orifice, the jet is broken into thousands of droplets. A frequency of 64 kHz (kilohertz) produces 64,000 droplets per second, but numbers up to a million have been reported. Each droplet must have identical size, or its trajectory will be distorted, resulting in misregister of the dot.

The resulting shear stresses on the ink are complex, and the formulation of the ink must meet complex interactions. The physical and chemical properties required to meet the requirements of inkjet printing are very demanding. Ink viscosity and rheology, surface tension, conductivity, volatility or vapor pressure, particle size, and insoluble matter are closely specified. Conductivity is important for CIJ inks. Formulation requires not only extensive laboratory work, but also extensive testing in the inkjet printer.

The formulation of the ink must meet these complex interactions. The inks contain solvent, dye (rarely pigment), binder, and additives.

The solvent is usually a blend of materials. Typical solvents for CIJ inks are MEK (methyl ethyl ketone), acetates, DMF (dimethyl formamide), N-methylpyrrolidone, glycol ethers, and ethanol, but water-based and UV-curable vehicles are also used. For DOD inks, glycol and water-based inks dry by absorption into the paper, but they must not cause feathering.

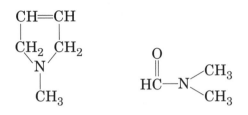

N-methylpyrrolidone   *Dimethyl formamide*

Dyes are preferred, but if pigment particles are less than one micron (1 μ) they may be used. Pigments tend to settle out, and they are often abrasive. New dyes with significant lightfastness are being continually developed.

Inkjet printing provides a rare opportunity for the use of black dyes in printing inks. For other processes, graphite provides excellent properties at low cost, but the graphite particles cannot be handled in jets and nozzles. One black dye used in black inks is food black 2. Its highly sulfonated structure makes it readily soluble in aqueous solution.

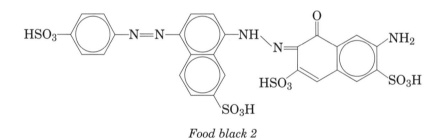

Food black 2

The binder must provide good gloss, adhesion, scuff resistance, and print definition in the dried ink and good flow properties in the liquid inks. Binders must promote good droplet formation. They must be soluble in the selected solvent. Cellulosics and block copolymers are replacing the phenol-formaldehyde prepolymers that were widely used in CIJ inks.

Ink additives promote flow and adhesion, electrical conductivity, stability to oxidation, and they reduce foam and plasticize the binder. Organic or inorganic salts (such as lithium nitrate, $LiNO_3$) are required in CIJ inks to promote conductivity. They must be noncorrosive.

Formulas include biocides to prevent the growth of mold, buffers to control the pH, crusting inhibitor (e.g., polyethylene glycol) to prevent formation of a crust or skin over the nozzle,and sequestering agents to suppress heavy metal contamination from the equipment.

Piezo impulse, thermal jet or bubble jet, hot melt, and valve jet are four types of DOD printers. Piezo printers eject a droplet from a channel on an electrical impulse. Thermal jet or bubble jet replaces the piezo crystal with a heater element. Hot-melt systems keep the ink above its melting point (typically 100–150°C or 210–300°F), and the droplet is discharged by an electrical impulse from a piezo-electric crystal. In the valve technique, the ink is held under pressure, and a droplet is ejected when the valve is electronically opened. Dyes must remain stable under these conditions. They must remain stable under storage and while in use in the printer. Dyes must produce bright, strong colors. Pigments are rarely used because they present nozzle wear, flow problems, clogging of nozzles, and stability problems in the ink.

Ink formulation varies depending on the application.

Piezo and bubble-jet inks must remain fluid without drying in the nozzle, but they must dry quickly when they are printed on the paper. Inks typically contain water and glycols or polymeric glycols.

Valve-jet printers can use inks of high volatility (e.g., MEK) because the nozzle is covered except when it is actually printing.

Toners

The word "toner" may apply to a pigment used to mask or alter an undesired color in printing inks such as the blue toner used to mask the brown tone in black inks, but in this section, a "toner" means the material used to develop an electrostatic image.

Two types of toners are used for electrostatic printers: powders (or dry toners) and liquid (wet toners). Dry toners are used on plain-paper copiers and laser printers. Dry toner particles are larger than liquid toner particles, but they avoid problems associated with an organic liquid. Liquid toners consist of a suspension of very fine particles of pigment (normally carbon black) and adhesive suspended in a hydrocarbon liquid. Because of the particle size, they give better print resolution than dry toners.

A powder toner typically consists of 90% resin and wax, 5% pigment (usually carbon black), and 5% of other materials such as flow promoters, charge modifiers, and drum cleansing agents. The additives give the toner the proper electrical properties to stick to the charged areas on the drum and to transfer cleanly to the paper when the charge is reversed. Toners for zinc-oxide-coated paper or printing plates have similar requirements, although they do not need to transfer from the original electrostatic image.

The toner receives the necessary charge by swirling with a selected friction partner, generally iron filings or magnetite, known as the carrier. The carrier may be incorporated with the toner in a one-package or mixed in the copier. Each type of copier or laser printer uses a specially formulated toner.

Coatings and Laminates

In addition to the overprint coatings discussed earlier in this chapter as an aid to ink drying, there are many materials applied both to paper and print to provide gloss, handling resistance, rub resistance, adhesion, chemical and product resistance, slip or slip resistance, and barriers to water vapor or air. They are applied to book and magazine covers, labels, annual reports, and all sorts of product brochures and packaging materials. These products are applied as liquids that are then dried or cured, or they are applied as film laminates.

Functional or barrier coatings are especially important to packaging materials, providing a barrier to oxygen, moisture, oils, soaps, and solvents, and protecting the packaged products. Coatings may even be used on the inside of packages. Moisture-vapor barriers formed by such polymers as poly(vinylidene chloride) and oxygen barriers from poly(vinyl alcohol) are often used. Moisture-vapor barriers applied to the inside of boxes of frozen foods prevent freezer burn, which occurs when the frozen product becomes dehydrated.

Overprint coatings provide resistance to handling of packages as well as magazines and other printed products. The term "squalene resistance" is used to express the resistance to a test oil called squalene that simulates the oils found in fats and human skin.

Coatings include varnishes, lacquers, laminates, UV/EB coatings, aqueous coatings, and primers. Varnishes are air-drying vegetable oil formulations that require hours to dry. They are usually applied on the printing press. Conventional drying-oil varnishes provide gloss, adhesion, product protection, and other properties, but they offer all the problems of a heavy ink film such as setoff and blocking, and they require the judicious application of spray powder, which sometimes reduces gloss.

Lacquers have higher gloss, but they are solvent-based and are often applied by an independent company with special equipment for applying the lacquer and for disposing of the solvent after the lacquer is applied. Nitrocellulose lacquers provide good gloss and protection, but they often lack good adhesion.

Laminates are also applied off press, usually not by the printer. These are thermoplastic polymers, such as polyethylene or polypropylene, applied by extruding the melted resin and laminating the film. They are used more commonly in packaging than in printing.

UV/EB coatings provide high gloss and good protection, and they require solvent. They are applied on the press (sheetfed or web), but they require special, expensive curing equipment. A UV varnish applied over a water-based primer coat gives magazine covers a brilliant, glossy finish.

Aqueous coatings are sometimes based on shellac or on urethane resins, but the most common are acrylic coatings. Ethylene-acrylic acid or styrene-acrylic acid copolymers are soluble or dispersible in aqueous ammonia or amines. Wax emulsions and polyethylene emulsions provide gloss and moisture resistance to cartons for frozen foods and milk. Nitrocellulose and shellac lacquers provide squalene resistance.

Aqueous coatings are applied on sheetfed or web offset presses, flexographic, or gravure presses. Offset presses can usually handle them without requiring special equipment, but flexo and gravure presses require special dryers to reach the 300–320°F (150–160°C) required to dry the aqueous coating.

Primer coats improve ink holdout and adhesion of overcoatings. Overprint coatings protect the print and the contents of the package. Poly(ethylene imine) is used on metal and polyester to promote an excellent bond to foils, paper, film, or composites.

Deinking

The removal of ink from printed paper grows increasingly important. A significant amount of paper is recycled without deinking: cuttings and trimmings from unprinted paper, and printed paper that is simply reprocessed without deinking. Cuttings and trimmings give good white paper or board, while printed paper that is not deinked (often mixed waste) gives a gray product that is made into paperboard and sold as tablet backing or chipboard. Often it is coated with a layer of white paper or pigment and printed to make boxes used for bottle carriers, shoes, suits or dresses, and many other products.

Deinking greatly increases the value of the waste paper. There are three processes for deinking: washing, flotation, and dispersion. Waste newspapers can be converted into newsprint by simple washing. The stock is beaten or refined, and surfactant or washing soda is added to the water. The news ink is removed much as soil is removed from dirty laundry. Flotation is more complicated. The printed paper is macerated (ground up) and the dried or cured ink (sheetfed, heatset, gravure, or UV) is separated from the fiber and floated away with foam. Dispersion is a combination of the other two processes. After cleaning by flotation, the fiber is collected and pressed. It is then dispersed in a disk refiner and washed to remove any remaining dirt or ink. Details of the process vary somewhat between heatset, sheetfed, or UV so that the value of the wastepaper depends somewhat on the printing process. Mixed waste is less valuable than a product all printed by the same process.

Further Reading

Leach, R. H., and R. J. Pierce, *The Printing Ink Manual.* 5th ed. 1993 (Kluwer Academic Publishers, Dordrecht, The Netherlands).

Thompson, B. *Printing Materials: Science and Technology.* 1998 (Pira International, Leatherhead, Surrey, UK).

9 Chemistry in the Pressroom

More chemistry is involved in printing by lithography than in any other printing process, but this chapter also applies to other processes.

When paper (chapter 7), ink (chapter 8), and printing plates (chapters 5 and 6) are brought together on the press (along with a dampening solution on lithographic presses), many chemical and physical changes occur, and they often interact. Some are helpful, others cause problems.

Chemical Formulations

The United States government, through OSHA (Occupational Safety and Health Administration), requires that suppliers of chemical formulations provide customers with Material Safety Data Sheets (MSDSs) that disclose the chemical contents of the materials. MSDSs are a ready source of information concerning products used in chemical process industries.

Products such as ink and dampening solutions used in the pressroom are chemical formulations covered by OSHA regulations. Some MSDSs specify the contents of chemical formulations more clearly than others, but if it becomes necessary to determine the contents of a chemical formulation or to confirm the formula supplied on the MSDS, modern instrumental methods of chemical analysis allow fast and accurate determination of the contents of any formulation.

Ink in the Pressroom

A chemical reaction may involve several materials. For example, when ink dries on paper, the rate of drying depends on the formulation of the ink, the properties of the paper, the temperature and relative humidity of the pressroom, and, with lithography, the pH of the dampening solution and the amount of dampening solution emulsified in the ink. Such reactions are discussed in this chapter.

Drying of Web Offset Inks

Nonheatset web offset inks dry by absorption. Absorbed solvents, high-molecular-weight hydrocarbons or water, usually pose no special health or environmental problems in the printed paper.

Heatset web offset inks dry by evaporation. It is neither legal nor desirable to discharge evaporated solvent, smoke, and odors into the air, and they must be disposed of properly. The most common methods of disposing of evaporated solvent, smoke, and odor are thermal incineration, catalytic incineration, and condensation. Incineration is the most common with heatset web offset inks, and condensation with flexo and gravure inks, although condensation is sometimes used with web offset inks.

Thermal incineration (figure 9-1) of evaporated web offset solvent requires afterburners that heat the exhaust gas from the dryer to 1,400–1,500°F (760–815°C), at which temperature the hydrocarbon solvents are burned to harmless carbon

Figure 9-1.
Thermal incinerator.
(Courtesy MEGTEC Systems)

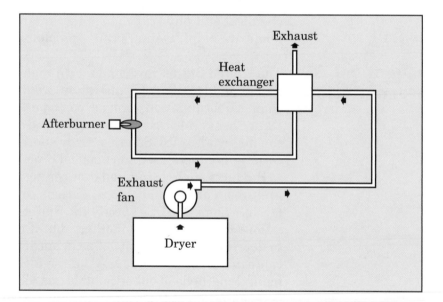

dioxide and water. Thermal incineration removes about 99% of the volatile organic compounds (VOCs) from the effluent air. At temperatures of 1,200°F, considerable amounts of carbon monoxide (CO) are formed, and at the higher temperatures, some air is converted to NO_x (oxides of nitrogen such as NO and NO_2.)

In some systems, the heat is recovered for use with the high-velocity, hot-air dryer. One system burns the solvents directly to heat the high-velocity hot air used to dry the web.

Catalytic incinerators (figure 9-2) oxidize the VOCs in the exhaust gas, removing about 95% of the VOCs. Since the incinerator operates at a lower temperature—600–800°F (325–425°C)—less energy is required, and very little CO and NO_x are produced. The exhaust gas passes through a chamber filled with the catalyst that induces oxidation of the hydrocarbon solvents. The most effective catalysts are the noble metals palladium (Pd) and platinum (Pt). Base metals, while cheaper, are not very effective. If properly maintained, the catalyst in a catalytic converter remains effective for several years. Like catalytic converters in automobiles, they are poisoned with lead, chlorinated hydrocarbons, and phosphorus. (Phosphorus in the dampening solution causes little problem unless it is used in excessive amounts.) Sometimes silicones in the paper or pressroom air get into the converter and reduce its effectiveness.

Figure 9-2.
Catalytic incinerator.
(Courtesy MEGTEC Systems)

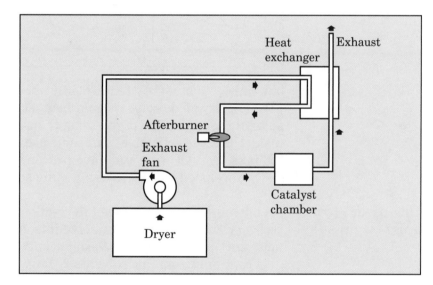

Heatset web offset solvents can also be recovered with a cooler-condenser that removes 60–85% of the VOCs (figure 9-3). Cooling the exhaust condenses the ink oils to a fine mist that is filtered out of the air. Because they boil at a higher temperature, the use of charcoal is not ordinarily required to remove heatset solvents from the air.

Technical advances in the design of the dryer and the chilling section have increased web speeds and operating efficiency, while reducing waste and improving solvent recovery.

Ultraviolet (UV) and electron beam (EB) inks and coatings are promoted as "environmentally friendly." They dry with-

Figure 9-3.
Cooler-condenser.
*(Courtesy MEGTEC
Systems)*

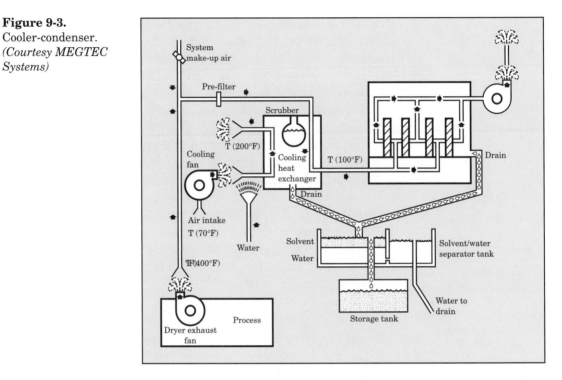

out emission, and they require only about 20–25% as much power to dry as do conventional heatset, web offset inks. In spite of the high price of the inks, their advantages often make them cost-effective, and their use is growing steadily. EB inks are VOC-free and have proven compatible with specially developed VOC-free dampening solutions.

Drying of Flexo and Gravure Inks

An advantage of water-based flexo and gravure inks is that recovery and disposal of the solvent is greatly simplified, but most gravure printing and flexo package printing still require inks containing organic solvents. (Many aqueous inks contain some organic solvents that must be properly disposed of.)

Organic solvents in the exhaust gas from gravure and flexographic press dryers are commonly removed by passing the gas through a chamber filled with activated carbon. The solvents are removed by adsorption onto the surface of the activated carbon particles. When the carbon becomes saturated with solvent, steam is used to strip the solvents from the carbon. While this operation is proceeding, the exhaust gas from the dryer is transferred to a second chamber, also filled with activated carbon. On cooling, the steam and solvents liquefy, separating into two layers. The recovered solvents can be either sold or used on the press as a dilution solvent.

Drying of Sheetfed Inks

Setting and drying of inks involve many physical and chemical processes. An ink has set on the sheet when it is ready for the next process (folding, cutting, or binding), but the chemical process that results in curing of sheetfed inks may continue for several hours longer.

Sheetfed inks containing drying-oil varnishes dry by a highly complicated oxidation-polymerization chemical reaction. The following discussion concerns the drying of these inks used in sheetfed litho, letterpress, and screen inks.

If a printed sheet cannot be handled in a reasonable length of time (say 2–4 hours), the ink is often blamed. However, not only the ink but also the paper, the press, and the atmosphere are involved. Ink characteristics are discussed in detail in chapter 8.

Ink components affect the setting and drying rate:
- **Pigments.** Some pigments accelerate drying while others consume drier. Lead pigments (no longer used) and iron blue pigments accelerate drying. Carbon black has a very high surface area that adsorbs drier and reduces its effectiveness.
- **Varnishes.** Some varnishes dry much faster than others. Linseed synthetic oils dry quickly, and chinawood oil dries even faster. Other oils dry more slowly.
- **Type of drier.** Cobalt, manganese, and zirconium compounds are used. They are not equally efficient as driers. Cobalt is most effective in promoting drying at the surface, while manganese and zirconium are most effective in promoting drying throughout the film. Since both are important, mixed driers are usually used.
- **Amount of drier.** The drying time is not directly proportional to the amount of drier. Doubling the drier content seldom doubles the rate of drying.
- **Drier dissipation.** Drier dissipation is explained in chapter 8. To avoid problems, the drying rate of old inks should be checked before they are used on a job.

Paper characteristics affecting ink drying are discussed in chapter 7. These characteristics are:
- **Absorbency of the sheet.** Absorbency of the paper does not affect real drying, but the more it is absorbed, the faster printed ink will set.
- **Paper or coating pH.** The more acid an uncoated paper is, the slower the printed ink dries. The more alkaline a

coating is, the faster the ink dries. The coating on coated paper can also be responsible for the chalking of inks.

- **Moisture in the paper.** The higher the moisture content of the paper, the slower the printed ink dries. Modern instrumental control of moisture during paper and board make wide variations rare.

Press control is particularly important in lithographic printing. Good ink/water balance is required: enough ink should be used to get full color, with only enough water to keep the plate running clean. Important **press factors** are:

- **Emulsification of water in ink.** The more water that becomes emulsified in the ink, the slower the ink dries. If a form with light ink coverage is printed, water has a greater chance of becoming emulsified in the ink.
- **Dampening solution acid.** As the dampening solution becomes increasingly acidic, inks with drying-oil varnishes take longer to dry. High acidity reduces the effectiveness of the drier.
- **Multicolor printing.** On a multicolor sheetfed press, the fourth-down ink may dry more slowly than the first-down ink (assuming that they dry at equal rates on a slab) because the sheets have picked up moisture from the first three units and have more moisture in the areas on which the fourth ink prints.

Temperature and relative humidity also affect the ink drying rate (see figures 9-4 and 9-5). Increasing the temperature at which sheets are stored after printing increases the drying rate. An ink that requires six hours to dry at 68°F (20°C) will dry in about three hours at 80°F (27°C). On the other hand, as the relative humidity of the air increases, inks dry more slowly.

Crystallization of First-Down Ink

When process inks are printed on a single-color press, the second-down ink sometimes does not trap well over the dried first-down ink if an extremely hard film has formed. It is often said that the first-down ink has crystallized or that crystallization has occurred. (Ink crystallization has nothing to do with the formation of crystals in the ink.) The use of a cobalt drier in the first-down ink is often blamed for this problem, although the use of waxes and hard-drying oils such as chinawood oil can also cause trapping problems. If the inkmaker has used large amounts of cobalt to get the ink

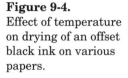

Figure 9-4.
Effect of temperature
on drying of an offset
black ink on various
papers.

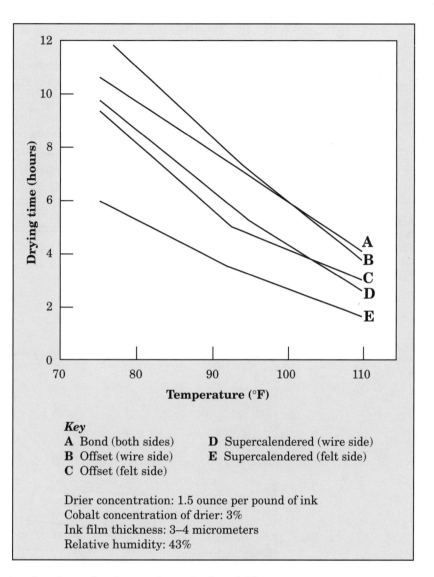

Key
A Bond (both sides)   **D** Supercalendered (wire side)
B Offset (wire side)   **E** Supercalendered (felt side)
C Offset (felt side)

Drier concentration: 1.5 ounce per pound of ink
Cobalt concentration of drier: 3%
Ink film thickness: 3–4 micrometers
Relative humidity: 43%

to dry fast, the formation of a hard film may cause a trapping problem.

Improperly formulated inks undoubtedly cause trouble with trapping of subsequent colors. The printer is urged to discuss the job with the ink supplier to be sure that inks will perform properly on the job. Inks that are altered or doctored in the pressroom have always caused problems.

**Drying of
Overprinted
Ink Films**

When a first-down ink is dried and promptly overprinted, the first-down ink exerts a catalytic effect that accelerates the drying of the second-down ink. The power of this effect decreases day by day. In fact, if the first-down ink contains drying-oil varnishes and has been aged a week or two, it retards the drying of the second-down ink.

The acceleration of drying of the second-down ink printed over recently dried ink is due to a volatile drying accelerator emitted by the first-down ink. The following experiment verified this explanation. The first-down ink was allowed to dry on a sheet of tinplate. Next, another sheet of tinplate was printed with a wet ink. The two ink films were brought within about 2 mm (0.08 in.) of each other. The wet ink film dried on the tinplate in one-fifth to one-third of the time

Figure 9-5.
Effect of relative humidity on drying of an offset black ink on various papers.

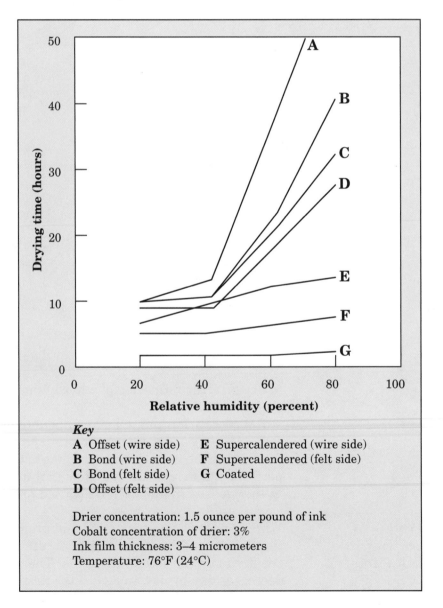

Key
A Offset (wire side)   **E** Supercalendered (wire side)
B Bond (wire side)   **F** Supercalendered (felt side)
C Bond (felt side)   **G** Coated
D Offset (felt side)

Drier concentration: 1.5 ounce per pound of ink
Cobalt concentration of drier: 3%
Ink film thickness: 3–4 micrometers
Temperature: 76°F (24°C)

(depending on the percentage of drier in the wet ink) required when the ink was exposed to room air.

This drying accelerator is probably a low-molecular-weight material that is a by-product of the drying of varnish. Many materials such as acrolein, acetaldehyde, propionaldehyde, and acetic acid and its esters have been isolated from drying ink and identified spectroscopically.

Gloss Ghosting and Poor Trapping on the Second Side

Suppose that a form with fairly light ink coverage is printed on the first side of a stack of press sheets and that the inks are allowed to dry briefly. Then the sheets are backed up with a heavy-coverage form. Occasionally, the ink on the second side shows more gloss, after drying, in those places where it is opposite inks printed on the first side. This effect is called gloss ghosting. Volatile drying accelerators from the ink on the first side appear to cause the ink opposite to dry much faster. Thus less varnish soaks into the paper, and the accelerated drying causes an area that has more gloss.

This strange drying effect has several causes. Evidence indicates that all of them may occur at one time or another.

Whether a ghost appears on the second side may depend on the amount of drier in the first-side ink (the more drier, the greater the tendency for ghosting), the pH of the paper (more ghosting with an alkaline coating), ink absorbency of the paper (less ghosting as ink absorbency increases), and phase separations in the ink (quickset inks printed on poorly absorbent paper may aggravate gloss ghosting). Glossy jobs have more problems: the ghost is more visible, and glossy jobs also are usually more expensive and are judged more closely.

A trapping problem sometimes occurs when four-color printing is done on two-color sheetfed presses. The four colors are printed on one side, two at a time, with no trapping problem. Then the first two colors are printed on the second side, again with no trapping problem. After the first two colors on the second side have dried, the sheets are printed with the last two colors. In such cases, these last two colors sometimes fail to trap properly over the first two colors in only those places where there is heavy ink coverage on the first side. This failure produces a ***trapping ghost*** on the second side. It appears that the volatile drying accelerator from the first-side inks helps to dry the first two inks on the second side, which become so hard that the last two inks will not trap properly on them.

Water Emulsified in Lithographic Inks

Some water always becomes emulsified in a lithographic ink when it is being run on a press. Excessive water can lead to slow drying, snowflaky prints, piling of ink on the rollers, pigment flocculation, and poor trapping in multicolor wet printing.

Tinting of Lithographic Inks

If sheets run on a lithographic press are covered with a more or less uniform tint of the ink, the press crew faces a *tinting* problem. One cause of tinting is formation of an ink-in-water emulsion (figure 9-6).

Figure 9-6.
Water-in-ink emulsion *(left)* and ink-in-water emulsion.

If an ink-in-water emulsion forms, the dampening solution is said to be "dirty," and it carries ink droplets on the plate's nonimage areas. These are transferred to the blanket and then to the paper where they tint the press sheets.

Another cause of tinting may be a surface-active agent in the coating of a coated paper. The agent dissolves in the dampening solution and reduces its surface tension, promoting an ink-in-water emulsion.

In other cases, a small amount of some material in the coating of coated paper may dissolve and deposit on the non-printing areas of the plate. These areas may gradually accept a small amount of ink, which will transfer to the paper and create tinting. If a coated paper causes it, the tinting will disappear when a different paper is put onto the press.

Fading of Inks

Occasionally an ink with iron blue (including reflex blue and Milori blue) or chrome green pigments fades or changes color on the printed sheets (while the ink is drying). This change usually happens in the center of the sheets and happens more often if a large solid is in that area. The vehicle in the ink is the cause in this case. The vehicle needs oxygen in order to dry, but oxygen is limited in the center of the delivery pile. Consequently, the vehicle becomes oxidized at the expense of the iron blue pigment or of the iron blue part of the chrome green pigment. As a result, the iron in the pigment is reduced from ferric iron to ferrous iron. This chemical reaction removes most of the color from the iron blue pigment.

Winding (or "airing") the sheets while the ink is drying prevents this problem. Winding allows the vehicle to take oxygen from the air instead of the iron blue pigment. The ink-fading problem arises only with inks containing a considerable percentage of drying-oil varnishes.

Antisetoff Powders

Powders to prevent wet-ink setoff are made of different kinds of starch, including starch from corn, arrowroot, tapioca, rice, potato, and sago. In general, cornstarch has the finest particles and potato starch the coarsest. Particle diameters vary from about 5–100 microns (0.2–4 mil). Blending two or three kinds of starch produces any desired range of particle sizes. Powders treated with either a silicone resin or a hydrocarbon wax are also available.

Spray powders are usually sold on the basis of their size in microns, but every antisetoff powder contains a range of particle sizes. The coarser the particles, the more effectively they deposit on the sheet, preventing setoff. Fine particles (about 5–20 microns in size) tend to become airborne. These deposit dust on presses and pressroom walls and pipes. Dust collectors can be installed to remove a large part of this "nuisance powder." A coarse powder, with the majority of particles 75–100 microns (3–4 mils) in size, protects ink on sheets very well, but the finished sheets have a rough, sandpaper feel. A compromise must be employed, using as coarse a powder as possible without producing a rough surface. Antisetoff powder also provides slip between sheets so they jog better. Furthermore, since sheets slip over one another more easily, less static electricity is generated.

The use of quickset inks or inks that set rapidly with infrared radiation reduces the amount of antisetoff powder needed.

It has been reported that antisetoff powders accelerate the drying of sheetfed inks. If this is true, it is owing to the fact that they allow more air between the sheets. Starch is inert and does not affect the drying rate of sheetfed inks.

These powders are sometimes used on nonheatset web offset presses equipped with sheeters for the same reasons they are used on sheetfed presses. If film or foil is printed on a flexographic press, and the web is rewound, antisetoff powder provides the slip that results in a better-formed roll. The powder is also employed in the converting of paper, board, and film, where jobs are laminated and plastic films are extruded. In these applications, the powder is used to achieve some purpose other than the prevention of ink setoff.

Silicone Fluids on Web Presses

A dilute emulsion (1.0–3.0%) of a silicone fluid in water is often applied to one side or both sides of a paper web. The emulsion is applied with a special roll applicator between the chill rollers and the former-folder on a web offset press. The thin film of silicone on the web helps to prevent smearing of the ink as the web passes over the former-folder. It also helps reduce static. Silicone liquids vary from very thin, easy-flowing liquids to very high-viscosity, slow-flowing syrups.

Silicon (Si), like carbon, has a valence of four. Unlike carbon, silicon does not bond to itself, and it does not form double bonds. In straight-chain aliphatic hydrocarbons, carbon atoms are attached to each other to create a chain of varying length. Long-chain compounds containing silicon must have oxygen atoms between the silicon atoms. One silicone fluid, poly(dimethyl siloxane), has the following formula:

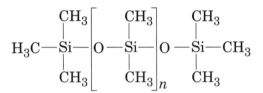

(The symbol "n" indicates that the section of the formula in the brackets is repeated many times.)

Silicones have the property of improving the slip of anything to which they are applied, such as chutes, collating machines, and the feedboards and back cylinders of presses. They minimize ink transfer from sheets to press parts such as delivery tapes and feedboard wheels.

The dimethyl silicones have been approved by the U. S. Food and Drug Administration for use in food packaging inks.

Control of Ink Feed

In modern printing plants, the amount of ink printed is controlled by a spectrophotometer or a reflection densitometer. Statistical data show the amount of variation to be expected and whether the level on any sheet is within tolerance. For this purpose, a spot on the print is monitored by the color control device, or a small solid-color patch printed along one edge of the sheet is monitored. Modern printing presses, equipped with remote control inkers, control the ink color by computer.

Instrumental control of color is essential if the printer is to remain competitive by keeping control of the process and minimizing waste. On the other hand, instrumental control must be supplemented by visual control. The eye cannot assign quantitative values to color, and the instrument cannot judge overall quality and freedom from defects.

Control of color on press depends on control of the ink-film thickness. An ink film thickness gauge (figure 9-7) has a 2-in.-diameter roller supported by a holder. To measure the ink-film thickness, it is held against the last vibrating steel roller of the offset press. The gauge reads ink film thickness between

Figure 9-7.
Ink film thickness gauge.
(Courtesy BYK-Gardner USA)

0 and 1.0 mil (thousandths of an inch). Metric gauges are also available. This instrument, like all measuring devices, should be calibrated before it is used. A good machine shop can help with the calibration. It is important to use the instrument carefully, because it is used around moving equipment. (It cannot be used on most new presses that come equipped with safety guards, and they must not be removed.)

When an ink is printing properly on a lithographic press, the ink film thickness is usually between about 0.20 and 0.60 mil (5 and 15 microns or 0.005 and 0.015 mm).

Paper in the Pressroom

Among the paper factors discussed in chapter 7 that relate to performance on the press are felt and wire sides, grain (the reason sheets are printed grain-long on sheetfed lithographic presses), appearance of inks on coated and uncoated paper, moisture in paper (the reason paper changes size with a change in moisture content and the reason this size change can lead to wavy-edged or tight-edged sheets), and the effect of the pH of paper or paper coating on the drying time of certain types of ink.

Air Conditioning in the Pressroom

Paper causes fewer problems if the pressroom is air-conditioned. Controlled humidity and temperature help not only to reduce waviness in the summer but also to reduce static and tight edges in the winter. Air conditioning includes the control of both temperature and relative humidity. If air coolers are used alone (notably in the summer), they remove humidity from the air, and the dry air causes paper problems. On the other hand, during humid weather, lowering the relative humidity by air conditioning helps to reduce the drying time of some types of ink.

See the section "Water in Paper" in chapter 7 for an extensive discussion of air conditioning, relative humidity, and paper conditioning.

Static Electricity and Its Elimination

Static electricity interferes with the slippage, separation, and normal travel of sheets of paper and plastic films. When paper moves over a metallic surface, the metal may give up some of its external electrons to the paper. When the paper suddenly moves away from the metal, the transferred electrons are trapped on the paper, giving it a negative charge. When one sheet of paper slides over another sheet, electron transfer may also occur, leaving one sheet negatively charged and the other positively charged. Therefore, they attract each other, interfering with sheet separation and causing poor jogging at the end of the press.

Static electricity is always of concern where volatile chemicals are used, because sparks can cause explosions and fires. Removing both the static and the flammable materials is the best prevention.

Avoiding low humidity in the pressroom is the best way to eliminate static. Paper falls somewhere between being a conductor and a nonconductor of electricity. Its conductivity increases as its moisture content increases. Generally, static electricity is not a problem if the relative humidity is above

35–40%. Since a coating on paper acts as an insulator, static charges remain longer and give more trouble with coated paper.

Because of the presence of moisture on a lithographic press, static electricity is usually less of a problem in lithography than in other printing processes. Water-based flexo and gravure inks also reduce static, but the standard organic solvents create both static and flammability hazards. Web presses, partly because of their higher speed, tend to build up more static in paper than do sheetfed presses. Another reason why webs are more susceptible to acquiring static charge is that paper loses moisture while passing through web press dryers.

If paper acquires electrons from parts of the press or bindery equipment, then the equipment becomes charged. Unless the equipment is well grounded to neutralize these charges, they can build up and become dangerous.

Static can be reduced by the use of antisetoff powders and also by the application of a silicone solution to a printed web. Antistatic materials, such as fabric softeners used in home laundry, reduce static when added to the silicone solution on the web press.

Static eliminators in one way or another neutralize the charge on paper or film. One method utilizes high-voltage electricity to generate a corona discharge that creates an ion cloud. Bits of copper foil, hung on a wire ("tinsel"), are not very effective.

Use of Static in Gravure Printing   Static electricity is not always detrimental. It is useful in *gravure electrostatic assist* (or *electroassist)*, where the electrical charge helps to pull ink out of gravure cells, thus reducing the number of missing dots in highlight areas. It is also used to create lithographic plates by electrostatic imaging.

Chalking of Inks on Coated Paper   Chalking is a problem in which inks that are apparently dry rub off the sheet much as chalk rubs off a blackboard. Inks may chalk a few hours after they are printed, and then dry to a nonchalking state after a day or two, or sometimes after two weeks or a month. Chalking seems to be caused by the inactivation of part of the drier in the ink by something present in the paper coating. Thus an ink that chalks temporarily is merely a slow-drying ink, made so by loss of part of the drier.

Work at GATF and elsewhere supports this explanation. Using the number of days required for an ink to become non-chalking as the criterion, experimenters have found that a particular ink dries in a shorter time as the percentage of drier in it is increased. Inks with drying-oil varnishes become nonchalking in a shorter time on papers with a high-pH coating, but a longer time is required as more water becomes emulsified in the ink. There is no correlation between the number of days to nonchalking of a particular ink and the absorbency of the paper on which it is printed. Anything that accelerates the drying rate of the ink helps to reduce chalking.

Moisture Sensitivity of Paper

Occasionally a paper that shows a good resistance to picking, judged by test results with one of the commercial pick testers, will pick badly on the third or fourth units of a multicolor lithographic press. This unexpected picking is due to the moisture sensitivity of the paper, which absorbs a certain amount of moisture as it passes through the first printing units. This additional moisture lowers the paper's resistance to picking. Some papers are more sensitive to moisture than others.

Blistering of Coated Papers on Web Presses

A coated paper sometimes blisters (figure 9-8) while it is in the web offset dryer. The heat of the dryer vaporizes some of the moisture in the paper. If this water vapor cannot escape fast enough through the coating, it creates blisters. Here are some conditions that make blistering more likely to occur:
- The moisture content of the paper is high.
- A high-basis-weight paper is being printed (more moisture needs to escape).
- The internal bond strength of the paper is low.
- The coating is not porous enough.
- Ink coverage is heavy. This condition is particularly important when the heavy coverage is in the same areas on both sides of the web, preventing the escape of moisture.
- The temperature of the paper web is raised rapidly in the dryer, rather than more slowly, as it is in a longer dryer or at lower press/dryer speeds.

Blistering of coated paper has become less of a problem partly owing to the production of better coated papers and partly to the better design of web press dryers.

Figure 9-8.
Blistering.

Dampening Solutions

In lithographic printing it is necessary to keep the nonimage areas of a plate moistened with water so that they will not accept ink. During platemaking, these nonimage areas are desensitized (usually with a thin adsorbed film of gum arabic or other hydrophilic material) so that they adsorb water instead of ink.

If the film of desensitizing gum remained on a plate indefinitely, it would be possible to run plates with nothing but water in the dampening fountain. Water alone can, in fact, be used on sheetfed presses for short runs. The film, however, wears off gradually as the plate runs on the press. The chemicals in the dampening solution replenish the desensitized film.

Dampening Solution Ingredients

The concentrated dampening solutions purchased by the printer are commonly referred to as ***fountain concentrates, fountain etches,*** or just ***etch.*** These usually contain a gum, but if an acid etch contains no gum, then gum arabic or a cellulose or starch derivative must be added in the pressroom. The term "dampening solution" is used for the diluted etch.

Usually, etches contain ingredients in addition to the desensitizer gum and acid. Magnesium nitrate, $Mg(NO_3)_2$, serves partly as a buffer and partly to reduce plate corrosion. Fungicides keep the dampening solution free from mold and mildew. Wetting agents, such as isopropyl alcohol or an alcohol substitute, assist in wetting the plate by decreasing the surface tension of water and water-based solutions.

When zinc plates were used, dampening solutions usually contained ammonium dichromate. It helped as a corrosion inhibitor to prevent the acid from reacting with the zinc. With modern plates, there is no need to use dichromate, and the danger of chromic poisoning is avoided.

Acid Dampening Solutions

The three main ingredients in an acid dampening solution are water, a desensitizing gum, and an acid (phosphoric acid, an acid phosphate compound, or citric or lactic acid).

Few manufacturers rely solely on the desensitizing ability of natural Sudanese gum arabic (chapter 5) because of its cost. Other desensitizing materials include carboxymethyl cellulose (CMC) and other cellulose derivatives, starch derivatives such as dextrin, hydrolyzed starches or chemically modified starches, and alginates.

The acid converts the desensitizing gum to its free acid form in which the molecules contain carboxyl groups, $-COOH$. These groups help the gum adsorb to the metal plate. Furthermore, phosphoric acid not only acts as an acid, but it also has desensitizing properties.

Enough acid should be used to convert most of the gum to its free-acid form. Beyond this point, the acid serves no useful purpose. The acidity of the dampening solution is measured by determining its pH. Modern acid dampening solutions usually have a pH in the range of 3.5–4.5. The printer should follow the recommendations of the manufacturer of the dampening solution regarding the proper pH for the type of ink and paper being used. The effect of fountain solution concentrate on the pH and conductivity of dampening solution is illustrated in figure 9-9.

Excess acid in the dampening solution is not only useless but may be a disadvantage. If such a solution is emulsified into inks containing varnishes with drying oils, the inks take longer to dry. Furthermore, high acidity promotes corrosion, and can actually cause sharpening of dots on a lithographic plate.

Alkaline Dampening Solutions

Alkaline dampening solutions that require no desensitizing gum are popular in newspaper printing and can be used for other work, too. These dampening solutions contain a mixture of sodium dihydrogen phosphate (NaH_2PO_4) and disodium hydrogen phosphate (Na_2HPO_4) at a pH around 10 or 11. At this pH, the phosphate ion reacts rapidly with the aluminum surface and functions as the desensitizing material. Since gum arabic or other gums are highly soluble at a pH greater

Figure 9-9.
Effect of fountain solution concentrate on pH and conductivity.

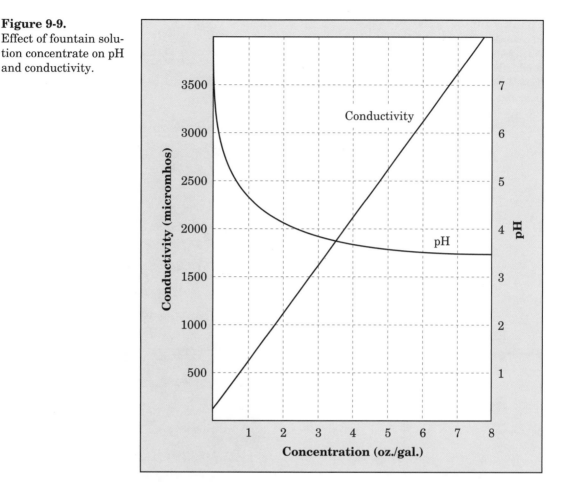

than 6 or 7, their presence in the alkaline dampening solution has little or no effect on the nonimage area.

Because these solutions are highly buffered, measurement of pH is not useful for determining the concentration. Concentration of alkaline dampening solutions is made by determining the conductance. Conductivity and pH can be determined with a single instrument (figure 9-10).

Several advantages are claimed for the use of alkaline dampening solutions on newspaper offset presses:
• Less linting occurs, probably because it usually requires a little more dampening solution to keep the plate clean. This promotes release of paper fibers from the plate and blanket.
• Ink rollers usually do not strip.
• Blankets do not become glazed, since the solution contains no gum.
• Fungus does not grow in the fountain pan.

Figure 9-10.
Instrument for measuring both pH and conductivity.
(Courtesy Cole-Parmer Instrument Company)

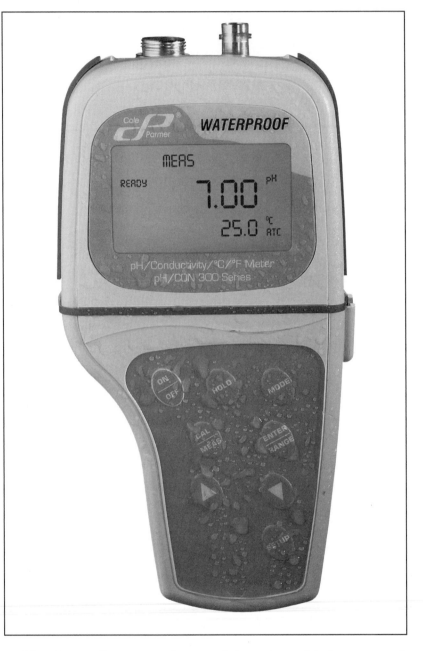

- Aluminum plates run clean and do not need to be gummed up, even for overnight.
- Plates do not need to be wiped down when starting up after an interruption in production.
- "Ink dot scum," which occasionally occurs on aluminum plates run with an acid dampening solution, has not been reported when an alkaline solution is used.

In spite of these advantages of alkaline solutions for newspaper presses, almost all commercial offset printers use acid dampening solution. When the use of an alkaline solution is attempted, commercial printers get foaming in the dampening solution, excess water emulsified in the ink, and bleeding of some pigments into the dampening solution, causing tinting on the sheets.

An alkaline solution is recommended when printing gold inks made with bronze powders. Alkaline dampening offers no special advantage when running a silver (aluminum powder) or an imitation gold ink (ink made with yellow pigment).

Successful printing of ultraviolet inks with an alkaline dampening solution has been achieved at the GATF research laboratory. It was reported that halftones printed sharp and that no roller stripping was encountered, probably because the special varnishes used in ultraviolet inks do not form soaps with an alkaline material.

Neutral Dampening Solutions

In addition to acid and alkaline dampening solutions, neutral fountain solutions with pH of 6.5–7.5 are also used. These may be neutral phosphates, or sometimes citrates or salts of other weak acids. Neutral dampening solutions, like alkaline dampening solutions, can be used on commercial sheetfed or web presses, but they are most frequently used in printing newspapers and business forms.

Alcohol or Alcohol Substitutes

With the introduction of dampening systems that transferred dampening solutions over one of the ink form rollers, it became necessary to use about 20% by volume of some kind of alcohol in the dampening solution. For a variety of reasons, isopropyl alcohol (isopropanol) was almost always used.

Alcohol has several important effects on dampening solutions. It reduces the surface tension, it increases their viscosity (it makes them syrupy), it helps dampening solution to spread over the inked form roller that carries it to the plate, and it alters the way that inks take up water or dampening solution, thus improving the behavior of marginal inks. In this way, alcohol makes it easier for the crew to maintain ink/water balance.

One reason that an alcohol-water solution is able to wet the surface of the ink and allow the solution to be transferred to the nonimage areas of the printing plate is that alcohol reduces the surface tension of a dampening solution. A 20% (by volume) aqueous solution of isopropyl alcohol has a surface tension only about half that of pure water. Alcohol also reduces the ink/water interfacial tension so that the alcohol/water solution wets the surface of the ink on the ink form roller, and the roller carries this solution around to the printing plate.

Isopropyl alcohol, a volatile organic compound (VOC), has some toxicity, and it causes air pollution when it is released to the atmosphere. Accordingly, government regulations restrict its use, and printers now use alcohol substitutes to reduce or replace alcohol on both web and sheetfed presses. Under closely controlled conditions it is possible to operate dampening systems with no isopropyl alcohol, using either a one-step solution or a two-step system of etch and alcohol substitute, but many printers prefer to use 2–5% isopropyl alcohol and use the alcohol substitute as an extender.

Addition of isopropyl alcohol has relatively little effect on the conductivity of dampening solutions (see figure 9-11). Addition of alcohol substitutes, which are used in smaller amounts, has even less effect on conductivity.

Figure 9-11.
Effect of alcohol on conductivity of dampening solution.

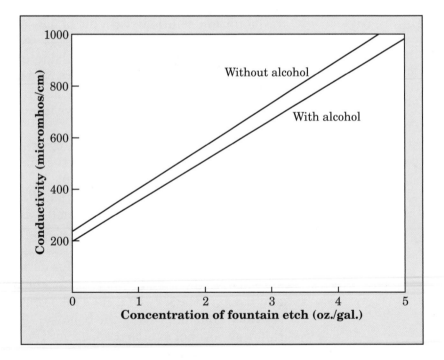

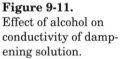

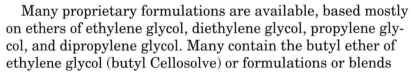

Many proprietary formulations are available, based mostly on ethers of ethylene glycol, diethylene glycol, propylene glycol, and dipropylene glycol. Many contain the butyl ether of ethylene glycol (butyl Cellosolve) or formulations or blends containing it or other glycol ethers combined with various surfactants. In addition, glycols such as 2-ethyl-1,3-hexanediol are also used in blends.

Chemical	Formula
Ethylene glycol	$HOCH_2CH_2OH$
Butyl ether of ethylene glycol	$C_4H_9OCH_2CH_2OH$
Diethylene glycol	$HOCH_2CH_2OCH_2CH_2OH$
Diethylene glycol ethyl ether	$CH_3CH_2OCH_2CH_2OCH_2CH_2OH$
Propylene glycol	$CH_3CH(OH)CH_2OH$
Dipropylene glycol	$HOCH_2CH(CH_3)OCH_2CH(CH_3)OH$
2-ethyl-1,3-hexanediol	$CH_3CH_2CH_2CH(OH)CH(C_2H_5)CH_2OH$

The alcohol substitutes reproduce many of the functions and properties of alcohol, but, of course, no substitute, no two materials, ever have exactly the same properties. (If the substitutes had the same toxicity and volatility, they would be useless.) On the press, alcohol still seems better able to reduce blanket piling and to give more ink and water uniformity across the press. Under carefully controlled conditions, the substitutes work very well.

Control of Dampening Solutions

Concentration and acidity, as measured by conductivity and pH (figures 9-10 and 9-10), are the most important factors to control in dampening solutions. Dampening solutions, acid, alkaline, and neutral, are buffered so that the pH does not change greatly as the concentration changes. The amount of fountain etch in the solution, accordingly, is controlled by monitoring the conductivity. Control of pH is also important for good printing, and it should also be monitored.

Any unusual conductivity readings justify rechecking the conductivity of the water and the fountain solution concentrate. It is normal for the conductivity to increase during the pressrun because materials from the ink and paper contaminate the dampening solution. Therefore, conductivity measurements should be made before the dampening solution is used on the press, and they should be checked at least once per shift to determine if contamination requires draining tanks.

Drying Stimulator

A drying stimulator consists of a water solution of a compound such as cobalt chloride ($CoCl_2$), cobalt acetate [$Co(CH_3COO)_2$], or manganese nitrate [$Mn(NO_3)_2$]. The proportion is 1–2 fl.oz./gal. (8–16 mL/L) of dampening solution. Cobalt and manganese accelerate the drying of inks that contain drying-oil varnishes. As the press runs, the dampening solution containing the salt compound becomes emulsified in the ink, adding more drier to the ink.

These compounds have been used by printers for a long time, and they sometimes help to overcome a drying problem. It is usually better to get at the root of the problem and solve it by calling the inkmaker or by using a better dampening solution.

Lithographic Plates in the Pressroom

Lithographic plates are subject to two main troubles: they scum and they go blind. *Scumming* occurs when the non-image area becomes sensitized to ink, and *blinding* occurs when the image area of the plate fails to accept ink.

Plate finishers are applied to the plate after it is developed in order to reduce the likelihood that the plate will scum or go blind. They are also applied in the pressroom when scumming or blinding is thought to be a problem.

There are four types of plate finishers:

- Clear finishers for subtractive plates contain a wetting agent and buffer plus a hydrophilic colloid such as gum arabic, a cellulosic resin, or poly(vinyl pyrrolidone).
- Finishers for additive plates include complex formulations such as asphaltum/gum emulsions. They are used as finishers in the plateroom and can be used on press as preservatives for plates that are to be stored and reused.
- Specialty finishers for subtractive plates contain gum and solvent but no asphaltum. They are mostly used for hand finishing of plates.
- A weak gum arabic (8° Baumé) is the oldest and least effective finisher.

Asphaltum emulsions are prepared from asphaltum, a solution of asphalt (residue from the refining of petroleum) in hydrocarbon. Emulsifying the solution in water makes an asphaltum emulsion or lithographic "asphaltum." Gum arabic is added to the aqueous phase. When the plate is wiped down with an asphaltum/gum emulsion *(AGE),* the emulsion breaks down easily as it is applied, depositing gum arabic on nonimage areas and asphaltum on the image areas. This gives a quick roll-up to the image area and desensitizes the nonimage area, protecting the plate from "oxidation."

Asphaltum is somewhat greasy and preserves the ink-receptivity of the image areas, even after a long period of plate storage. When the plate is to be run again, the printer goes over the plate with a sponge soaked with water, and the plate is again ready to print.

Asphaltum finds only limited application, and with some types of plates, it may cause gum blinding in the image areas. The printer should use only plate finishers recommended by the plate manufacturer.

Scumming

Scumming (figure 9-12) occurs when ink adheres to part of the nonimage areas of a plate. The term "nonimage areas" applies not only to the large areas where there is no image but also to the open areas between halftone dots. Thus, when a halftone begins to fill in, it is said that the plate is beginning to scum.

Figure 9-12.
Scumming due to insufficient acid, gum, or both in the dampening solution.

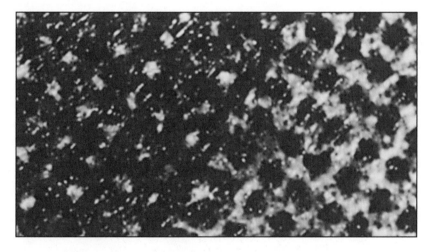

The mechanism of scumming. The nonimage areas are desensitized during the platemaking operation: they are made water-receptive by applying a thin adsorbed film of a hydrophilic gum, such as gum arabic.

When plates are running clean, the gum film may gradually wear off, but it is replaced with gum from the dampening solution. If for some reason this process does not occur smoothly, ink varnish may become adsorbed onto the nonprinting areas. The result is a scummy plate.

Ink and paper constituents may contribute to scumming. Some variable in pigment manufacture may occasionally be responsible for plate scumming, but scumming is seen in any color.

Prevention of scumming. Widespread use of anodized aluminum plates has reduced the frequency of scumming problems, but scumming remains troublesome.

To prevent scumming, the plate must be properly desensitized when it is made. If a plate on the press is to stand for an hour or more, it should be gummed to protect the nonimage areas. If these procedures are followed, a plate should run clean, unless a very greasy ink is being used.

When a plate begins to scum, something should be done at once. If ink becomes firmly attached to parts of the nonimage areas of the plate, it is very difficult to reverse the process and make these areas water-receptive again. The plate should be treated with a good desensitizing plate cleaner or etch. Some pressure must be used on the scummed areas to remove the ink so that it can be replaced with the desensitizing gum. Such a plate etch should finally be dried on the plate, since any desensitizing etch protects better if it is dried. A procedure such as this should be tried at least twice before resorting to the addition of more etch or fountain concentrate to the dampening solution.

Ink dot scum on aluminum plates. Sometimes aluminum lithographic plates develop a peculiar type of scum, called "ink dot scum." Such scum consists of thousands of tiny, sharp dots of ink (figure 9-13). The areas between the ink dots are still well desensitized.

Ink dot scum is associated with the pit corrosion of aluminum. When aluminum corrodes, the corrosion occurs in many little spots, which become pits. When the desensitizing gum is removed from these pits, they can hold ink.

Figure 9-13.
Ink dot scum.

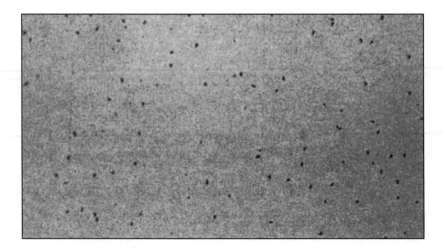

This type of scum can often be produced by covering a plate with water and allowing it to evaporate slowly. The scum may appear in a band on a plate opposite a wet dampener roller.

If ink dot scum has not progressed too far, it can be eliminated by treating the plate with a desensitizing solution consisting of phosphoric acid and gum arabic.

Anodized aluminum plates are covered with a hard, adherent layer of aluminum oxide. Such plates are far less subject to pit corrosion than unanodized aluminum plates, and ink dot scum is less of a problem with them.

Plate Blinding

A plate is blind if it will not produce an image. This may be caused by the erosion of the image area from the plate or by the covering of the image area with gum or another hydrophilic material. If the image is worn off the plate, the problem may simply be called wearing out. Blinding (figure 9-14) usually refers to the presence of an image on the plate, an image that will not accept ink. A plate goes blind when the image areas do not accept ink from the ink form rollers as well as they should. With modern plates and plate processors, blinding is greatly reduced and plate life is greatly extended.

Plate blinding has five principal causes:
- Abrasion of image areas
- Original image areas that are not fully ink-receptive
- Poor adhesion of image areas to the metal
- Partial desensitization of image areas
- Solvents in the ink or plate cleaner used

Figure 9-14.
Plate blinding caused by excessive gum in the fountain solution or improper pH of fountain solution.

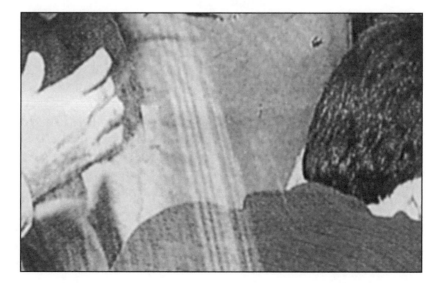

Abrasion of image areas. If parts of the image are abraded from the plate, the plate will be blind in these areas. Little can be done to restore a plate ruined by abrasion. It is often caused by hard, improperly set form rollers. Abrasion of the image areas can also be due to excessive pressure between the plate and the blanket. It can also be caused by abrasive particles in the ink or paper. Some pigments are more abrasive than others. In general, inorganic pigments are more apt to be abrasive than others. Opaque white ink with titanium dioxide (TiO_2) and metallic inks are especially bad offenders.

Sometimes the pigment particles used in the coating on a coated paper are hard and abrasive in nature. This possibility can be tested by passing a clean, cold electric iron over the paper several times, using moderate pressure. If the coating is abrasive, it will become discolored by particles of iron removed from the bottom of the electric iron. Such a coated paper can cause wear of the image areas of a plate if the coating is not securely bound to the paper.

Paper fibers, too, can be somewhat abrasive. When newsprint is run on a lithographic press, the fibers often accumulate on the rubber blanket. Such accumulation (called "linting") often leads to blinding of the image areas, due to the abrasive action of these fibers.

Another source of abrasion is dry antisetoff spray. When sheets treated with this spray for the first color are run through the press again for the second color, the dry spray particles on the sheets can exert an abrasive action on the second color plate.

The increased thickness of a swollen blanket increases the pressure between the blanket and plate. This excessive pressure may lead to abrasion of the plate and ultimately to image blinding. The causes of blanket swelling are outlined in the subsequent section on "Blankets for Offset Presses."

Original image areas that are not fully ink-receptive. The platemaker aims to make plates that are ink-receptive in the image areas. But sometimes things happen such that a plate does not roll up properly at the start of a run. (The press operator drops the ink rollers, but not the dampeners, and applies ink to the entire plate. When the dampeners are then dropped onto the plate, the nonimage areas should clean up promptly.) For example, if the chemicals in the plate processor are old or dilute, or if the machine is not properly functioning, a troublesome plate may result. Plate processors give far fewer problems than the old hand-processing methods did.

Poor adhesion of image areas to the metal. Sometimes a plate starts to print satisfactorily and then begins to go blind as the run proceeds. One reason for this development can be poor adhesion. Depending on the type of plate, more than one material-to-material adhesion can be involved. The light-hardened coating must adhere to the metal; the lacquer, if one is used, must adhere to the coating; and the developing ink must adhere to the lacquer.

If the chemical or physical adhesive bond is poor, the result is a partial blinding of the image areas. Sometimes the crew must have another plate made, and either the procedure or the chemicals must be changed to avoid a recurrence of the problem.

Partial desensitization of image areas. During development, lithographic plates are treated with a solution of a water-soluble gum. A desensitizing gum such as gum arabic is also used in the water dampening solution. Occasionally, gum adheres to part of the image areas of the plate. If this happens, these areas accept water instead of ink. The result is that the plate is blind: the plate will not print in these areas. Any change in press conditions that favor the gum may lead to a partial blinding of the image areas. Some of these conditions are as follows:

- **Running the ink film too thin.** This condition makes it easier for the gum to break through the ink film and become attached to the material underneath. If the color must be reduced, it is much better to add a transparent extender to the ink and then to print a thicker ink film.
- **Too much water in the ink.** If the ink becomes water-logged, it is much easier for the gum to replace it on the image areas of the plate.
- **Too much gum in the dampening solution.** The more gum there is in the dampening solution, the more gum will be emulsified in the ink. This gives the gum a better chance to adhere to the image areas.
- **Too much acid in the dampening solution.** The acid makes the gum a better desensitizing agent so that it can adhere more easily to the image areas if it gets a chance.

Plates are often gummed with a solution of gum or an asphaltum/gum emulsion if the plates are to be left overnight on the press. The aim is to leave a film of gum on all of the nonimage areas of the plate and to leave no gum on any of the image areas. If a film of gum remains on part of the image areas of a plate for several hours, the plate will print with gum streaks when the press is started. The gum adheres to part of the image area over which it is dried and makes that part of the image area water-receptive. Thus, gum streaks can be described as one type of blinding of the image areas.

Gum streaks usually occur if the ink film is too thin on the image areas when the plate is gummed up, or if the ink is badly waterlogged, or if the gumming solution is not buffed down enough over the plate to remove it from the image areas. Gum streaks are unnecessary; they can be avoided if a good plate finisher is properly applied to a plate that is to be left for any length of time.

There is one more way in which the image areas of a plate can be partially desensitized to ink. If printing ink is allowed to dry on the image areas of the plate, these areas often do not accept ink properly when the plate is run again. If a plate is to be left overnight or stored for a rerun, the plate should be rubbed with a good finisher before shutting down the press.

Solvents in the ink or plate cleaner. Most light-hardened coatings and the lacquers used over them are highly resistant to the solvents in most lithographic inks and plate cleaners. Such cleaners must not be used on unbaked, positive-working presensitized plates. If these plates have been exposed to light for any length of time, an alkaline cleaner can attack the image areas, causing the plate to go blind.

The light-sensitive materials in ultraviolet inks are polar in nature and will attack the coating on some types of lithographic plates. For this reason, diazo-type presensitized and wipe-on plates cannot be used with ultraviolet inks. Many photopolymer plates are satisfactory for printing ultraviolet inks.

Treatment of Sponges

Even sponges and their use involve some chemistry. To avoid contaminating lithographic solutions with chemicals from new sponges, they should be washed carefully before use.

Synthetic sponges often contain a detergent material that must not get onto the plate or into dampening solutions.

Natural sponges may contain alkaline materials that affect dampening solutions and plate chemistry. They may also contain sand or small bits of shells that scratch the plates. These shells of marine animals consist largely of calcium carbonate, which, although not soluble in water, changes the pH of dampening solutions, neutralizing the acid and reducing the effectiveness of the etch. It is recommended that a new natural sponge be soaked for 10–15 minutes in a solution containing 1 fl. oz. of concentrated hydrochloric acid (40% HCl) per gallon of water (8 mL/L). The diluted HCl solution neutralizes the alkalinity of the sponge. (Some sponges are so treated before they are sold.)

Press Rollers

Every printing process uses rollers in some way. Even screen and inkjet printers use rollers to transport paper or film when printing on webs. Offset, letterpress, and collotype printing use rollers to transport the ink from the ink fountain to the form rollers. A simple inking system involving one or two rubber rollers is used in flexographic printing, and a special rubber impression roller is used in the conventional and electroassist gravure processes. Electrostatic printing uses rollers to transport the paper and distribute the toner.

The material used to cover rollers depends on the printing process and the function of the roller. Ink rollers are usually covered with an elastomeric material, but chromed steel or ceramic is used as an anilox roller to carry ink on flexo presses. Water rollers on offset presses are covered with rubber, chrome, or ceramic.

For ink rollers, the roller compound must be formulated so the rollers will not swell and grow tacky when they come in contact with the materials in the ink. For example, ultraviolet inks swell the rollers ordinarily used on offset presses, and special rollers must be ordered.

It is common to speak of "rubber" rollers, but rollers covered with natural rubber are used today only on some flexographic presses with certain inks. For other applications, several organic elastomers (or synthetic rubbers) are employed. Here is a list of the principal ones and the printing processes in which they are used:
- Nitrile rubber (a copolymer of butadiene and acrylonitrile)—offset and letterpress
- Butyl (a copolymer of isobutylene and butadiene)—with ultraviolet inks; flexography
- Vinyl [poly(vinyl chloride)]—offset and letterpress

- Neoprene [poly(chlorobutadiene)]—flexography
- EPDM (ethylene propylene diene monomer)—with ultraviolet inks; flexography
- Thiokol (trade name for a polysulfide rubber)—flexography

Besides these, rollers covered with polyurethane (polyester cross-linked with an isocyanate) are used in offset and letterpress. They are tough, and they are not affected by the solvents used in offset and letterpress inks. They generally are not suitable for printing with ultraviolet inks.

Offset press rollers are often covered with a mixture of nitrile rubber and vinyl. Some rollers are covered only with nitrile rubber, and others only with vinyl.

Rollers for Ultraviolet (UV) Inks

The vehicles and the photoinitiators used in UV inks cause conventional rollers to swell. The synthetic rubber materials used in conventional rollers are affected very little by the hydrocarbon solvents used in heatset and quickset inks, but synthetic rubber generally will not resist the polar vehicles present in UV inks.

To make rollers that are satisfactory for use with UV inks, other elastomers, such as butyl or blends of synthetic elastomers, are employed. Some of these blends give fairly good results with both regular and UV inks, but for best results, rollers should be designed either for regular inks or for UV inks.

Dampening Rollers

Dampening rollers must be hydrophilic: they must prefer water rather than ink. To hold water, rollers are commonly covered with molleton or a paperlike sleeve as shown in figure 9-15.

Figure 9-15. Press operator installing a dampening sleeve on a roller *(left)* and using warm water to shrink the sleeve *(right). (Courtesy Rogers Corporation)*

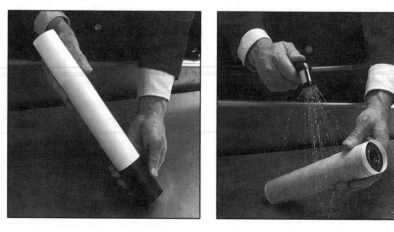

Most rubber rollers are oleophilic (oil-loving) instead of hydrophilic. Treatments are available to make rubber bareback water form rollers hydrophilic.

Most metal water rollers are chrome-plated. Aluminum or stainless steel can also be used. These are metals relatively easy to desensitize with an acidified solution of gum arabic. Even so, as they continue to run, the metal rollers often begin to pick up a layer of ink. When this happens, all of the ink should be removed with a good solvent. The rollers should then be treated for about two minutes with a mixture of 1 fl. oz. of 85% phosphoric acid in 32 fl. oz. of 14° Baumé gum arabic solution (30 mL of the acid per liter of gum). Then the etch should be wiped down and allowed to dry.

Rollers for Flexographic Presses

The flexographic roller material and plate material must be compatible, and both depend on the type of ink being used. Three examples follow:

- For water- or alcohol-based inks, the plates can be natural rubber, and the rollers can be natural rubber or neoprene.
- For inks containing ketones or esters, a butyl plate is used and the rollers can be either butyl or EPDM.
- If the ink contains aliphatic hydrocarbon solvents, then nitrile rubber rollers are used.

Many different types of ink are used in flexography since the process is used to print on a wide variety of substrates. The ink must be formulated to work well with and adhere well to whatever substrate is being printed. Rollers must be compatible with the ink being run.

The anilox roller (the metering roller) on flexo presses is engraved with fine cells whose volume determines the amount of ink that is fed to the plate. The most familiar rollers are chromed steel, but ceramic rollers, which are expensive but resist wear, are becoming popular. Chromed rollers must be changed more frequently because they wear rapidly, and, as they wear, the amount of ink they carry—and therefore the color of the print—changes. This is especially troublesome on repeat runs. Ceramic anilox rollers have largely replaced the chrome-plated steel anilox rollers on narrow webs and in the printing of corrugated board.

Rollers for Gravure Electrostatic Assist

Gravure is an intaglio process: the ink is held in recessed cells below the surface of the printing cylinder. With rare exceptions, gravure printing is done on webs, webs of paper, board, or plastic. The web passes between the printing cylinder and the impression roller. Before the electrostatic-assist process also called electroassist) was developed by the Gravure Association of America, the only way that the ink was transferred to the paper was by pressure and capillary forces. This often resulted in missing dots, particularly in the highlight areas. Paper for gravure must be exceptionally smooth and compressible in order to contact the metal cylinder that carries the image. Electrostatic assist improves printability on a smooth paper, but it will not overcome the problems of a rough sheet.

The creation of an electrical field between the printing cylinder and the impression roller helps to lift the ink from the cells, greatly reducing the number of missing dots. Generating the electrical field requires use of a semiconductive impression roller. This roller is covered with a rubber coating, 0.5–1.0 in. (12–25 mm) thick. It is made semiconductive by the inclusion of special carbon blacks or a long-chain amino or hydroxyl polymer.

**Rollers for
Hickey Removal**

Sometimes matter being printed develops defects of a kind that can be roughly divided into the categories called "hickeys" and "spots." **Hickeys** are small areas that print almost solid in the center and are surrounded by a white ring or halo. They are caused by small pieces of ink skin or other particles that become attached to the plate or, on a lithographic press,

Figure 9-16.
Hickey (enlarged 20×).

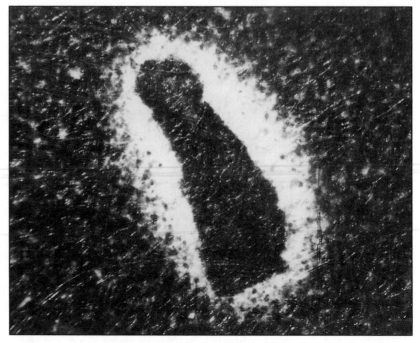

to the blanket. They can also be caused by a deteriorating rubber roller. They print dark because ink skin and rubber are ink-receptive. The ink can't print around the edges of the thick pieces, creating the white ring (figure 9-16). Also, coating particles may occasionally accept ink and produce hickeys.

Other printing defects *(spots)* can be caused by paper fibers, slitter dust, and fibers from molleton dampener rollers (figure 9-17). The resulting spots usually print white or light gray, because such materials are more water-receptive than ink-receptive.

"Hickey rollers" are designed to remove hickeys from lithographic printing plates. Several types have been used. One of the most popular has a nap or suede finish that brushes or wipes debris off the plate. Another device that virtually eliminates hickeys is a system that causes one of the water form rollers to run a few percent slower than the plate. The resulting drag removes hickeys without causing serious damage to the plate.

Modern presses are equipped with any of a variety of hickey removal systems.

Figure 9-17.
Spots caused by lint from paper (enlarged 20×).

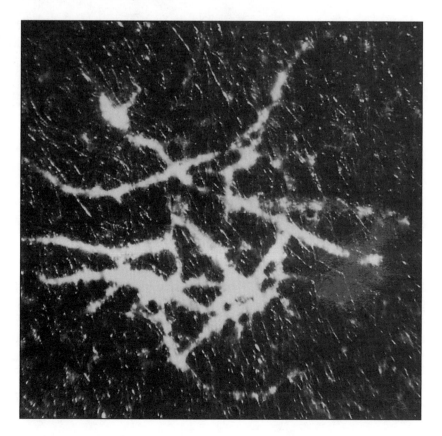

Steel Rollers

On offset inking systems, the rollers alternate between elastomeric (synthetic rubber) and steel. Steel rollers accept ink only if the ink is applied before any desensitizing material, such as gum arabic, has had a chance to come in contact with the steel.

While steel accepts ink if ink is applied first, steel is more water-receptive than ink-receptive. Under certain conditions, the gum in the lithographic dampening solution can become adsorbed onto the steel, causing stripping of the ink film (figure 9-18). The steel does not accept ink in these areas.

Figure 9-18.
A metal roller that is stripping.

Ink stripping of steel rollers is prevented by the use of a roller that is covered with a highly cross-linked hard rubber or with a plastic such as nylon. Both are ink-receptive materials. Stripping is also prevented by covering the steel with a film of copper, which is more ink-receptive than steel. The best way to apply the copper film is to remove the steel rollers from the press and have them electroplated with copper.

A thin film of copper can be deposited chemically on steel rollers without removing them from the press. The rollers must first be thoroughly cleaned. Then they are treated with a solution that commonly contains cuprous chloride, Cu_2Cl_2, and hydrochloric acid, HCl, dissolved in a mixture of ethylene glycol and isopropyl alcohol. The cuprous ions of the cuprous chloride react with the iron atoms of the steel and are reduced to metallic copper. The reaction is:

$$2Cu^+ + Fe^\circ \rightarrow 2Cu^\circ + Fe^{2+}$$

The chemically deposited copper is a very thin film because the reaction stops when the steel is completely covered with copper and the cuprous ions can find no more iron atoms to react with. Such a film may wear off in a week or two, and the treatment must then be repeated.

Some of the conditions that cause stripping of steel rollers are as follows:

- **Using too much dampening solution.** Excess dampening solution causes the ink to become waterlogged, making it easier for the gum in the dampening solution to penetrate the ink and become adsorbed to the surface of the steel.
- **Too much gum in the dampening solution.** A dampening solution usually should contain no more than 1 fl. oz. of 14° Baumé gum arabic per gallon of dampening solution (8 mL/L). If much more than this is used, it is likely to reach and coat the steel.
- **Too much acid in the dampening solution.** Excess acid may cause stripping for two reasons. First, extra acid improves the desensitizing power of gum arabic. Second, the acid is usually phosphoric acid, which itself has the capability of desensitizing steel.

Prevention or Cure of Roller Stripping

If copper plating or a covering of hard rubber is not used on steel rollers to prevent stripping, avoiding the main causes of stripping helps to prevent it. It also helps to clean the steel rollers occasionally by rubbing them with a mixture of pumice and an acid such as acetic, hydrochloric, or nitric (not phosphoric). This pumice-and-acid mixture removes any dried ink or film of desensitizing gum, making the steel ink-receptive again. All pumice must be *carefully cleaned* away or it will abrade the image on the plate.

Steel rollers are not the only ones that suffer ink stripping. Sometimes there is partial stripping of ink from rubber rollers as well. A thin film of ink often remains after each washup. In time, the roller surface becomes hard and glazed and does not accept the ink as well as it should.

To prevent rubber rollers from stripping, the roller train should be cleaned daily with one of the special two-step roller cleaners. Such cleaning helps to prevent the formation of dried ink films and gum glaze.

A small area that has become glazed and stripped can often be made ink-receptive again by treating it with a greasy material in which a fine abrasive is incorporated. (Such preparations are available commercially.) A little of it is applied to the area that is stripping and the press is allowed to idle for a few minutes. The abrasive grinds off the dried ink or desensitizing gum; the greasy material makes the surface ink-receptive again.

Ceramic Rollers

Ceramic rollers are available for use as water rollers on offset presses. These smooth rollers are hydrophilic (water-loving) and, on press, they carry a very thin film of water. They resist ink, so that sensitivity to ink is no problem. They are helpful in running alcohol replacements. Unlike the ceramic anilox rollers used in flexography, these rollers are smooth and glossy. Ceramic rollers replace chrome dampening rollers and help reduce or eliminate streaking. Being extremely hard, they resist scoring and grooving.

Blankets for Offset Presses

A lithographic blanket usually consists of two, three, or four plies of a textile fabric covered with an elastomer (a synthetic rubber) compound. The fabric is woven to strict specifications from either long-staple cotton or synthetic yarn. The textile plies are then coated with about sixty to eighty very thin coats of a rubber compound containing a solvent. Each coat is dried by evaporating the solvent with heat.

Blankets, like rubber rollers, must be made of materials that will not swell unduly when they are in contact with the ink being used. For example, blankets and rollers must both be made with special elastomers that will resist swelling when ultraviolet inks are printed.

Most lithographic rubber blankets are made with nitrile rubber. It is mixed with pigments, softeners or plasticizers, accelerators, and curing or vulcanizing agents (usually sulfur, sometimes peroxides). When all coats have been applied, the blanket roll is heated for a length of time sufficient to convert the uncured plastic mass into the final tough, elastic "rubber," a polymerization process referred to as ***vulcanization.***

Blankets made with nitrile rubber are inert to inks made with drying-oil and heatset varnishes. Blankets must not swell in contact with ink, since a swollen blanket increases the pressure between blanket and plate and can be one cause of plate blinding. Furthermore, if the image areas swell with one job, these areas can create a ghost image when the next job is printed due to the increase in pressure in the embossed areas.

Sometimes coating from coated paper accumulates or "piles" on the blanket. In addition to the blanket, ***piling*** (figure 9-19) can be caused by paper, ink, and dampening solution. Piling caused by a tacky blanket that pulls pigment particles out of the paper can be reduced by adding materials such as ethylene glycol to the dampening solution to reduce blanket tack.

Figure 9-19.
Blanket piling.

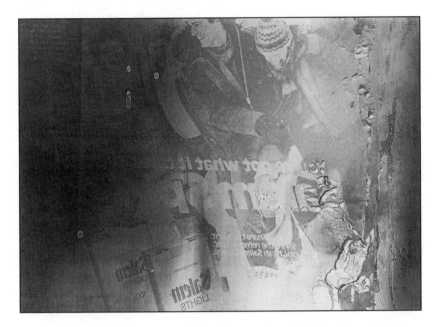

It is important to store blankets properly. Heat, sunlight, or blue fluorescent light causes the rubber surface to harden, glaze, and crack. Storing blankets in a dark, cool place, preferably in the tubes supplied by the manufacturer, will keep them in good condition. If they are stacked on shelves, they should be placed rubber against rubber and fabric against fabric to prevent the transfer of the fabric pattern to the rubber.

Lithographic blankets can be classified as conventional or compressible. Because of their advantages, ***compressible blankets*** are favored by most printers (figure 9-20).

Rubber is not truly compressible, but it flows under pressure. When a conventional blanket revolves against a metal cylinder, a bulge is formed (figure 9-21). Blankets are made compressible to reduce this rolling bulge by introducing a compressible layer between two fabric layers or under the rubber surface. This layer is a spongy material containing thousands of tiny cells filled with air. Besides being compressible, the layer must also be elastic; it must have the ability to spring back after compression. When a blanket has such a layer, the pressure applied is transmitted vertically.

Different manufacturers use different methods to produce the compressible material, and different brands of blankets have properties that differ enough that only one brand should be used at one time on a press.

Usually about 0.002–0.004 in. (0.05–0.10 mm) more packing must be used under compressible blankets than under conventional ones to obtain good printing. With compressible blankets, plate life is increased, streaks due to worn cylinder gears are minimized, and there is less chance of blanket smashup caused by web breaks or doubling sheets. Compressible blankets sometimes recover from blanket smashes.

Blanket Release   Work in the GATF laboratories shows that it requires more force to peel a printed sheet from a smooth blanket than from one with a rough surface. The surprising result was that the actual hardness of the rubber on the surface had little effect on the ability of the blanket to release paper. It is generally observed that soft blankets have a greater tendency to cause paper to pick while it is being printed. Apparently, when the rubber on the surface of the blanket is softened by solvent, it swells and becomes very smooth from contact with the plate and the paper. The hardness of the elastomer itself does not affect release.

Figure 9-20.
A photomicrograph showing the printing surface, compressible layer, and carcass of a compressible blanket. *(Courtesy Day International, Inc.)*

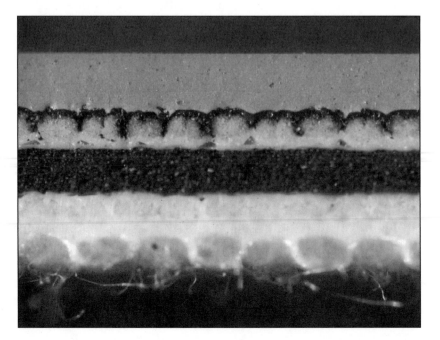

Figure 9-21.
Deformation of compressible blanket *(top)* and conventional blanket *(bottom)* at the nip between the plate and blanket.

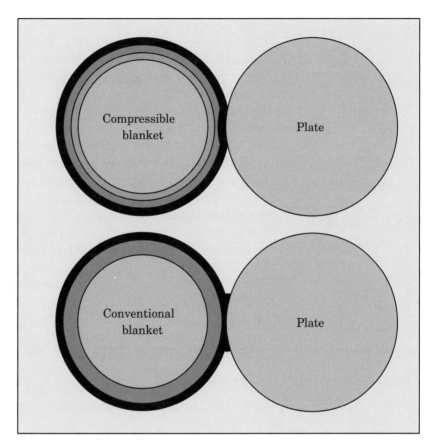

On the other hand, a very smooth surface produces a better halftone dot than a blanket with a rough surface. Special techniques for finishing the rubber surface now produce blankets with a smooth surface that can print a fine halftone dot and yet give good paper release.

The GATF study showed no difference in release between conventional and compressible blankets (wide variations existed in each group).

Press Washup Solutions

It is a general rule of chemistry that *like dissolves like.* Solvents that are powerful in dissolving dried oil-based inks also dissolve or soften rubber rollers and blankets, which also contain polymers of organic chemicals. Water and alcohol, on the other hand, with very different chemical properties, are poor solvents for dried ink and do not soften rollers or blankets. To remove deposits from the rollers without swelling them, a press washup solution requires careful formulation and compounding.

Various solutions or solvents are used to wash ink from rollers, plates, and litho blankets. The wash solution must do the desired cleaning job without swelling rollers or blankets or damaging the image areas of plates. It must have low toxicity and a high flash point. It must not evaporate too rapidly or too slowly. The toxicity and flash points of many organic solvents are discussed in chapter 10. Because of the complicated technical requirements and because of the valuable materials being run on and through the press, not to say the loss of time when things fail, the printer should always buy approved press materials and washes from manufacturers and dealers of these products.

It is convenient to divide press cleaners into two categories: one to remove wet ink and the other to remove glaze. Naphtha is a good cleaner for wet ink. It has low toxicity, little tendency to swell rollers or blankets, and a flash point of 52–53°F (11–12°C). If fire or insurance regulations demand a solvent with a flash point over 100°F (38°C), a good grade of mineral spirits can be used that evaporates without leaving an oily residue. The flash point of mineral spirits is about 100–110°F (38–43°C). Higher-boiling materials such as kerosene should not be used. In the other direction, gasoline presents an intolerable fire hazard, and it contains many materials that should not be used on press.

Special commercial two-solution cleaners are also used for the cleaning of ink rollers. When such cleaners are used for the first time on a particular press, they not only remove the wet ink but also begin to remove ink of other colors that were run previously. After two or three washups, it is possible to change from black to yellow with one washup.

The first solution usually is an aqueous emulsion containing a detergent and a hydrocarbon solvent. The hydrocarbon will be mostly aliphatic, but some aromatic solvent may be incorporated to increase its solvent power.

After most of the ink has been removed by the first solution, the second solution is applied. It is a hydrocarbon such as naphtha or mineral spirits. These solutions will do a good job of cleaning ink rollers, but it is important that enough of the final solution be used to remove the previous solution completely. Otherwise, the rollers will not ink up properly.

Deglazers

The word "glaze" refers to many different types of materials that give a shiny surface to the rubber rollers and cause stripping or streaking by changing the roller's ability to carry ink or dampening solution. Materials from inks are precipitated onto the rollers during the complicated interactions of ink and dampening solution. These may even include dried linseed alkyd or phenolic materials. Most of them are soluble in strong organic solvents. Dampening solutions also contribute to glaze. Gum glaze (from precipitated gum arabic) can be removed with water, but calcium citrate (from citric acid or citrates in dampening solution formulations) requires a fairly strong acid.

Any glaze remover can be used with or without pumice. The mixture should be left on the rubber as short a time as possible, since it has a tendency to swell the rubber. It is best used at shift end so that the solvents can evaporate from the rubber over a period of time, allowing the rubber to return to its normal thickness. Since these materials dissolve grease, they are hard on the hands and should be used only when wearing gloves.

Commercial glaze removers are formulated to be effective without containing any materials that are highly toxic. Highly toxic materials that should never be used include benzene, carbon tetrachloride, and turpentine.

Rubber Rejuvenators

A rubber rejuvenator is primarily a deglazer, but it may perform other functions as well, depending on its formulation. Rejuvenators are based on aggressive, aromatic solvents designed to remove tenacious layers of organic matter that become attached to rubber rollers. Formulations may include xylenes and toluene, a surfactant, and perhaps a plasticizer. Chlorinated hydrocarbons, ketones, and glycol ethers also serve to deglaze or "rejuvenate" rubber rollers.

Roller Conditioners

Like rejuvenators, roller conditioners perform different functions, depending on how they are formulated. They are usually applied to a clean, dry roller system, and the press is allowed to run for a few minutes. The product may be formulated with water to remove gum or other water-soluble materials, surfactants to promote dissolving or removal of hardened dirt, and they may contain dispersed plasticizers to replace those that are removed during washup.

Pressroom Safety

Although it is covered at length in chapter 10, safety is so important that it must be mentioned here. It is management's responsibility to provide safe equipment and training in the safe use of that equipment. The press crew must be instructed in the safe handling of chemicals. Among other things, federal and other laws, enacted to prevent hazardous use of chemicals, require that hazardous chemicals used in the pressroom (or anywhere else) must be described in Material Safety Data Sheets (MSDSs) that present the nature of the material, its hazards, and safe handling procedures. Laws also require that these MSDSs be made available to any employee who requests them.

Skin Sensitivity to Chemicals

Some people are more sensitive than others to chemicals that contact their hands or arms. Rubber gloves should be worn in the plateroom, the pressroom, or anywhere that toxic, corrosive, or irritating materials are handled. Hands must be clean to avoid contaminating the inside of gloves when putting them on. The gloves should be washed well with water before taking them off. It is wise to wash the hands with soap and water as soon as possible after handling irritating or corrosive chemicals, even if none of the material is known to be on the hands or arms.

Further Reading

Destree, Thomas M., ed. *Lithographers Manual.* 9th ed. (426 pp) 1994 (Graphic Arts Technical Foundation, Sewickley, Pa.).

GATF Staff. *Solving Sheetfed Offset Press Problems.* 3rd ed. (144 pp) 1994 (Graphic Arts Technical Foundation, Sewickley, Pa.).

GATF Staff. *Solving Web Offset Press Problems.* 5th ed. (184 pp) 1997 (Graphic Arts Technical Foundation, Sewickley, Pa.).

10 Safety, Health, and the Environment

This chapter has two purposes: to explain some fundamentals of safe handling of chemicals and to discuss some of the laws and regulations relating to health, safety, and the environment.

Accidents can happen when people are unaware of a hazard or are not properly trained and equipped to work with hazardous materials. Chemicals present many hazards, and it is helpful to know what materials present what hazards and how to handle materials safely. To protect workers and the environment from chemical hazards, federal, state, and local agencies have established rules and regulations that help to assure safety, health, and a clean environment.

General Chemical Safety

There is no such thing as a "safe" chemical, but safe ways can be found to handle any material. All substances are chemicals. Even pure water and clean air are dangerous under certain conditions. Some materials that require unusual or extra care are designated as "hazardous." Although most materials used in printing contain no chemicals commonly designated as hazardous, they must be handled in a safe manner.

Under certain conditions, inks, solvents, fountain solutions, and other chemicals used in printing can cause safety and health problems for the people exposed to them. If any material contains a chemical that is designated as a health hazard by the Occupational Safety and Health Administration (OSHA), its manufacturer or importer is required by law to provide a Material Safety Data Sheet (MSDS) for that material. The MSDS contains information on the chemical and physical characteristics of the product and on safe handling and proper disposal procedures. OSHA requires employers to make MSDSs available to employees and to maintain them for thirty years.

Figure 10-1.
The Hazardous Materials Identification System

A completed label is affixed to the chemical container so that employees can be aware of the hazards associated with the chemical and wear proper personal protection gear.

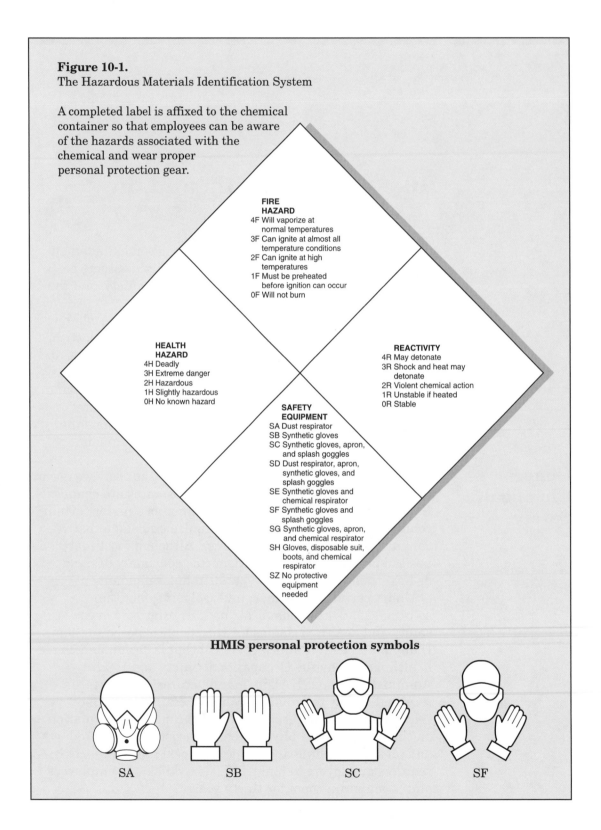

FIRE HAZARD
4F Will vaporize at normal temperatures
3F Can ignite at almost all temperature conditions
2F Can ignite at high temperatures
1F Must be preheated before ignition can occur
0F Will not burn

HEALTH HAZARD
4H Deadly
3H Extreme danger
2H Hazardous
1H Slightly hazardous
0H No known hazard

REACTIVITY
4R May detonate
3R Shock and heat may detonate
2R Violent chemical action
1R Unstable if heated
0R Stable

SAFETY EQUIPMENT
SA Dust respirator
SB Synthetic gloves
SC Synthetic gloves, apron, and splash goggles
SD Dust respirator, apron, synthetic gloves, and splash goggles
SE Synthetic gloves and chemical respirator
SF Synthetic gloves and splash goggles
SG Synthetic gloves, apron, and chemical respirator
SH Gloves, disposable suit, boots, and chemical respirator
SZ No protective equipment needed

HMIS personal protection symbols

SA　　　SB　　　SC　　　SF

A completed Hazardous Materials Identification System (HMIS) label, shown in Figure 10-1, should be affixed to all chemical containers. The label identifies the hazards associated with the product and lists the personal protection required when handling the chemical.

Irritants

While they are not the most dangerous products, irritants are probably the most common health hazard. Some people are more sensitive than others to chemicals that contact their hands or arms. Some products cause rashes, itching, and inflamed eyes in most people. It is a good idea to wear gloves when handling toxic, corrosive, or irritating inks, solvents, or other materials in prepress, press, or other operations.

Lightweight, long-sleeve rubber gloves are inexpensive, but they must be resistant to the material being used. The plastic gloves worn by doctors, nurses, and health workers are useful for protection from some chemicals, but they have low solvent resistance; therefore, they are *not* recommended for use in the printing plant. Hands must be clean to avoid contaminating the inside of gloves when putting them on. The gloves should be washed well with water before removal. It is wise to wash the hands well with soap and water as soon as possible after handling chemicals, even if none of the materials is known to be on the hands or arms.

The eyes are more sensitive to irritants than the skin. Many chemicals cause sore, red, itching, or burning eyes. Goggles or other appropriate eye protection should be readily available to anyone who works with irritating materials. Liquids are especially apt to be sprayed or wiped into the eyes. The solvents in flexographic and rotogravure inks, for example, are highly irritating in the eyes. Eyewash fountains (figure 10-2A) should be readily available for everyone working near irritating chemicals.

If any material is splashed or sprayed into eyes, the eyes should be washed for at least 15 minutes in an eyewash fountain with large quantities of clean water, and a doctor should be seen if there is any redness or irritation after washing.

Corrosive Materials

Corrosive materials are more dangerous than irritants. A corrosive material is any solid, liquid, or gaseous substance that burns or attacks organic tissues, most notably the skin, and, when taken internally, the mouth, stomach, and lungs.

Figure 10-2.
A. Eyewash fountain.
B. Safety goggles for use when working with materials hazardous to the eyes.
C. Safety shower for use when handling flammable or corrosive material in the plant or laboratory.
(Courtesy Cole-Parmer Instrument Company)

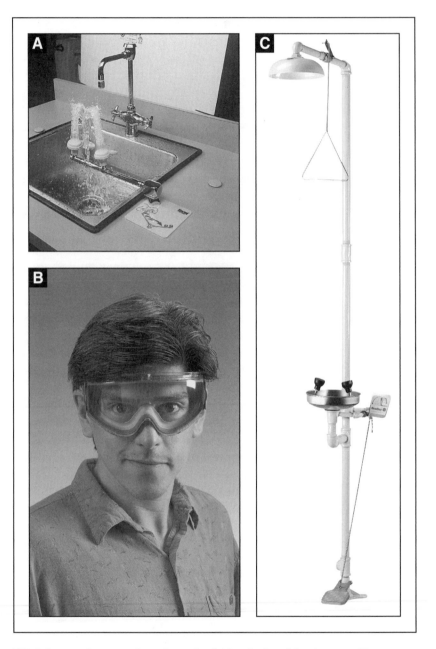

Widely used corrosive chemicals include chlorine, sodium hydroxide (lye), ammonia, phosphoric acid, hydrochloric acid, sulfuric acid, and nitric acid.

When handling such products, the worker should wear rubber or plastic gloves, goggles, or a face shield (figure 10-2B), and a rubber splash apron. If any corrosive material is splashed onto the skin, it should be washed off immediately under a stream of water. If any get into the eyes, they should be washed for 15 minutes in an eyewash fountain, and a doctor should always be seen. Safety showers are recommended in areas where these materials are commonly handled.

Toxic Materials

A toxic material causes damage to living tissue, impairment of the central nervous system, severe illness or, in extreme cases, death when ingested, inhaled, or absorbed by the skin. The amounts required to produce these results vary widely with the nature of the substance and the time of exposure to it.

The Toxic Substances Control Act (TOSCA) provides the legal basis for regulating all aspects of the manufacture of highly toxic products. The U.S. Environmental Protection Agency (USEPA) establishes and enforces such regulations.

Heavy metals such as lead and chromium are toxic when taken into the body in the form of soluble salts. On the other hand, some lead and chrome compounds are so insoluble that they do not enter the blood stream or cause toxic reactions when they are taken into the stomach. Nevertheless, because of the high toxicity of their soluble salts, these metals are largely banned from commerce, and chemists are finding ways to replace lead, barium, chromium, and other chemicals that are frequently hazardous.

Almost every volatile organic compound (VOC), a chemical that readily evaporates at room temperature, has some degree of toxicity. In fact, the word "intoxicated" has become associated with the drinking of ethyl alcohol. Most organic liquids are far more toxic than ethyl alcohol.

It is a good idea to keep the workplace well ventilated and to avoid using high concentrations of solvents. Where necessary, such as in gravure and heatset web offset inks, the solvents are either captured and recycled or burned.

It is not possible to list solvents in order of their relative danger. Although the allowable concentration in the air is related to the acute toxicity, some solvents cause only temporary body dysfunction; continued exposure to others causes permanent damage.

Threshold limit value. The American Conference of Governmental Industrial Hygienists (ACGIH) is concerned with the relative toxicity of a wide variety of chemicals. The organization has established threshold limit values (TLVs) for these chemicals. The values change from time to time as ACGIH obtains additional information about toxic effects of chemicals.

The TLV is the concentration in the air to which most people can be exposed for eight hours a day, year after year, without adverse effects. The averages are time-weighted, permitting higher concentrations when exposures are only for short intervals. A few substances are assigned ceiling concentrations; the concentration of vapor in the air should never exceed the ceiling concentration even for a short length of time.

Most TLV values are expressed in "parts per million" (ppm) based on volumes—the parts (by volume) of the vapor of the chemical that are present in one million parts (by volume) of air—the more toxic a material is, the lower its TLV.

Choice of a solvent or a mixture of solvents therefore requires not only attention to the useful properties of the materials but also to the health and safety hazards involved. A cleaner must do the job well but must not contain solvents that have low TLVs or low flash points.* On the other hand, if solvents evaporate too slowly, they may leave a residue on the press.

Permissible exposure level. Permissible exposure levels (PEL), based on an 8-hour time-weighted average exposure, are set by OSHA and therefore have the force of law. A short-term exposure limit (STEL) is based on a 15-minute exposure. Since OSHA may consider information in addition to that considered by ACGIH, the numbers are not always the same. PELs and TLVs of some solvents used, one way or another, in the graphic arts are listed in table 10-I.

Flammable Materials

The terms "flammable" and "combustible" do not mean the same thing, and the definition varies. According to a definition used by the U.S. Department of Transportation, a flammable substance has a flash point below 100°F (38°C), as determined by the closed-cup method, and a combustible liquid has a flash point of 100–200°F (38–93°C). OSHA and the National Fire Prevention Association define "flammable" as a combustible liquid with a flash point below 80°F, while the USEPA hazardous waste regulations specify 140°F. Special care is required when using such materials.

---

*The ***flash point*** of a liquid is the lowest temperature at which its vapor will ignite when a small flame is passed over the surface. At this temperature and within a range of several degrees above it, the vapor will ignite, but it will not continue to burn. The ***flame point*** or fire point is the lowest temperature at which the liquid will continue to burn in an open container once it has been ignited.

Table 10-I.
Permissible exposure levels (PELs) and flash points of selected solvents.

Name	PEL[1] TWA[3]	PEL[1] STEL	Flash Point[2] °F	Flash Point[2] °C
Saturated hydrocarbons				
Cyclohexane	300[4]	—	−4	−15
n-Heptane	400	500	39	−4
n-Hexane	50	—	−7	−21
Kerosene	—	—	100–165	38–74
VM&P naphtha	300	400	20–50	−7 to 13
Aromatic hydrocarbons				
Benzene	10	—	12	−11
Toluene	100	150	40	4
Xylene (mixed)	100	150	80	27
Alcohols and glycols				
Methanol	200	250	52	11
Ethanol (95%)	1000	—	63	17
n-Propanol	200	250	77	25
iso-Propanol (anh)	400	500	53	12
Ethylene glycol	50[5]	—	232	111
iso-Butanol	50		82	28
Esters				
Methyl acetate	200	250	14	−10
Ethyl acetate	400	—	24	−4
n-Propyl acetate	200	250	58	14
iso-Propyl acetate	250	310	40	4
n-Butyl acetate	150	200	72	22
Ketones				
Acetone	750	1000	0	−15
Cyclohexanone	25	—	111	44
Methyl ethyl ketone (MEK)	200	300	20	−7
Methyl iso-butyl ketone (Hexone)	50	75	73	23
Miscellaneous				
Carbon tetrachloride	2	—	none	none
n-Butoxy ethanol (Butyl Cellosolve)	25	—	155	68
Methoxy ethanol (Methyl Cellosolve)	0.1	0.5	105	41
Ethoxy ethanol (Cellosolve)	0.5	2.5	110	44
1,1,1-Trichloromethane (Methyl chloroform)	350	450	none	none
Turpentine	100	—	91	33

[1]PEL, the permissible exposure level, is enforced by OSHA. TLV, the threshold limit value, is recommended by the American Conference of Governmental and Industrial Hygienists. TLVs are usually but not always the same as PELs. Values in the table are those in force in the year 2000.

[2]Closed cup method

[3]TWA is the time-weighted average for an 8-hr. day, 40-hr. week. STEL is the short-term exposure level (15-min.).

[4]Exposure is in parts/million or mg/m^3.

[5]Ceiling concentration

Two methods are used to determine the flash point of a liquid: the closed-cup method and the open-cup method. In general, the closed-cup method is used for liquids with low flash points; the open-cup method is used for liquids with higher flash points. Flash-point readings of the same solvent with the open-cup method are about 5–10°F (3–6°C) higher than readings made using the closed cup.

Such familiar materials as isopropanol, ink solvents, and washup solvents like naphtha ignite easily and explode under the proper conditions. Smoking and open flames must be prohibited when using them. Fire extinguishers are required by law in factories and offices, and they are especially important in the printing plant. They must be clearly labeled and readily accessible. They must be checked at regular intervals to be sure that they are fully charged and in operating condition.

Areas in the plant or laboratory where flammable or corrosive chemicals are handled should be equipped with safety showers (figure 10-2C).

Different types of fires require different methods of extinguishing:

- **Type A**—paper, wood, solids. Water is a satisfactory extinguishing agent. (Carbon dioxide will also work.)
- **Type B**—liquids, solvents, gasoline. Carbon dioxide is satisfactory. Water may cause the flaming solvent to spread.
- **Type C**—electrical fires. Special extinguishers are required. Water will aggravate the fire.
- **Type D**—igniting metals (Mg, Al, Na). Although these are not usually found in printing plants, carbon dioxide and water do not extinguish the flame.

Spontaneous ignition or combustion. The oxidative-polymerization reaction by which sheetfed inks dry is exothermic (i.e., it gives off heat). Waste, wiping cloths, and rags that contain drying oils generate heat as the oil slowly reacts with oxygen in the air. As the temperature increases, the reaction speeds up. The problem is aggravated by the presence of an easily ignited solvent used to remove the ink. If a pile of wiping cloths or rags soaked in ink or varnish is allowed to stand, it can generate enough heat to ignite the solvent and the cloths. Spontaneous ignition or spontaneous combustion has caused many serious fires.

Cans containing cloths used to clean the press or to wipe up ink must be emptied regularly, preferably once a day.

Electrical ignition: sparks, grounding, and bonding.

The printing plant usually has solvent vapors present in the air. The concentration can be high, close to solvent cans or the press. A spark can ignite the solvent, creating a fire or explosion. All bulk containers of flammables should be both grounded and bonded to prevent static electricity sparks (figure 10-3). The terms "bonding" and "grounding" are often used interchangeably, but bonding is the elimination of a

Figure 10-3.
Grounding and bonding of containers for flammable liquids.

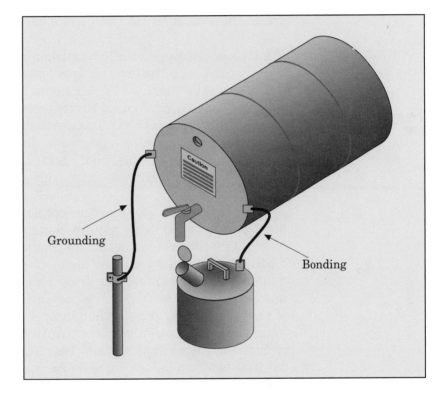

difference in electrical potential between objects; grounding eliminates a difference in potential between an object and ground. Both are effective only when the connections are conductive. When large volumes of solvent are being handled, a ground wire should connect the solvent drum with the container into which the solvent is poured.

Metal safety cans with a spring-closed lid and metal screen (flame arrester) must be used for storing small amounts of flammable liquids in the pressroom. The screen inside the mouth of the can helps to prevent sparks from entering the can and causing an explosion. They act by diffusing the heat of the spark, in the same way that the screen inside a miner's lamp prevents explosions.

Carcinogens

A carcinogen is any substance that causes the development of a cancerous growth in living tissues. A wide variety of materials can cause cancer; OSHA periodically updates its list of substances known to be carcinogenic. Materials that cause cancer in small quantities are usually banned from commerce, or they are tightly controlled. Other materials are banned as their ability to cause cancer is proven.

Some materials commonly used in today's printing plant are known to be carcinogenic. Therefore, the proper safety precautions must be observed when handling them.

General Safety in the Printing Plant

Printing is not an especially hazardous occupation, but hazards do exist, and the law requires management to make the workplace safe. The U.S. Occupational Safety and Health Administration (OSHA) used surveys conducted by the Bureau of Labor Statistics of the U.S. Department of Labor as a guide to help establish inspection priorities. According to the Bureau of Labor Statistics in 1999, injury and illness rates in the printing industry were below the average for all manufacturing industries.

To provide a safe workplace for employees in every printing plant and to meet OSHA requirements, printers (regardless of size) must have a formal safety program, with responsibility formally assigned (even on a part-time basis) to one manager or department. The program must include all areas and activities in the plant. The program includes all aspects of safety, including the safe handling of chemicals.

All areas and activities of the printing plant must be included in the safety program. This includes such activities as office work and shipping or delivery, in addition to maintenance and manufacturing. In planning the construction or expansion of a plant, management should consider safety factors such as equipment configuration, maintenance, chemical storage, material flow, ventilation, fire hazard, and waste disposal.

Every employer must comply with OSHA's Hazard Communication Standard. Container labeling and MSDSs are cornerstones of this standard. The MSDS provides basic information to help employers and employees assess chemical hazards posed by materials they use.

Elements of a General Safety Program

The OSHA Hazard Communication Standard requires that a viable safety program include a written hazard communication program. The program must include:

- A commitment from management to support the program
- An organization with specific responsibilities and authority for safety
- Warnings or cautionary signs on cans, containers, and equipment

In addition, a satisfactory safety program must include proper training of operators as well as accident and illness prevention programs that include personnel protective equipment, work practices, accident investigation, and record keeping.

The printer must also see that employees:

- Observe and follow all warnings or cautionary signs on cans, containers, and equipment
- Do not smoke in the press area or in other areas where flammable/combustible chemicals or materials are used or stored
- Use proper (metal) solvent safety cans
- Use grounding straps when transferring solvents to and from other containers
- Store solvent wipes or rags in tightly covered containers
- Keep fire doors clear and free from obstruction
- Maintain adequate housekeeping

Safety engineers and inspectors now pay strict attention to repetitive motion injuries, lifting hazards, and **_ergonomics,_** the design of equipment to improve convenience and operator comfort and safety.

Chemical Safety in the Printing Plant

As a general rule, prepress areas have more chemical hazards than physical hazards, while the pressroom has more physical hazards than chemical. Photo-processing and platemaking chemicals require special safety training and special protective equipment.

Prepress Chemical Safety

Engraving is replacing etching for preparing gravure cylinders, but etching chemicals (see chapter 6) still pose a health hazard and require protective clothing and equipment, as well as training in handling of chemicals and equipment.

Photography and photo processing. Although digital management of prepress operations is replacing film and wet processing of plates, wet processing is still widely used in prepress operations. Most of the chemicals used in photography and the processing of photographic images are toxic. Photographic developers contain sodium hydroxide or sodium carbonate, which are toxic and very hazardous to the eyes. Materials such as silver nitrate, hydroquinone, formaldehyde, and hypo are extremely toxic if taken internally. Formaldehyde is a known carcinogen. Many of these materials are affected by the materials on human skin, so that photographers keep their hands out of processing chemicals both for self-protection and to avoid contaminating the photographic bath.

Photographers are used to handling wet paper and film with tongs, or using gloves when fingers are more convenient. Automatic processors reduce the chances for getting chemicals onto the hands, or into the mouth or eyes.

Platemaking. Solvents should never be handled without gloves. The mild alkali used to develop positive lithographic plates, like any alkali, is hard on the skin and dangerous if splashed into the eyes. Automatic lithographic plate processors contain hazardous chemicals that should be handled according to directions. The same applies to processors for flexographic photopolymer plates.

Chemicals used to etch gravure cylinders are especially corrosive and must be kept off the skin and out of the eyes.

Prepress environmental concerns. Local sewer authorities are concerned about heavy metal contamination. Photographic film usually contains silver, and when film is developed or discarded, silver at concentrations greater than 5 ppm becomes a hazardous waste. Silver recovery is currently required for printers throughout the United States. Recovery-disposal systems for silver include electrolytic precipitation and ion exchange. The cartridge-type silver recovery unit is widely used (figure 4-8).

Aqueous-developable lithographic plates (chapter 5) are designed to reduce the amount of material discharged to the sewer.

Pressroom Chemical Safety

Even if physical hazards in the pressroom are greater, there are still many chemical hazards in the pressroom.

Some materials cause rashes and skin problems. Many people are sensitive to solutions of ammonium dichromate or potassium dichromate (usually called "bichromate" by printers). Ammonium dichromate has been used in some fountain solution etches. Potassium dichromate was used as a sensitizer for carbon tissues used to make gravure cylinders and screen process printing stencils. Fortunately, these materials are rarely used now, but they remain dangerous.

Skin sensitivity to dichromate is commonly called "chromic poisoning." It is characterized by skin sores and severe itching (rather like poison ivy.) Sometimes people suddenly become sensitive to dichromate after handling it with no trouble for years. People who become sensitive usually remain sensitive, and it is advisable to use preventive measures.

Solvent handling has been treated casually in many printing plants. Ink solvents present fire, health, and environmental hazards. Training programs together with proper equipment are required to reduce accidents and injuries.

Printers are required to know the level of exposure to hazardous chemicals that workers experience in the pressroom. Printers should contact consultants or local universities that provide air quality tests and analysis of pressroom areas to determine the concentrations of hazardous materials in the air. Establishing a history of monitoring the concentrations of organic vapors will go a long way in complying with OSHA right-to-know regulations.

Before a new printing ink or press wash is used, the technical manager should review the MSDS and product label to evaluate any potential hazards. Some MSDSs contain warnings not present on a product label. The printer must have trained personnel and written emergency handling procedures to deal with accidental exposure and injury caused by hazardous chemicals. For personal injuries, the printer needs to file proper documentation as required by OSHA.

Toxicity of common solvents is summarized in table 10-I, which lists PELs and TLVs.

Printing inks. Different printing processes use different kinds of printing inks. Liquid inks used in gravure and flexography often contain flammable solvents. (Even the water-based inks popular in flexography contain some volatile organic materials to assist in dissolving the binder and to improve performance.) Web offset inks contain large amounts of VOCs. Paste inks—used for offset, screen, and letterpress printing—may or may not be classified as hazardous, depending on the solvent or ink oil, pigments, and additives used in them.

Printing inks rarely cause health or environmental problems when they are used as recommended by the manufacturer. However, MSDSs should be consulted to be sure that inks are properly handled. Traditionally, printing ink manufacturers and chemical suppliers have been responsive to the printing industry's desire to reduce or eliminate the use of highly hazardous materials. In addition, irritating materials have been largely eliminated from inks.

The ink oils in printing inks are VOCs that can be emitted during the printing process. Other VOCs in the printing plant include isopropyl alcohol and cleaning solvents. To prevent the amount of emissions from exceeding the amount specified by law, these must be controlled or reduced.

Storage of printing inks. Small quantities of printing inks should be stored in tightly closed cans in a cool, clean room (figure 10-4). Large quantities, stored in tote tanks, drums, or pails, require a secondary containment such as dikes, drip pans, or absorbent material in order to limit any spill to a manageable size, reduce employee exposure to a minimum, and prevent the spill from becoming uncontrollable during an emergency such as fire or explosion.

Dampening solutions. Dampening solutions usually contain phosphoric acid and other additives. The MSDS tells which of the components may be regarded as hazardous. Adding isopropyl alcohol (IPA), or alcohol substitutes, poses

Figure 10-4.
Ink storage room with laboratory three-roll mill.
(Courtesy National Association of Printing Ink Manufacturers)

additional hazards. IPA vapors can be intoxicating, and, when discharged to the outside air, IPA can react to cause air pollution contaminants.

Bindery Chemical Safety

Few chemical hazards exist in the bindery. Adhesives now used are mostly aqueous suspensions or hot melts. Hot melts can cause thermal burns. If solvent-based adhesives are used, the solvents must be treated like any other solvents in the printing plant.

Environmental Hazards

Many air contaminants, including oxides of nitrogen, ozone (O_3), and other materials, are extremely irritating to the eyes and lungs. Severe pollution of the atmosphere is known to have caused large numbers of deaths, particularly among the aged and the sick. Air pollution causes many other environmental problems. Many of the solvents in the printing plant contribute to air pollution, and they must be captured, destroyed, or avoided. Control of these materials is mandated by USEPA and many state and local agencies.

Handling of Chemical Wastes

Hazardous wastes are waste materials specified by the USEPA under the Resource Conservation and Recovery Act (RCRA). Hazardous waste may be ignitable, corrosive, explosive, or TCLP toxic. (TCLP is an acronym for toxic characteristic leaching procedure.) Under RCRA, printers must seek new permanent ways to dispose of hazardous waste besides landfilling.

The major waste streams for a printing operation consist of paper, spent solvent, solvent-laden cleaning cloths or rags, and spent ink. Means of disposal of these materials are limited, but fortunately, many printers are classified as small-quantity generators and have less stringent requirements.

Waste inks. The printer has several options for disposing of waste inks. First, ink may be recycled or recovered in-house or the printer may contract for this service with a supplier. Many flexo and gravure inks, whether they are solvent- or water-based, may be considered hazardous waste and should be incinerated if recycling or recovery is not available. Under federal regulations, waste lithographic ink is considered hazardous only if it contains high concentrations of heavy metals such as barium, lead, or chromium. Whether or not a specific lot of ink is classified as hazardous, few waste transporters are willing to remove ink without a hazardous waste designation and manifest. Lithographic, flexo, and gravure ink containers, if properly cleaned, can be discarded as non-hazardous waste containers.

Solvents. Solvents and other liquid hazardous wastes cannot be disposed of in a landfill without prior treatment (e.g., solidification or incineration), and the printer is liable for all future disposal costs if a landfill becomes designated as an abandoned hazardous waste site. Incineration is the most reliable but often the most expensive alternative. Keeping the amount of waste generated to a minimum effectively reduces the amount of hazardous waste as well as any potential liability.

Most local sewer codes forbid discharging anything that may discolor the wastewater flowing in the sewer system. This poses special problems for users of water-based inks.

The printer needs to be very careful about washing ink residue from empty ink containers, cleaning cloths, rags, etc. Printers classified as small-quantity generators can often contract with an approved launderer to pick up used solvent-laden wiping cloths, launder them, and return them. Some states and EPA regions are pressuring industrial launderers to reduce their own emissions and effluents and are requiring them to have air and wastewater disposal permits. As a result, launderers may refuse to accept solvent-laden wiping cloths.

Waste minimization. Waste minimization is a regulatory requirement. Printers must separate waste streams, reduce the volume of solvent-bearing cleaning materials, adjust their cleaning procedures, and establish waste minimization programs, along with reuse, recovery, and recycling. Many states and cities have stringent regulations concerning the disposal of nonhazardous industrial waste. The printer must be aware of such regulations, because disposal of empty printing ink pails and cans is subject to those regulations.

Regulations Affecting Printers

The public's awareness and concern have increased legislative activity. Federal and state environmental regulations have been amended and strengthened since they were first enacted in the 1970s. The impact of these issues on the printer has increased, and the very existence of a company may depend on how it deals with health, safety, and the environment. A combination of increased industrial activity, increased automobile usage, and increasing public demand for safe working conditions and clean environment has stimulated much legislation and new regulations.

Printers are required to comply with the laws and regulations of federal, state, and local governments. Some major U.S. federal laws are summarized in table 10-II.

Hazard Communication Standard. The Hazard Communication Standard is not an act but an OSHA standard (29 CFR 1910.1200). It requires communication of hazardous material risks to workers in regulated facilities. To comply with the standard, printers must meet the following five requirements:

• Compile a list of all chemicals used in the plant.
• Obtain a Material Safety Data Sheet (MSDS) from the manufacturer or distributor for each chemical that appears on the list.
• Properly label each container that contains a hazardous chemical.
• Provide ongoing employee training, including the identification and proper use of all chemicals used in the plant.
• Develop a written hazard communication program. (See the section "Elements of a General Safety Program" on page 370.)

Clean Air Act. Amendments to the Clean Air Act require the USEPA to pressure state and local governments to adopt and enforce more stringent rules designed to meet the National Ambient Air Quality Standard for ozone. Emissions of VOCs from lithographic printing plants consist mainly of dampening solution additives (isopropyl alcohol and alcohol substitutes), ink oils, and cleaning solvents. These contribute to the formation of ozone in the lower atmosphere. Ozone is formed through a series of complex chemical reactions involving VOCs and nitrogen oxides in the presence of sunlight. The EPA has set stringent limitations on the acceptable amount of ozone in the workplace or the lower atmosphere. Ozone concentration exceeds established limits in many urban areas.

The ink oil or solvent in printing inks contains volatile organic compounds (VOCs) that are emitted during the printing process (see chapter 9). Other VOCs in the printing plant include isopropyl alcohol and cleaning solvents. In the presence of sunlight, these VOCs react with nitrogen oxides in the air to form ozone. To prevent the amount of emissions

Table 10-II.
Some important federal health, safety, and environmental acts.

The Clean Air Act (CAA): 1970, 1977, 1990
The predecessor to these acts appeared in 1955. The 1977 amendments mandated and enforced toxic emissions standards for stationary sources and motor vehicles. The 1990 amendments reauthorized the Clean Air Act. They contain four sections pertaining to healthy air standards: control of acid rain; improved ozone control; reduced contamination from mobile sources; and reduced air emissions of 189 toxic compounds. Reducing emissions from small business is the target of the 1990 amendments.

Occupational Safety and Health Act (OSHA): 1970
This law created the Occupational Safety and Health Administration, which develops and enforces regulations for workplace health and safety.

Clean Water Act (CWA): 1972, 1977, 1987
The Federal Water Pollution Control Act was amended in 1977 to become the Clean Water Act. It regulates the discharge of pollutants into surface waters. The Water Quality Act Amendment (1987) focuses on tighter limits and stormwater discharges.

Resource Conservation and Recovery Act (RCRA): 1976, 1984
The act regulates management and disposal of hazardous materials and wastes currently being generated, treated, stored, disposed of, or distributed. The Hazardous and Solid Waste Amendments (1984) establish a timetable for banning landfills and more stringent requirements for underground storage tanks (USTs).

Toxic Substances Control Act (TOSCA or TSCA): 1976
This law authorizes the U.S. Environmental Protection Agency (USEPA) to gather information on chemical risks, order manufacturers to test, and under certain conditions ban the distribution and use of chemicals.

Comprehensive Environmental Responsibility, Compensation, and Liability Act (CERCLA): 1976
Referred to as the "Superfund," this law provides the authorization to identify and correct abandoned hazardous waste sites and to seek reimbursement from contributors of the costs associated with cleanup. The 1986 Superfund Amendments and Reauthorization Act (SARA) reauthorizes and expands the jurisdiction of CERCLA.

Emergency Planning and Community Right-to-Know Act (EPCRA): 1986
Title III of SARA, commonly called "Right-to-Know," mandates public disclosure of chemical information and the storage and release of certain chemicals and mandates development of emergency response plans.

Pollution Prevention Act: 1990
Any company completing Form R under EPCRA must now include information on waste minimization activities.

from exceeding that specified by law, they must be collected and/or disposed of.

Air pollution control devices, such as incinerators, cooler-condensers, and carbon absorption devices can be used to reduce VOCs generated by flexo, gravure, or heatset web offset lithography. These devices are expensive to install and to operate, and they pose a major economic burden, one that may be prohibitive to small printers. Reducing the VOC content of inks and solvents is usually less expensive.

Press and plate washes containing little VOCs are currently available. Although these cleaning products are not yet fully evaluated, the newspaper industry has found cleaning solvents that are virtually free of VOCs. The use of exempt chlorinated hydrocarbon solvents reduces VOCs, but many are toxic and require special handling.

Other amendments to the Clean Air Act require further reductions of VOC emissions from all sources. Sheetfed and nonheatset web printers—like heatset web, flexographic, and gravure printers—now are required to provide some form of VOC emission control. Permits for the installation and/or operation of all printing presses will be required by most states.

Groundwater and soil contamination. Leaks from underground storage tanks or spills from outdoor storage drums that contain chemicals or solvent-bearing hazardous waste may contaminate the soil and/or groundwater. Regulations for underground storage tanks require printers to use tanks and underground piping with corrosion protection, leak protection, and secondary containment, and to carry at least $500,000 in insurance. Hazardous waste and solvent drums should be stored inside prior to appropriate disposal.

Storage and transfer will be inhibited by increasingly severe and expensive regulations. Real estate sales will require environmental audits and/or certification. The printer should avoid the possibility of groundwater or soil contamination.

Clean Water Act. Neutralization of waste waters and the elimination of silver is required by all public systems accepting waste water. As a result of the Water Quality Act of 1987, additional process water pretreatment beyond silver recovery and neutralization is required. Organic pollutants must be controlled, and, to treat activated sludge, the printer may have to install costly treatment systems, including flocculators and precipitators.

Printers who do not discharge effluent into public sewer systems must provide approved disposal sites or ship waste-water streams for disposal elsewhere. On-site treatment is extremely expensive except for very great volumes, and obtaining permits is also costly.

Emergency Planning and Community Right-to-Know. Title III of Superfund Amendment and Reauthorization Act (SARA) or Emergency Planning and Community Right-to-Know Act (EPCRA) adds another administrative burden to the printer. The purpose of this amendment, aside from making the local areas aware of hazardous chemicals on-site in case of accidents or spills, is to provide a database for the USEPA in determining sources of emissions from specific industries. Printers are required to report all accidental releases of materials, storage, and release of certain other materials.

Hazardous chemical release. USEPA requires plant managers to examine their facilities to determine whether there is any potential for an uncontrollable release of hazardous material. Chemicals subject to inventory reporting include any materials that have an MSDS and those classified as extremely hazardous. These inventory reports are due March 1 of each year. On July 1 of each year, an emission report is required for any listed chemicals that exceed specified thresholds. The reports are required by state agencies and the USEPA. Lists include pigments and solvents that are sometimes found in printing inks.

A Chemical Elements and the Periodic Table

Chemical elements.

Name	Symbol	Atomic Number	Atomic Weight
Actinium	Ac	89	227.028
Aluminum	Al	13	26.9815
Americium	Am	95	243.061
Antimony (Stibium)	Sb	51	121.760
Argon	Ar	18	39.948
Arsenic	As	33	74.9216
Astatine	At	85	209.987
Barium	Ba	56	137.327
Berkelium	Bk	97	247.070
Beryllium	Be	4	9.012
Bismuth	Bi	83	208.980
Bohrium	Bh	107	262.12
Boron	B	5	10.811
Bromine	Br	35	79.904
Cadmium	Cd	48	112.411
Calcium	Ca	20	40.078
Californium	Cf	98	251.080
Carbon	C	6	12.011
Cerium	Ce	58	140.115
Cesium	Cs	55	132.905
Chlorine	Cl	17	35.453
Chromium	Cr	24	51.996
Cobalt	Co	27	58.933
Copper	Cu	29	63.546
Curium	Cm	96	247.070
Dubnium	Db	105	262.114
Dysprosium	Dy	66	162.50
Einsteinium	Es	99	252.083

Chemical elements
(continued).

Name	Symbol	Atomic Number	Atomic Weight
Erbium	Er	68	167.26
Europium	Eu	63	151.965
Fermium	Fm	100	257.095
Fluorine	Fl	9	18.9984
Francium	Fr	87	223.020
Gadolinium	Gd	64	157.25
Gallium	Ga	31	69.723
Germanium	Ge	32	72.61
Gold (Aurum)	Au	79	196.967
Hafnium	Hf	72	178.49
Hassium	Hs	108	255
Helium	He	2	4.0026
Holmium	Ho	67	164.930
Hydrogen	H	1	1.00794
Indium	In	49	114.818
Iodine	I	53	126.904
Iridium	Ir	77	192.217
Iron (Ferrum)	Fe	26	55.845
Krypton	Kr	36	83.80
Lanathanum	La	57	138.906
Lawrencium	Lr	103	262.11
Lead (Plumbum)	Pb	82	207.2
Lithium	Li	3	6.941
Lutecium	Lu	71	174.967
Magnesium	Mg	12	24.305
Manganese	Mn	25	54.938
Meitnerium	Mt	109	256
Mendelevium	Md	101	258.10
Mercury (Hydrargyrum)	Hg	80	200.59
Molybdenum	Mo	42	95.94
Neodymium	Nd	60	144.24
Neon	Ne	10	20.180
Neptunium	Np	93	237.048
Nickel	Ni	28	58.693
Niobium (Columbium)	Nb	41	92.906
Nitrogen	N	7	14.0067
Nobelium	No	102	259.101
Osmium	Os	76	190.23
Oxygen	O	8	15.9994
Palladium	Pd	46	106.42

Chemical elements
(continued).

Name	Symbol	Atomic Number	Atomic Weight
Phosphorus	P	15	30.9738
Platinum	Pt	78	195.08
Plutonium	Pu	94	244.064
Polonium	Po	84	208.982
Potassium (Kalium)	K	19	39.098
Praseodymium	Pr	59	140.908
Promethium	Pm	61	144.913
Protactinium	Pa	91	231.036
Radium	Ra	88	226.025
Radon	Rn	86	222.018
Rhenium	Re	75	186.207
Rhodium	Rh	45	102.906
Rubidium	Rb	37	85.468
Ruthenium	Ru	44	101.07
Rutherfordium	Rf	104	261.11
Samarium	Sm	62	150.36
Scandium	Sc	21	44.956
Seaborgium	Sg	106	263.118
Selenium	Se	34	78.96
Silicon	Si	14	28.086
Silver (Argentum)	Ag	47	107.868
Sodium (Natrium)	Na	11	22.9898
Strontium	Sr	38	87.62
Sulfur	S	16	32.066
Tantalum	Ta	73	180.948
Technetium	Tc	43	98.906
Tellurium	Te	52	127.60
Terbium	Tb	65	158.925
Thallium	Tl	81	204.383
Thorium	Th	90	232.038
Thulium	Tm	69	168.934
Tin (Stannum)	Sn	50	118.710
Titanium	Ti	22	47.867
Tungsten (Wolfram)	W	74	183.84
Uranium	U	92	238.029
Vanadium	V	23	50.942
Xenon	Xe	54	131.29
Ytterbium	Yb	70	173.04
Yttrium	Y	39	88.906
Zinc	Zn	30	65.39
Zirconium	Zr	40	91.224

Periodic Table

Active metals — Transition metals — Nonmetals

1A	2A	3B	4B	5B	6B	7B	8B	8B	8B	1B	2B	3A	4A	5A	6A	7A	8A
3 Li 6.941	4 Be 9.012											5 B 10.811	6 C 12.011	7 N 14.0067	8 O 15.9994	9 F 18.9984	2 He 4.0026
11 Na 22.9898	12 Mg 24.305											13 Al 26.9815	14 Si 28.0855	15 P 30.9738	16 S 32.066	17 Cl 35.453	10 Ne 20.180
19 K 39.098	20 Ca 40.078	21 Sc 44.956	22 Ti 47.867	23 V 50.942	24 Cr 51.996	25 Mn 54.938	26 Fe 55.845	27 Co 58.933	28 Ni 58.693	29 Cu 63.546	30 Zn 65.39	31 Ga 69.723	32 Ge 72.61	33 As 74.9216	34 Se 78.96	35 Br 79.904	18 Ar 39.948
37 Rb 85.468	38 Sr 87.62	39 Y 88.906	40 Zr 91.224	41 Nb 92.906	42 Mo 95.94	43 Tc 98.906	44 Ru 101.07	45 Rh 102.906	46 Pd 106.42	47 Ag 107.868	48 Cd 112.411	49 In 114.818	50 Sn 118.710	51 Sb 121.760	52 Te 127.60	53 I 126.904	36 Kr 83.80
55 Cs 132.905	56 Ba 137.327	*	72 Hf 178.49	73 Ta 180.948	74 W 183.84	75 Re 186.207	76 Os 190.23	77 Ir 192.217	78 Pt 195.08	79 Au 196.967	80 Hg 200.59	81 Tl 204.383	82 Pb 207.2	83 Bi 208.980	84 Po 208.982	85 At 209.987	54 Xe 131.29
87 Fr 223.020	88 Ra 226.025	†	104 Rf 261.11	105 Db 262.114	106 Sg 263.118	107 Bh 262.12	108 Hs 255	109 Mt 256								86 Rn 222.018	71 Lu 174.967

* Lanthanide series

57 La 138.906	58 Ce 140.115	59 Pr 140.908	60 Nd 144.24	61 Pm 144.913	62 Sm 150.36	63 Eu 151.965	64 Gd 157.25	65 Tb 158.925	66 Dy 162.50	67 Ho 164.930	68 Er 167.26	69 Tm 168.934	70 Yb 173.04	71 Lu 174.967

† Actinide series

89 Ac 227.028	90 Th 232.038	91 Pa 231.036	92 U 238.029	93 Np 237.048	94 Pu 244.064	95 Am 243.061	96 Cm 247.070	97 Bk 247.070	98 Cf 251.080	99 Es 252.083	100 Fm 257.095	101 Md 258.10	102 No 259.101	103 Lr 262.11

The boldface number in each box is the atomic number, and the lightface number is the atomic weight. Those elements in gray are semimetals.

B The SI or Metric System and Conversion Tables

The Metric System

Precise measurement is the first requirement of all scientific disciplines. Accordingly, a convenient scale for measuring is most important. Chemists throughout the world have adopted a system of measurements, the International System (SI), that modifies and expands the metric system. The older metric system was proving inadequate to new requirements, and the 11th General Conference on Weights and Measures formulated the new system in 1960, correctly called the SI, but still usually referred to as the metric system.

All major countries of the world except the United States have now adopted the metric system (now succeeded by the more precise SI). Even in the United States, the metric system is used almost universally in the scientific and medical professions and extensively in the chemical industry, the electric power industry, photography, optometry, and the electronics industry. Official U.S. weights and measures are defined in terms of the metric system. For example, an inch is legally defined as a length equal to exactly 25.4 mm.

Increasing commerce and communications between countries must make a universal measuring system inevitable. Already, American machinists use metric tools to work on imported automobiles.

The metric system came out of the French Revolution, and it spread throughout the world as revolutions came to Latin America and China, and as the emperors were restored to nineteenth-century Japan. The meter was defined as one ten-millionth of the distance from the equator to the North Pole, but this proved to be insufficiently precise for some scientific measurements, and it is now defined as the distance traveled by light in a vacuum during the time interval of 1/299,792,458 of a second. This definition means that matter itself is the standard of measure, and the length of the meter can be established in any properly equipped laboratory.

The metric system is easy to learn and to remember. It is a decimal system; the units differ by factors of ten:

- 1000 millimeters (mm) = 100 centimeters (cm) = 1 meter
- 1000 meters = 1 kilometer (km)
- 1/1000 liter (L) = 1 milliliter (mL)

The various multiples are all given prefixes:

Multiple	Prefix	Symbol
10^{12}	tera	T
10^{9}	giga	G
10^{6}	mega	M
10^{3}	kilo	k
10^{2}	hecto	h
10	deka	da
10^{-1}	deci	d
10^{-2}	centi	c
10^{-3}	milli	m
10^{-6}	micro	μ
10^{-9}	nano	n
10^{-12}	pico	p
10^{-15}	femto	f
10^{-18}	atto	a

The SI defines six basic measurements, and all others are derived from these six: the meter (length), the kilogram (mass), the second (time), the ampere (electric current), the Kelvin scale of temperature, and the candela (light intensity).

The kilogram is derived from the mass of one liter of water. The temperature scale is set at 0 for absolute zero and at 273.16 as the triple point of water (the temperature at which water, ice, and water vapor are at equilibrium). The Celsius scale is derived from the Kelvin scale.

The following tables in appendix B permit conversion from the U.S. to metric equivalents, and vice versa.

Length equivalents:
feet/meters.

1 ft. = 0.3048 m
1 m = 3.28 ft.

Feet	Foot or meter measurement to be converted	Meters
3.28	1	0.305
6.56	2	0.610
9.84	3	0.914
13.12	4	1.219
16.40	5	1.524
19.68	6	1.829
22.96	7	2.134
26.24	8	2.438
29.52	9	2.743
32.80	10	3.048

Length equivalents:
inches/millimeters.

1 in. = 25.4 mm
1 mm = 0.03937 in.

Inches	Inch or millimeter measurement to be converted	Milli-meters
0.039	1	25.4
0.079	2	50.8
0.118	3	76.2
0.157	4	101.6
0.197	5	127.0
0.236	6	152.4
0.276	7	177.8
0.315	8	203.2
0.354	9	228.6
0.394	10	254.0

Mass equivalents: ounces/grams.

1 oz. = 28.3495 g
1 g = 0.03527 oz.

Ounces	Ounce or gram mass to be converted	Grams
0.035	1	28.3
0.071	2	56.7
0.106	3	85.0
0.141	4	113.4
0.176	5	141.7
0.212	6	170.1
0.247	7	198.4
0.282	8	226.8
0.317	9	255.1
0.353	10	283.5

Mass equivalents: pounds/kilograms.

1 lb. = 0.4536 kg
1 kg = 2.2046 lb.

Pounds	Pound or kilogram mass to be converted	Kilo-grams
2.20	1	0.454
4.41	2	0.907
6.61	3	1.361
8.82	4	1.814
11.02	5	2.268
13.23	6	2.722
15.43	7	3.175
17.64	8	3.629
19.84	9	4.082
22.05	10	4.536

Liquid capacity equivalents: fluid ounces/milliliters.

1 fl. oz. = 29.573 mL
1mL = 0.0338 fl. oz.

Fluid ounces	Fluid ounce or milliliter capacity to be converted	Milli-liters
0.034	1	29.6
0.068	2	59.1
0.101	3	88.6
0.135	4	118.3
0.169	5	147.9
0.203	6	177.4
0.237	7	207.0
0.271	8	236.6
0.304	9	266.2
0.338	10	295.7

Liquid capacity equivalents: quarts/liters.

1 qt.. = 0.9463 L.
1 L = 1.0567 qt.

Quarts	Quart or liter capacity to be converted	Liters
1.06	1	0.946
2.11	2	1.893
3.17	3	2.839
4.23	4	3.785
5.28	5	4.732
6.34	6	5.678
7.40	7	6.624
8.45	8	7.571
9.51	9	8.517
10.57	10	9.463

Temperature
equivalents:
Fahrenheit/Celsius.

$°F = 1.8 \,(°C) + 32$
$°C = 5/9 \,(°F - 32)$

°F	°F or °C reading to be converted	°C	°F	°F or °C reading to be converted	°C
−40	−40	−40	78.8	26	−3.3
−36.4	−38	−38.9	80.6	27	−2.8
−32.8	−36	−37.8	82.4	28	−2.2
−29.2	−34	−36.7	84.2	29	−1.7
−25.6	−32	−35.6	86	30	−1.1
−22	−30	−34.4	87.8	31	−0.6
−18.4	−28	−33.3	89.6	32	0
−14.8	−26	−32.2	91.4	33	0.6
−11.2	−24	−31.1	93.2	34	1.1
−7.6	−22	−30	95	35	1.7
−4	−20	−28.9	96.8	36	2.2
−0.4	−18	−27.8	98.6	37	2.8
3.2	−16	−26.7	100.4	38	3.3
6.8	−14	−25.6	102.2	39	3.9
10.4	−12	−24.4	104	40	4.4
14	−10	−23.3	105.8	41	5
17.6	−8	−22.2	107.6	42	5.6
21.2	−6	−21.1	109.4	43	6.1
24.8	−4	−20	111.2	44	6.7
28.4	−2	−18.9	113	45	7.2
32	0	−17.8	114.8	46	7.8
33.8	1	−17.2	116.6	47	8.3
35.6	2	−16.7	118.4	48	8.9
37.4	3	−16.1	120.2	49	9.4
39.2	4	−15.6	122	50	10
41	5	−15	123.8	51	10.6
42.8	6	−14.4	125.6	52	11.1
44.6	7	−13.9	127.4	53	11.7
46.4	8	−13.3	129.2	54	12.2
48.2	9	−12.8	131	55	12.8
50	10	−12.2	132.8	56	13.3
51.8	11	−11.7	134.6	57	13.9
53.6	12	−11.1	136.4	58	14.4
55.4	13	−10.6	138.2	59	15
57.2	14	−10	140	60	15.6
59	15	−9.4	141.8	61	16.1
60.8	16	−8.9	143.6	62	16.7
62.6	17	−8.3	145.4	63	17.2
64.4	18	−7.8	147.2	64	17.8
66.2	19	−7.2	149	65	18.3
68	20	−6.7	150.8	66	18.8
69.8	21	−6.1	152.6	67	19.4
71.6	22	−5.6	154.4	68	20
73.4	23	−5	156.2	69	20.6
75.2	24	−4.4	158	70	21.1
77	25	−3.9	159.8	71	21.7

Temperature
equivalents:
Fahrenheit/Celsius
(continued).

°F	°F or °C reading to be converted	°C	°F	°F or °C reading to be converted	°C
161.6	72	22.2	266	130	54.4
163.4	73	22.8	269.6	132	55.6
165.2	74	23.3	273.2	134	56.7
167	75	23.9	276.8	136	57.8
168.8	76	24.4	280.4	138	58.9
170.6	77	25	284	140	60
172.4	78	25.6	287.6	142	61.1
174.2	79	26.1	291.2	144	62.2
176	80	26.7	294.8	146	63.3
177.8	81	27.2	298.4	148	64.4
179.6	82	27.8	302	150	65.6
181.4	83	28.3	305.6	152	66.7
183.2	84	28.9	309.2	154	67.8
185	85	29.4	312.8	156	68.9
186.8	86	30	316.4	158	70
188.6	87	30.6	320	160	71.1
190.4	88	31.1	323.6	162	72.2
192.2	89	31.7	327.2	164	73.3
194	90	32.2	330.8	166	74.4
195.8	91	32.8	334.4	168	75.6
197.6	92	33.3	338	170	76.7
199.4	93	33.9	341.6	172	77.8
201.2	94	34.4	345.2	174	78.9
203	95	35	348.8	176	80
204.8	96	35.6	352.4	178	81.1
206.6	97	36.1	356	180	82.2
208.4	98	36.7	359.6	182	83.3
210.2	99	37.2	363.2	184	84.4
212	100	37.8	366.8	186	85.6
215.6	102	38.9	370.4	188	86.7
219.2	104	40	374	190	87.8
222.8	106	41.1	377.6	192	88.9
226.4	108	42.2	381.2	194	90
230	110	43.3	384.8	196	91.1
233.6	112	44.4	388.4	198	92.2
237.2	114	45.6	399.2	200	93.3
240.8	116	46.7	395.6	202	94.4
244.4	118	47.8	399.2	204	95.6
248	120	48.9	402.8	206	96.7
251.6	122	50	406.4	208	97.8
255.2	124	51.1	410	210	98.9
258.8	126	52.2	413.6	212	100
262.4	128	53.3			

C   Relative Humidity and Moisture-Saturated Air

Measuring Relative Humidity

For a fixed-position hygrometer, fan the wet-bulb vigorously for 1 min. or more before taking readings; then read at once. A sling pyschrometer must be whirled until two wet-bulb readings agree. This usually takes 1–2 min.

The boldface numbers in the charts on the next two pages represent temperature, and the lightface numbers represent the relative humidity.

Dry-bulb temperature (°F)

Wet-bulb temperature (°F)	60	61	62	63	64	65	66	67	68	69	70	71	72	73	74	75	76	77	78	79	80
45	26	22	20	17	15																
46	30	27	24	21	18	16															
47	34	31	28	25	22	20	17	15													
39	35	32	29	26	24	21	19	16	14												
49	43	40	36	33	30	27	25	22	20	18	15										
50	48	44	41	37	34	31	29	26	23	21	19	17	15								
51	53	49	45	42	38	35	32	30	25	24	22	20	18	16							
52	58	54	50	46	43	39	36	33	31	28	25	23	21	19	17	15					
53	63	58	54	50	47	44	40	37	34	32	29	27	24	22	20	18	16	14			
54	68	63	59	55	51	48	44	41	38	35	33	30	28	25	23	21	19	17	16	14	
55	73	68	64	60	56	52	48	45	42	39	36	33	31	29	26	24	22	20	18	17	15
56	78	73	69	64	60	56	53	49	46	43	40	37	34	32	29	27	25	23	21	19	18
57	83	78	74	69	65	61	57	53	50	46	44	41	38	35	33	30	28	26	24	22	20
58	89	84	79	74	70	66	61	58	54	51	48	45	42	39	36	34	31	29	27	25	23
59	94	89	84	79	74	70	66	62	58	55	51	48	45	42	39	37	34	32	30	28	26
60	100	94	89	84	79	75	71	66	62	59	55	52	49	46	43	40	38	35	33	31	29
61		100	94	89	84	80	75	71	67	63	59	56	53	50	47	44	41	39	36	34	32
62			100	95	90	85	80	75	71	67	64	60	57	53	50	47	44	42	39	37	35
63				100	95	90	85	80	76	72	68	64	61	57	54	51	48	45	43	40	38
64					100	95	90	85	80	76	72	68	65	61	58	54	51	48	46	43	41
65						100	95	90	85	81	77	72	69	65	61	58	55	52	49	46	44
66							100	95	90	85	81	77	73	69	65	62	59	56	53	50	47
67								100	95	90	86	81	77	73	69	66	62	59	56	53	50
68									100	95	90	86	82	78	74	70	66	63	60	57	54
69										100	95	90	86	82	78	74	70	67	63	60	57
70											100	95	91	86	82	78	74	71	67	64	61
71												100	95	91	86	82	78	74	71	68	64
72													100	95	91	86	82	79	75	71	68
73														100	95	91	87	83	79	75	72
74															100	96	91	87	83	79	75
75																100	96	91	87	83	79
76																	100	96	91	87	83
77																		100	96	91	87
78																			100	96	91
79																				100	96

						Dry-bulb temperature (°F)									
	82	**84**	**86**	**88**	**90**	**92**	**94**	**96**	**98**	**100**	**102**	**104**	**106**	**108**	**110**
56	14	12	9												
58	20	16	14	11											
60	25	21	18	15	13	11									
62	30	26	23	20	17	15	12								
64	36	32	28	25	22	19	16	14							
66	42	37	33	30	26	23	20	18	15	13					
68	48	43	39	35	31	28	24	22	19	17	15	13			
70	55	49	44	40	36	32	29	26	23	21	18	16	14	12	
72	61	56	50	46	41	37	33	30	27	24	22	20	17	16	14
74	69	62	57	51	47	42	38	35	32	28	26	23	21	19	17
76	76	69	63	57	52	48	43	39	36	33	30	27	24	22	20
78	84	76	70	64	58	53	49	44	40	37	34	31	28	25	23
80	92	84	77	70	65	59	54	50	45	41	38	35	32	29	26
82	100	92	84	77	71	65	60	55	50	46	42	39	36	33	30
84		100	92	85	78	72	66	61	55	51	47	43	40	37	34
86			100	92	85	78	72	66	61	56	52	48	44	41	38
88				100	92	85	79	73	67	62	57	53	49	45	42
90					100	92	85	79	73	68	62	58	53	49	46
92						100	93	86	79	73	68	63	58	54	50
94							100	93	86	80	74	69	64	59	55
96								100	93	86	80	74	69	64	60
98									100	93	86	80	75	70	65
100										100	93	87	81	75	70
102											100	93	87	81	75
104												100	93	87	81
106													100	93	87
108														100	94
110															100

Wet-bulb temperature (°F)

Mass of water vapor in saturated air at various temperatures.

1 gr./ft.^{3} = 2.288 g/m^3
1 g/m^3 = 0.437 gr./ft.^{3}

Temperature °F	°C	Grains/ cubic foot	Grams cubic meter	Temperature °F	°C	Grains/ cubic foot	Grams cubic meter
32	0.0	2.119	4.849	78	25.6	10.39	23.78
34	1.1	2.287	5.234	80	26.7	11.06	25.31
36	2.2	2.466	5.643	82	27.8	11.76	26.91
38	3.3	2.657	6.080	84	28.9	12.50	28.60
40	4.4	2.861	6.547	86	30.0	13.28	30.39
42	5.6	3.080	7.048	88	31.1	14.10	32.27
44	6.7	3.313	7.581	90	32.2	14.96	34.23
46	7.8	3.561	8.149	92	33.3	15.86	36.29
48	8.9	3.826	8.755	94	34.4	16.82	38.49
50	10.0	4.108	9.401	96	35.6	17.82	40.78
52	11.1	4.407	10.08	98	36.7	18.88	43.21
54	12.2	4.723	10.81	100	37.8	19.99	45.74
56	13.3	5.063	11.59	102	38.9	21.15	48.40
58	14.4	5.419	12.40	104	40.0	22.38	51.21
60	15.6	5.798	13.27	106	41.1	23.65	54.12
62	16.7	6.201	14.19	108	42.2	24.98	57.16
64	17.8	6.627	15.17	110	43.3	26.39	60.39
66	18.9	7.080	16.20	112	44.4	27.86	63.75
68	20.0	7.561	17.30	114	45.6	29.40	67.28
70	21.1	8.064	18.45	116	46.7	31.00	70.94
72	22.2	8.605	19.69	118	47.8	32.70	74.83
74	23.3	9.169	20.98	120	48.9	34.46	78.86
76	24.4	9.763	22.34				

Index

About the Author

After receiving his doctorate in organic chemistry from Pennsylvania State University, Dr. Nelson R. Eldred set up a research project at the Union Carbide Corp. on chemicals and resins for manufacture of paper and ink. He later served as assistant manager of development for Buckman Laboratories, Inc., before joining the Graphic Arts Technical Foundation as supervisor of the Chemistry Division. Later, Dr. Eldred became manager of GATF's Techno-Economic Forecasting Division. He is now an independent graphic arts consultant headquartered in Tampa, Florida. Dr. Eldred has written extensively for both technical and trade journals, holds several patents, and presented numerous seminars and workshops on paper and ink. He holds degrees from Oberlin College, Wayne State University, and the Pennsylvania State University.

About GATF

The Graphic Arts Technical Foundation is a nonprofit, scientific, technical, and educational organization dedicated to the advancement of the graphic communications industries worldwide. Its mission is to serve the field as the leading resource for technical information and services through research and education. GATF is a partner of the Printing Industries of America (PIA), the world's largest graphic arts trade association, and its regional affiliates.

For 76 years the Foundation has developed leading edge technologies and practices for printing. GATF's staff of researchers, educators, and technical specialists partner with nearly 14,000 corporate members in over 80 countries to help them maintain their competitive edge by increasing productivity, print quality, process control, and environmental compliance, and by implementing new techniques and technologies. Through conferences, satellite symposia, workshops, consulting, technical support, laboratory services, and publications, GATF strives to advance a global graphic communications community.

The GATF*Press* publishes books on nearly every aspect of the field; learning modules (step-by-step instruction booklets); audio-visuals (CD-ROMs and videotapes); and research and technology reports. It also publishes *GATFWorld,* a bimonthly magazine of technical articles, industry news, and reviews of specific products.

For more information on GATF products and services, please visit our website *http://www.gatf.org* or *http://www.gain.net,* or write to us at 200 Deer Run Road, Sewickley, PA 15143-2600 (phone: 412/741-6860).

Orders to:
GATF Orders
P.O. Box 1020
Sewickley, PA 15143-1020
Phone (U.S. and Canada only): 800/662-3916
Phone (all other countries): 412/741-5733
Fax: 412/741-0609
Internet: www.gain.net

About PIA

In continuous operation since 1887 and headquartered in Alexandria, Virginia, Printing Industries of America, Inc. (PIA), is the world's largest graphic arts trade association representing an industry with more than 1 million employees and $156 billion in sales annually. PIA promotes the interests of over 14,000 member companies. Companies become members in PIA by joining one of 30 regional affiliate organizations throughout the United States or by joining the Canadian Printing Industries Association. International companies outside North America may join PIA directly.

Printing Industries of America, Inc. is in the business of promoting programs, services, and an environment that helps its members operate profitably. Many of PIA's members are commercial printers, allied graphic arts firms such as electronic imaging companies, equipment manufacturers, and suppliers.

PIA has developed several special industry groups to meet the unique needs of specific market segments. Each special industry group provides members with current information on their specific market and helps members stay ahead of the competition. PIA's special industry groups are the Graphic Communications Association (GCA), Web Offset Association (WOA), Web Printing Association (WPA), Graphic Arts Marketing Information Service (GAMIS), International Thermographers Association (ITA), Label Printing Industries of America (LPIA), and Binding Industries of America International (BIA).

For more detailed information on PIA products and services, please visit PIA at *http://www.gain.net* or write to 100 Daingerfield Road, Alexandria, VA 22314 (phone: 703/519-8100).

GATF*Press*: Selected Titles

Colophon

Chemistry for the Graphic Arts was edited, designed, and printed at the Graphic Arts Technical Foundation, headquartered in Sewickley, Pennsylvania. The manuscript was written using Microsoft Word, and the files were emailed to GATF. The edited files were imported into QuarkXPress for page layout. The text is set in 11-pt. New Century Schoolbook. Line drawings were created in Adobe Illustrator and Macromedia FreeHand, and photographs were scaled and cropped in Adobe Photoshop. Page proofs for author approval were produced on a Xerox Regal color copier with Splash RIP.

The author of the book wanted to reuse many of the illustrations that appeared in the second edition of the book. Although the second edition was published in 1992, the archived Adobe PageMaker 4.0 files were still readable using PageMaker 6.5. In addition, all of the line drawings had been embedded directly in the PageMaker 4.0 files. This enabled the GATF editor to open the archived PageMaker documents, save individual pages as EPS files, and then open and manipulate those images in Adobe Illustrator. This saved considerable production time, because these drawings did not have to be redrawn.

After the editorial/page layout process was completed, the images were transmitted to GATF's Center for Imaging Excellence, where all images were adjusted for the printing parameters of GATF's in-house printing department and proofed.

Next, the entire book was preflighted using a Power Macintosh, digitally imposed using DK&A INposition, and then output to a Creo Trendsetter 3244 platesetter. The cover was printed two-up on a 20×28-in., six-color Komori Lithrone 28 sheetfed press with tower coater. The interior was printed on a 26×40-in., four-color Heidelberg Speedmaster Model 102-4P sheetfed perfecting press. Finally, the book was sent to a trade bindery for case binding.